Wireless Networks

Series Editor
Xuemin Sherman Shen, University of Waterloo, Waterloo, Canada

The purpose of Springer's Wireless Networks book series is to establish the state of the art and set the course for future research and development in wireless communication networks. The scope of this series includes not only all aspects of wireless networks (including cellular networks, WiFi, sensor networks, and vehicular networks), but related areas such as cloud computing and big data. The series serves as a central source of references for wireless networks research and development. It aims to publish thorough and cohesive overviews on specific topics in wireless networks, as well as works that are larger in scope than survey articles and that contain more detailed background information. The series also provides coverage of advanced and timely topics worthy of monographs, contributed volumes, textbooks and handbooks.

Haipeng Yao • Tianle Mai

UAV Swarm Cooperation

A Networking Perspective

Springer

Haipeng Yao
Beijing, China

Tianle Mai
Beijing, China

ISSN 2366-1186 ISSN 2366-1445 (electronic)
Wireless Networks
ISBN 978-3-031-96443-5 ISBN 978-3-031-96444-2 (eBook)
https://doi.org/10.1007/978-3-031-96444-2

This Springer imprint is published by the registered company Springer Nature Switzerland AG
The registered company address is: Gewerbestrasse 11, 6330 Cham, Switzerland

Declarations

Competing Interests The authors have no competing interests to declare that are relevant to the content of this manuscript.

Contents

Acronyms

UAV	Unmanned Aerial Vehicle
AI	Artificial Intelligence
ML	Machine Learning
DL	Deep Learning
RL	Reinforcement Learning
DNN	Deep Neural Network
RSS	Received Signal Strength
CNN	Convolutional Neural Network
RNN	Recurrent Neural Network
LSTM	Long Short-Term Memory Network
MF-HPPO	Mean Field Hybrid Proximal Policy Optimization
MADDPG	Multi-Agent Deep Deterministic Policy Gradient
MARL	Multi-Agent Reinforcement Learning
SVM	Support Vector Machine
HSFL	Hybrid Split and Federated Learning
MCKP	Multiple Choice Knapsack Problem
DetNet	Deterministic Networking
TSN	Time-Sensitive Networking
ILP	Integer Linear Programming
3GPP	The 3rd Generation Partnership Project
ITU	The International Telecommunication Union
IEEE	The Institute of Electrical and Electronics Engineers
FANET	Flying Ad-Hoc Network
CH	Cluster Head
CMs	Cluster Members
LOWID	Lowest ID Algorithm
HIGHD	Highest Node Degree Algorithm
WCA	Weight-Based Clustering Algorithm
SC	Spectral Clustering
FSC	Fission Spectral Clustering
CRP	Channel and Relevant Properties

OPCP	Optimization Problem of Communication Performance
SINR	Signal to Noise Ratio
GT	Graph Theory
SF	Shannon Formula
LC	Link Connection
TC	Throughput
LCD	Link Connection Duration
TI	Time Indicator
GMM	Gaussian Mixture Model
AP	Affinity Propagation
AC	Agglomerative Clustering
DQN	Deep Q-Network
LoS	Line-of-Sight
NLoS	Non-Line-of-Sight
Dec-POMDP	Decentralized Partially Observable Markov Decision Process
GRU	Gate Recurrent Unit
MLP	Multilayer Perceptron
ReLU	Rectified Linear Unit
TD	Temporal-Difference
QMIX	Q-Function Mixing Network
AC	Actor-Critic
DF	Degrees of Freedom
ICT	Information and Communication Technology
ESS	Evolutionary Stable Strategy
EGT	Evolutionary Game Theoretic
OLSR	Optimized Link-State Routing
GCSR	Ground Control System Routing
AODV	Ad-Hoc On-Demand Distance Vector
GBR	Geolocation-based Routing
IoT	Internet of Things
SD	Source-Destination
MSE	Mean Square Error
SDN	Software-Defined Networking
MDP	Markov Decision Process
UNDN	UAV Named Data Networking
GS	Gale-Shapley
NDN	Named Data Networking
IC	Incentive Compatibility
IR	Individual Rationality
DT	Digital Twin
MEC	Mobile Edge Computin
VNE	Virtual Network Embedding
VNR	Virtual Network Request
IoV	Internet of Vehicle
DAO	Decentralized Autonomous Organizations

PoW	Proof-of-Work
QoS	Quality of Service
QoE	Quality of Experience
MBS	Macro Base Station
SBS	Small-cell Base Stations
VCG	Vickrey-clarke-Grove
FCFS	First-Come-First-Serve
LPO	Local Processing Option
LCO	Lowest Cost Option
V2V	Vehicle-To-Vehicle
V2I	Vehicle-To-Infrastructure
V2X	Vehicle To Everything
HFL	Hierarchical Federated Learning
FL	Federated Learning
CFL	Centralized FL
DFL	Decentralized FL
SGD	Stochastic Gradient Descent
AWGN	Additive White Gaussian Noise
SNR	Signal-To-Noise Ratio
PSNR	Peak Signal-To-Noise Ratio
VSP	Virtual Service Provider
MAB	Multi-Armed Bandit
UCB	Upper Confidence Bound
FPA	First-Price Auction
SPA	Second-Price Auction
SPA-0	SPA with Zero Reserve Price
Non-IID	non-Independent and Identically Distributed
SE	Stackelberg Equilibrium
OFL	Original Federated Learning
BS	Base Station
EDC	Euclidean Distance of Decomposed Cosine
SVD	Singular Value Decomposition
MARL	Multi-Agent Reinforcement Learning
FEMNIST	Federated Extended MNSIT
HDPG	Hierarchical Deep Policy Gradient
SAC	Soft Actor-Critic
PPO	Proximal Policy Optimization
AIoT	Artificial Intelligent and Internet of Things
IDA	Iterative Double Auction
EWA-DA	Experience Weighted Attraction Learning-Double Auction
DPDA	Dynamic Pricing-Based Double Auction
KKT	Karush-Kuhn-Tucker
DA	Double Auction
EWA	Experience-Weighted Attraction Learning
PG	Policy Gradient

GA	Genetic Algorithm
TSPG	Two-Stage Population Generation
BE	Best-Effort
GCL	Gate Control List
SMT	Satisfiability Modulo Theories
PSO	Particle Swarm Optimization
IIoT	Industrial Internet of Things
AVB	Audio Video Bridging
TT	Time-Triggered
TAS	Time-Aware Shaper
TSSDN	Time-Sensitive Software-Defined Networking

Chapter 1
Introduction

1.1 Background

In recent years, Unmanned Aerial Vehicle (UAV) swarms have been widely used in military and civilian fields. UAV swarms achieve mission objectives through distributed collaboration and communication, and can adapt to complex environmental changes and diversified mission requirements. Based on communication networking technology, UAV swarm networks can achieve efficient communication guarantee and coverage, and information interaction and synchronization between nodes further improve mission effectiveness. This chapter first outlines the concept and connotation of UAV swarm networks, and explains its research hotspots and applications in academia and industry. In addition, this chapter also discusses emerging technologies enabling UAV swarm networks and related research and industry standards of UAV swarm networks.

1.2 Overview of UAV Swarm Networks

1.2.1 Overview of UAV Swarm Networks

The UAV swarm acts as an operating unit composed of multiple UAVs controlled through collaborative control. It adopts a coordinated architecture that can effectively share the workload of a single node while significantly reducing the degree of dependence on centralized control. In complex application scenarios, UAV swarms can effectively disperse the risk of single point failures through a dynamic task allocation mechanism, thereby improving the reliability and safety of task execution. It has the following characteristics:

H. Yao, T. Mai, *UAV Swarm Cooperation*, Wireless Networks,
https://doi.org/10.1007/978-3-031-96444-2_1

(A) **High Degree of Autonomy and Collaboration**: Each UAV in the UAV swarm can act as an independent unit to perform tasks and make real-time decisions. And exchange information through real-time interaction with other UAVs to promote task collaboration. The above collaborative characteristics make UAV swarms suitable for highly uncertain and complex environments.
(B) **Self-Organization and Adaptability**: The topological structure of the UAV swarm will be adaptively adjusted according to the mission characteristics and environmental situation. For example, the UAV controls its flight trajectory according to the surrounding environment to achieve cluster trajectory optimization. The self-organizing characteristics of the UAV swarm enable it to achieve autonomous optimization without a centralized control unit and realize cluster collaboration in a dynamic environment.
(C) **Robustness**: When a node in the UAV swarm fails, it will not affect the overall mission execution, because the tasks will be redistributed and dynamically optimized. The redundant design of the UAV swarm for dynamic environments and mission conditions enables it to operate normally even when some nodes fail, thereby ensuring the robustness of the UAV swarm in the event of damage or emergencies.

Based on the above characteristics of UAV swarms, it has been widely used in military activities, emergency scenarios, disaster relief, low-altitude economy and other fields. For example, in the military field, UAV swarms can efficiently perform reconnaissance and rescue tasks, while reducing casualties and improving mission efficiency and reducing costs. In the field of low-altitude economy, UAV swarms can perform applications such as UAV logistics, showing great economic value. In emergency scenarios, UAV swarms can perform environmental perception, information feedback, and autonomous decision-making operations through their highly autonomous and collaborative characteristics.

As a new generation of intelligent collaborative system, UAV swarm network has become a research hotspot. Compared to the UAV swarm, UAV swarm network refers to a collaborative system with a networked control architecture. Based on communication networking technology and intelligent collaboration technology, UAV swarm network has greatly expanded its communication interaction, large-scale coverage and intelligent collaboration capabilities. In the UAV swarm network, each UAV can be equipped with computing loads and communication modules, etc., to interact and collaborate while performing environmental situation awareness. Different from the fully distributed architecture in the UAV swarm, the UAV swarm network can adopt a hybrid architecture to achieve hierarchical collaboration, that is, some nodes are used as backbone nodes and the remaining nodes are used as edge nodes to collaborate to achieve global optimization. The UAV swarm network has the following characteristics:

(A) **Communication guarantee**: When the ground network is damaged, the UAV swarm network can quickly cover to ensure communication capabilities. Unlike traditional ground networks, UAV swarm networks face a dynamic and complex environment. Its network topology, link status, and node status will be constantly adjusted and changed according to the environment. In order to ensure the robust communication capability in the UAV swarm network, it optimizes the topology structure through topology control and adjusts the transmission path through routing strategies. When the mission situation changes, it can quickly adjust and optimize to achieve rapid response.

(B) Dynamic Reconstruction: When a single point failure occurs in the UAV swarm network, it will not affect the overall task execution efficiency. At the same time, when the task is adjusted online, the UAV swarm network can realize dynamic network reconstruction, and adapt to complex environmental changes through operations such as failed nodes leaving the network and new nodes rejoining the network.

(C) Intelligent Collaboration: The UAV swarm network can emerge complex cooperative behaviors based on swarm intelligence. The UAV swarm network can realize intelligent collaborative behaviors such as collaborative perception and collaborative computing to achieve multi-objective optimization tasks. In addition, the communication resources and computing resources in the UAV swarm network can be optimized through joint allocation to improve the task efficiency and intelligent collaboration capabilities.

Although the UAV swarm network can achieve communication guarantee, dynamic reconstruction, intelligent collaboration, etc., the UAV swarm network still needs a lot of optimization to ensure its performance in highly dynamic and complex environments. First of all, in complex environments such as strong interference environments and urban environments, the communication between UAVs faces natural conditions and obstacles, which makes the performance optimization of UAV swarm networks face challenges. Diversified mission requirements also put forward higher requirements for UAV swarm networks. In order to improve the mission performance and adaptability in different mission environments, the UAV swarm network can achieve optimizations such as trajectory control, resource allocation, and task offloading through the joint optimization of communication, computing, energy and other resources. With the increase in the scale of UAV swarm networks and the increase in the complexity of tasks, UAV swarm networks need to achieve collaborative optimization of nodes through intelligent optimization methods and by characterizing game behaviors such as cooperation and competition between nodes. In addition, when the UAV swarm network performs collaborative perception of the environment, the collected sensor data needs to be further analyzed and processed, and advanced data fusion algorithms can be combined to achieve efficient information perception.

To this end, researchers have proposed many innovative solutions to enhance the intelligent collaboration capabilities of UAV swarm networks. With the continuous development of technologies such as machine learning and edge computing, the combination of UAV swarm networks and intelligent algorithms has been widely studied. By deploying intelligent algorithms, UAV swarm networks can achieve efficient information perception, processing, analysis and decision-making, thereby improving the intelligent writing capabilities and mission performance of UAV swarm networks. In the next section, the key technologies that promote the development of intelligent collaboration in UAV swarm networks will be introduced.

1.2.2 Emerging Technologies Enabling UAV Swarm Networks

This section introduces the emerging technologies enabling UAV swarm networks. In recent years, the emerging technologies commonly used in UAV swarm networks mainly include machine learning, distributed learning, semantic communication, deterministic technology, and so on. The following will introduce the application of these technologies in UAV swarm networks in detail.

1.2.2.1 Machine Learning for UAV Swarm Network

Based on its powerful data processing, predictive analysis and adaptive capabilities, machine learning (ML) algorithms can effectively solve challenges such as multivariable optimization, complex network prediction and environmental perception. The combination of ML and UAV swarm networks has shown significant advantages in improving the intelligence, autonomy and efficiency.

Deep learning (DL) is a machine learning method that simulates the complex neural network inside the human brain by building a multi-level neural network structure, thereby achieving learning and pattern recognition of complex data. In highly dynamic and complex scenarios such as communication, networking, path planning and environmental perception, UAV swarm networks learn complex patterns from real-time and historical data through DL, and can accurately perceive the environment, predict future states, and achieve optimal decisions. Reference [1] proposed a channel quality prediction method based on deep learning. By analyzing the received signal strength (RSS) data, it significantly improved the accuracy of channel prediction, reduced the probability of communication interruption, and provided stable communication guarantee for UAV swarm networks. Reference [2] proposed an obstacle avoidance path learning algorithm based on deep learning. It used convolutional neural networks (CNN) to process images, detect obstacles in real time, and efficiently plan flight paths in UAV swarm networks, effectively improving the autonomous obstacle avoidance capabilities of UAVs in unknown environments.

Reinforcement learning (RL) aims to enable intelligent agents to learn optimal behavior strategies through interaction with the environment to maximize cumulative rewards. In scenarios with high-speed dynamic changes and frequent changes in network topology, UAV swarms can learn the best cooperation strategy through real-time interaction with the environment through reinforcement learning, thereby optimizing overall performance. Reference [3] proposed a mean field hybrid proximal policy optimization (MF-HPPO) algorithm based on reinforcement learning, which optimizes resource allocation by dynamically adjusting communication resources (such as spectrum and power) to maximize network performance. Reference [4] introduced an algorithm based on multi-agent reinforcement learning (MARL), namely multi-agent deep deterministic policy gradient (MADDPG), to coordinate the communication behavior between multiple UAVs and reduce interference. Reference [5] combined distributed reinforcement learning and multi-agent reinforcement learning for the cooperative control scenario of UAV swarms, learned the optimal cluster control strategy, and achieved higher cooperative control performance.

Supervised learning uses labeled data sets to train models, learn the mapping relationship between input and output, and achieve accurate prediction and classification of new data. In mission scenarios that rely on a large amount of historical training data, UAV swarm networks use supervised learning to learn from past experience to accurately identify target objects, evaluate network conditions, and select optimal paths, thereby improving their collaboration and autonomy. Reference [6] uses support vector machines (SVMs) to quickly identify UAVs with abnormal behavior and ensures stable operation of the swarm by analyzing the flight parameters and energy consumption data of UAVs.

Unlike supervised learning, unsupervised learning discovers hidden patterns and structures from unlabeled data. In the scenarios that lack labeled data, UAV swarm networks automatically extract information from the environment through unsupervised learning, identify data patterns, and optimize group behavior and decision-making. Reference [7] uses the K-means algorithm to cluster UAV swarms, divides UAV nodes into multiple clusters, and selects a cluster head in each cluster to manage communication and data forwarding, which not only significantly reduces communication overhead but also improves overall network efficiency.

With the improvement of computing power and further optimization of algorithms, ML technology is expected to provide smarter and more reliable solutions for UAV swarm networks in the future, driving UAV technology towards greater intelligence and autonomy.

1.2.2.2 Distributed Learning for UAV Swarm Networks

With the development of AI technologies, distributed learning has found widespread applications in UAV swarm networks. In traditional architectures, UAVs send data to a central server in a centralized manner. However, due to the highly dynamic nature of UAV swarms and the rapid changes in communication links, this centralized architecture faces significant issues, including single points of failure. To address this challenge, distributed learning has been introduced into UAV swarm networks. Distributed learning allows individual UAVs in the swarm to function as computational nodes, working collaboratively to complete learning tasks.

Distributed learning enables UAVs in the swarm to jointly train AI models. Within the framework of distributed learning, UAVs can process and analyze the data they collect locally, meaning that the collaborative nodes do not receive the complete set of data gathered or generated by the entire swarm. Typically, UAVs only share model parameters with collaborative nodes to protect data privacy. Furthermore, homomorphic encryption can be applied to UAV swarm networks utilizing distributed learning, allowing UAVs to perform computations on encrypted data without the need for decryption. In this way, even if data is shared among UAVs, attackers cannot understand the content of the data.

As shown in Fig. 1.1, each UAV can continually optimize its local model based on the local model and interactions with other UAVs. Since there is no central control node, model parameters are distributed across UAV nodes, and thus, the allocation of computational resources must be considered. For example, optimization algorithms can be used to dynamically allocate resources to maximize the performance of the UAV swarm. By appropriately distributing computational resources, the UAV swarm can complete tasks more efficiently, avoid resource waste, and improve overall operational efficiency. Moreover, due to the decentralized nature of the system, the failure of a single UAV does not impact the training of the other UAVs, ensuring the swarm's continued operation and significantly enhancing the robustness of the UAV swarm network. Additionally, when new UAVs join the swarm, distributed learning avoids the need for frequent network reconfiguration, enabling the swarm to scale and adapt to different task sizes. For large-scale data

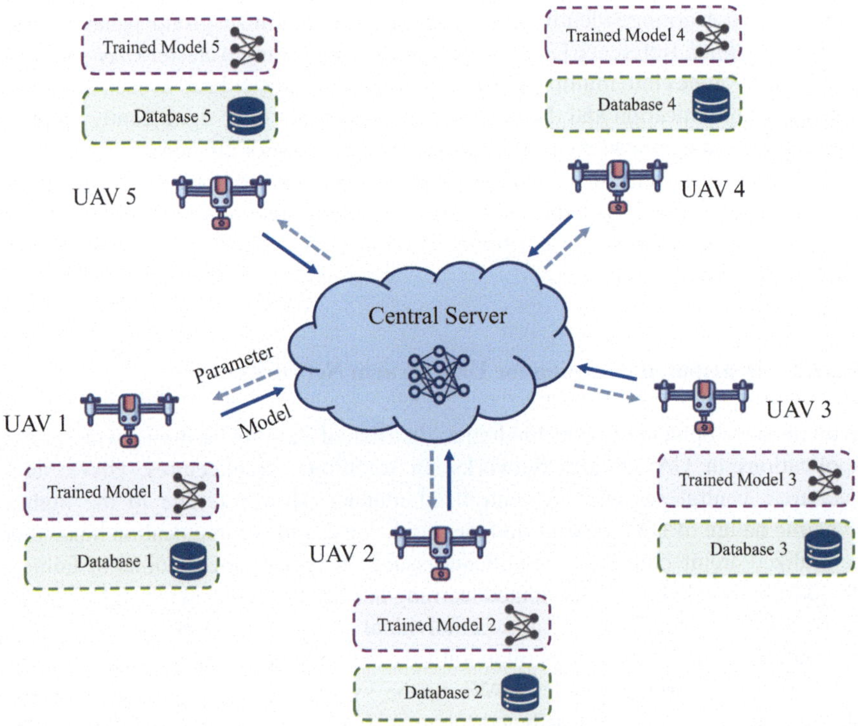

Fig. 1.1 Distributed learning architecture in UAV swarm networks

processing tasks, the workload can be distributed across multiple UAV nodes, alleviating data processing pressure.

Federated learning is a widely used distributed machine learning technique, first introduced by Google, to address personal data privacy concerns. As a privacy-preserving learning paradigm, federated learning can securely extract data from distributed devices and achieve collaborative training. In UAV swarm networks, different training times can affect the convergence performance of the system. In addition, the heterogeneity of UAV swarm data requires different operations such as weighted aggregation, gradient compression, semi-synchronous aggregation, and model splitting. Based on this, many researchers have proposed UAV swarm optimization methods based on federated learning. For example, the authors in [8] proposed a federated learning architecture for UAV swarms, which used low-rank technology to decompose the fully connected layer of CNN to compress model parameters, and introduced a filtering strategy to discard data from certain devices to achieve model compression, greatly improving the model training efficiency. The authors in [9] proposed an asynchronous federated learning framework and designed a device selection strategy to prevent the low-quality learning of certain UAVs from affecting the overall learning efficiency and accuracy of the system. In addition, they also designed a joint device selection, UAV layout, and resource management algorithm based on asynchronous dominant behavior critics to further improve the convergence speed and accuracy of the federated learning framework.

In addition, MARL is another distributed architecture that can be applied to UAV swarm networks. MARL involves multiple agents learning collaboratively in an environment, where each agent must consider the actions of other agents to optimize its own strategy. However, MARL faces challenges such as environmental non-stationarity, multi-agent training mode, and multi-agent trust allocation. To address these challenges, many researchers have proposed solutions. The authors in [10] proposed a multi-objective, independent prediction-based multi-agent actor-critic approach to achieve joint trajectory planning and task offloading for UAV swarms. This approach not only provides ultra-low latency services for mobile devices in UAV-assisted mobile edge computing, but also reduces the communication cost of the system.

Split learning is another technology related to UAV swarm networks, in which part of the model is trained on the client side and the other part is trained on the server side. Hybrid split and federated learning (HSFL) can be applied to UAV swarm networks to overcome the risks of UAV nodes accessing the global model and the high resource requirements of computing nodes, while also reducing the time loss caused by relay model training in traditional split learning. In this architecture, UAV nodes use split learning to train different parts of the network and use federated learning to aggregate the model. The authors in [11] proposed a HSFL architecture, which enhances the original framework by expressing user scheduling and training method selection as a multiple choice knapsack problem (MCKP), thereby improving accuracy while consuming less energy.

In the future, distributed learning will provide strong technical support and innovative solutions for UAV swarm networks. Training and learning methods for

UAV swarm networks, with a focus on intelligence, automation, and safety, will become a key research direction for distributed learning technology.

1.2.2.3 Semantic Communication for UAV Swarm Networks

In recent years, UAV swarms have become increasingly important in tasks such as military operations, rescue missions, and surveillance. Given the highly dynamic nature of UAV swarms, the positions of the UAVs and the states of the nodes continuously change throughout the execution of tasks. Moreover, communication links between swarm nodes are prone to disruptions [12]. In large-scale UAV swarms, individual UAVs need to exchange information with other UAVs. However, the time-varying nature of the environment, the dynamic relative positions of nodes, and frequent changes in communication channels often result in traditional communication systems struggling to maintain reliable communication quality. Traditional communication systems have limited bandwidth. When the node scale expands exponentially, it will bring about problems such as increased data transmission delay and increased packet loss rate. Ensuring efficient data transmission in a highly dynamic environment has become a core bottleneck restricting UAV swarm communication.

As an emerging communication method, semantic communication has shown its unique value. Traditional communication method only relies on increasing the dimension of information transmission and changing the way of resource mining and utilization to improve the capacity and performance of the communication system. Semantic communication is a task-based communication method of "understanding first, then transmission". It optimizes communication efficiency by transmitting the "semantics" behind the data, that is, whether the information can be effectively understood and used by the receiver [13]. Semantic communication actively uses information redundancy to match the channel state, and co-designs the source compression and channel transmission to improve bandwidth utilization.

Semantic communication usually includes semantic encoding and semantic decoding [14]. Semantic encoding compresses data as much as possible at the source and extracts the meaning and semantic features of the data to reduce redundancy during transmission. Semantic decoding involves the receiver interpreting the received signals, using prior knowledge to recover information that is meaningful to the receiver. In particular, in a semantic communication system, the primary objective of the encoding process is to encode the semantic meaning of the data, rather than encoding the data's specific bit stream, thus facilitating transmission. For example, in image or video transmission, the sender might only transmit crucial semantic information based on predictions of the receiver's needs, rather than transmitting all pixel data or video frames. Semantic encoding techniques often incorporate methods like data compression and feature extraction, ensuring that only the necessary semantic information is transmitted, minimizing redundancy and irrelevant data. Additionally, semantic communication can be enhanced by artificial intelligence techniques. For example, during the transmission of images, videos, or

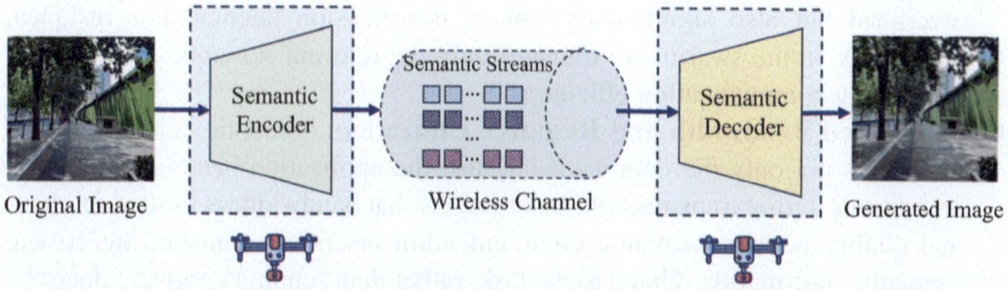

Fig. 1.2 Architecture of semantic communication

sensor data, deep learning models can automatically extract task-relevant features and generate compressed semantic representations.

For high-dynamic UAV swarms, the semantic communication system must exhibit adaptability, including adaptive modulation and demodulation, adaptive frequency allocation, and resource scheduling mechanisms based on task requirements. As shown in Fig. 1.2, the sending UAV encodes the original image into semantic streams through a semantic encoder and sends it in a wireless channel. After receiving the image, the receiving UAV restores the semantic information through a semantic decoder to generate a new image. It should be capable of adjusting its strategies in real-time based on network status, task types, and the dynamic changes within the UAV swarm. This adaptability can be powered by machine learning and deep learning algorithms, which continuously analyze network states and contextual information, learning and adjusting optimization strategies to adapt to increasingly complex and dynamic communication environments. For instance, when network bandwidth is limited, the system can prioritize the transmission of critical semantic information, while secondary data may be selectively discarded or delayed. Semantic communication is not limited to optimization at the network layer. It also involves cross-layer design. By optimizing across the physical layer, link layer, network layer, and application layer, the system can maximize bandwidth utilization and information transmission efficiency, ensuring high-quality communication even under limited bandwidth and resource constraints.

The advantages of semantic communication are as follows:

(A) **Reduction of Redundant Information**: Traditional communication methods often require transmitting large amounts of redundant data, which places a significant strain on communication bandwidth and increases delay in UAV swarms. Semantic communication allows UAVs to selectively transmit valuable information based on current task needs, rather than each UAV transmitting all the data. For example, in disaster rescue operations, UAV swarms only need to transmit data related to the location of affected individuals, rather than transmitting all environmental monitoring data.

(B) **Improved Communication Efficiency**: By preprocessing, compressing, and semantically encoding data at the sender's end, semantic communication reduces the amount of data transmitted. This not only reduces communication overhead but also significantly lowers transmission latency. For instance, each UAV in the swarm can transmit only the relevant semantic information, enhancing communication efficiency.

(C) **Optimized Bandwidth and Resource Utilization**: Semantic communication considers not only the data itself but also the application needs and network conditions during transmission. In situations that bandwidth is limited or channel quality is poor, semantic communication prioritizes transmitting critical semantic information related to the task, rather than sending complete datasets. This approach optimizes the use of bandwidth and computational resources, preventing unnecessary resource wastage.

Semantic communication holds great potential for enhancing the communication performance, and collaborative efficiency in high-dynamics, resource-constrained UAV swarm networks. However, several challenges remain in the application of semantic communication to UAV swarms. One of the most complex tasks is accurately understanding and extracting the semantic information, particularly in multi-modal data transmissions (e.g., images, videos, sensor data). Moreover, semantic communication relies on artificial intelligence technologies such as deep learning, which will incur high computing overhead. Achieving efficient semantic processing still needs further research.

1.2.2.4 Deterministic Networking

In recent years, the rapid advancements in UAV technology have given rise to novel application scenarios, such as remote control, remote sensing monitoring, and collaborative operations. These emerging applications place increasingly stringent demands on network performance, particularly in terms of real-time capabilities, reliability, and so on. Specifically, UAV networks supporting these applications require strict Quality of Service (QoS) guarantees, including deterministic data transmission, low latency with bounded delays, and minimal jitter. These network performance metrics are crucial for the effective operation of UAV systems. For instance, in disaster rescue missions, UAV swarms must dynamically adjust their flight paths and execute tasks, necessitating that network delays be constrained to milliseconds to ensure precision and timely responsiveness. Similarly, in collaborative UAV operations such as formation flying, precise spatiotemporal synchronization of multiple UAVs is critical, as even minor network delays or jitter can lead to loss of formation or mission failure.

Although the "best effort" transmission mode adopted by traditional networks can prioritize data transmission efficiency, it is difficult to meet the strict demands of emerging applications for real-time. In this mode, the network only guarantees the basic accessibility of data packets, but lacks effective constraints on key performance indicators such as transmission delay, jitter and packet loss rate, resulting in unpredictable latency and bandwidth changes. This variability undermines the network's ability to provide time-critical services requiring deterministic and stable performance. Consequently, the non-deterministic nature of "best-effort" networks presents significant limitations, rendering them unsuitable for high-reliability and real-time applications that demand precise and consistent network characteristics.

Deterministic networking (DetNet) has attracted widespread attention in both industry and academia as an innovative solution to break through the bottlenecks of traditional network architecture. DetNet is a collection of technologies that aim to provide deterministic QoS, thereby ensuring dependable and predictable end-to-end data transmission across heterogeneous services [15]. IEEE 802.1 Time-Sensitive Networking (TSN) task group and the IETF DetNet working group has achieved remarkable advancements in the development of technologies pertaining to the data link and network layers [16]. For example, TSN, which is an extension of Ethernet standards, combines important mechanisms such as time synchronization (IEEE 802.1AS), time-aware scheduling (IEEE 802.1Qbv), frame preemption (IEEE 802.1Qbu), and traffic shaping (IEEE 802.1Qav) to support low-latency, low-jitter, and high-reliability communications. These capabilities are best applied in industrial automation, automotive, and audio-video transport applications. Similarly, the IETF's DetNet framework aims to provide deterministic transmission services within IP and MPLS networks [17]. The core components of DetNet include resource reservation, traffic isolation, and redundant path transmission, which collectively enhance the reliability and real-time performance of wide-area networks through mechanisms such as traffic engineering and packet replication [18].

In addition to advancements in wired deterministic networking, recent innovations in wireless deterministic networking have been driven by developments in 5G and Wi-Fi technologies [19]. For instance, 5G Deterministic Networking (5GDN) leverages 5G network resources to create predictable, programmable, and verifiable mobile private networks, offering differentiated and deterministic service experiences. The integration and innovation of these technologies have facilitated end-to-end deterministic transmission capabilities for UAV control. As illustrated in the Fig. 1.3, the UAV swarm controller can achieve end-to-end remote deterministic control of UAVs through the use of deterministic wired networks and wireless access networks.

Besides, deterministic networking has also gained attention in the academic community for its application in remote control. In [19], the authors addressed the challenges of traffic bursts and severe jitter of UAV data transmission in 6G

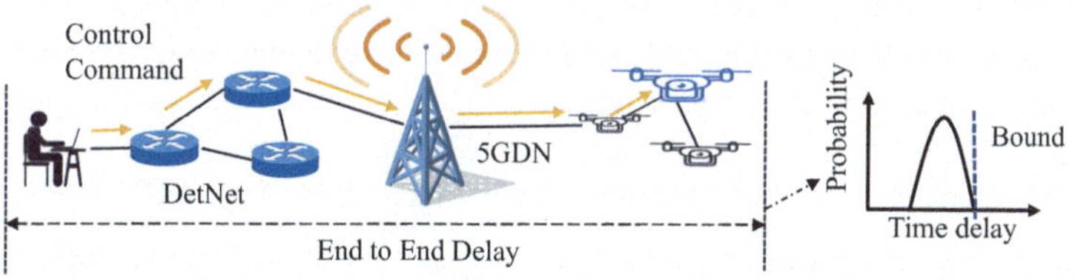

Fig. 1.3 Deterministic network technology for UAV swarm networks

network by constructing a dynamic scheduling model based on deep reinforcement learning (DRL) to dynamically allocate time slots. They designed a scheduling algorithm based on historical information evaluation and prediction, achieving high scheduling success rates and stable latency performance. The authors in [20] proposed a deterministic network transmission model for cyber-physical systems, introducing a geometric multi-path generation method based on topology views. This method improves reliability by reducing the joint-degree of multi-paths and promoting convergence. The authors in [21] proposed a network traffic deterministic scheduling scheme that accounts for dynamic channel variations. By predicting link quality using a channel change model, the network is divided into stable segments. They employ a TDMA/CSMA mechanism to provide guaranteed service flows that are scheduled based on priority, optimizing the objective function in order to minimize transmission jitter and ensure real-time delivery for time-sensitive services. This scheme ensures a deterministic transmission jitter of less than 60 μs, effectively guaranteeing the certainty of transmission.

1.3 Related Research and Industry Standards

In recent years, UAV swarm networks have experienced unprecedented development, with specialized networks featuring high-speed mobility, instant networking, and automated management—such as FANETs and VANETs—being widely used in scenarios like disaster detection and humanitarian rescue [22, 23]. In order to have reliable and widely available wireless connections, UAV swarm networks are required to achieve uniform technical specifications in various aspects such as communication protocols, spectrum allocation, and network architecture. Currently, UAV swarm networks, as an emerging technology, are receiving attention from several international standardization organizations [24, 25].

3GPP

The 3rd Generation Partnership Project (3GPP) is a global communication standardization organization responsible for developing technical specifications and standards for cellular communications. In recent years, the organization has addressed the communication demands of UAVs. It actively identifies communication protocols, network technologies, and standards required for UAV operations [26]. In TR 22.829, 3GPP highlights the technical requirements and standardization needs for UAV swarm networks. In TR 23.754, it proposes enhanced support for network infrastructure and procedures to enable UAV connectivity, identification, and tracking, ensuring secure and efficient UAV operations within cellular networks. Specifically, 3GPP supports a wide range of UAV applications and use cases in terms of network capacity, communication requirements and security management.

(A) **Network capabilities**: Cellular networks are optimized for the flight characteristics of UAV devices, including extended signal coverage, enhanced uplink performance, and effective management of high-altitude cellular interference problems, thus enabling support for airspace devices such as UAVs. At the same time, the dynamics of UAVs and devices are managed. Identifying that the speed and trajectory of UAVs are significantly faster than that of ground-based user devices, the standard defines a network-optimized switching mechanism to reduce dropout problems caused by frequent switching. Moreover, UAV communication is dominated by uplink data transmission, such as high-definition video streaming and telemetry data transmission, and the cellular network under this standard meets the application requirements of UAV swarm networks by improving the bandwidth and reliability of the uplink.

(B) Communication requirements: The standard flexibly regulates communication resources to meet different communication demand tasks oriented by UAV swarm networks. The standard supports real-time control and monitoring, such as remote flight control and mission collaboration, to ensure low-latency communication between UAVs. At the same time, highly reliable communication links are provided for mission scenarios with high reliability requirements, such as emergency rescue, to ensure successful mission execution. For high-bandwidth-demanding tasks such as high-definition video stream transmission or sensor data upload, high transmission bandwidth is provided to meet the mission requirements. Multicast and broadcast capabilities are improved in order to support group mission distribution and command in group missions using UAVs.

(C) Security management: The standard supplies identity authentication and authorization services for the UAVs, providing a distinct authentication for each UAV in order to keep network access legitimate and secure. It ensures dynamic updates as well as real-time monitoring of the trajectories of the flights of the UAVs, providing accuracy in scheduling missions. In addition, the standard secures data transfer between the UAVs, preventing against malicious attacks as well as risks of information theft.

ITU

As the central organization in global telecommunication standardization, the International Telecommunication Union (ITU), promotes the growth of UAV technology, takes on significant responsibility in research on drone communication standards. In response to the fast growth of drone technology, the increasing diversification of its communications requirements raises higher requirements for global standardization. The relevant work of ITU aims to establish a unified communication framework to enhance the interoperability and performance of UAV communications, as well as to provide technical guidance for future deployment of UAV swarm networks [27].

The ITU-T F.749.10 standard focuses on the study of system intelligence in UAV communications, particularly in communication optimization and intelligent design for critical mission services [28]. These studies not only promote technological advancements in UAV network systems but also provide important references for the future direction of network development. In response to applications in areas such as emergency rescue, logistics delivery, and environmental monitoring, the ITU emphasizes the necessity of intelligent network systems and explores how intelligent technologies can improve task execution efficiency and communication reliability. Research on artificial intelligence techniques to optimize network resource allocation, such as using machine learning algorithms to improve spectrum efficiency, reduce communication latency, and enhance UAV collaboration, aims to optimize the communication performance of UAV swarm networks. On the other hand, ITU actively promotes standard collaboration and explores mechanisms with international standardization organizations like 3GPP to ensure global consistency and broad applicability of UAV communication standards. By combining with existing mobile communication standards, it promotes the smooth deployment of UAV communication in different network environments.

The ITU-T Y.UAV.arch standard defines the overall architecture of UAV swarm networks and aims to provide a unified framework for UAV communications and network services [29]. It specifies the layered architecture of UAV swarm networks, including the UAV network layer, the ground control layer and the user layer, to ensure effective communication and collaboration between the layers. Communication protocols between UAVs, as well as between UAVs and ground stations, are specified to ensure that equipment from different manufacturers are interoperable and to improve the compatibility and scalability of the network. On the other hand, security measures such as authentication, encrypted communication and data integrity verification are introduced to safeguard the security of UAV swarm networks against unauthorized access and data tampering. And dynamic management strategies for network resources, including spectrum allocation, bandwidth management and power control, are provided to optimize network performance and meet the needs of different application scenarios.

IEEE

The Institute of Electrical and Electronics Engineers (IEEE) is committed to developing air-to-air communication standards for UAV swarm networks and UAV applications, including network structure, communication protocols, spectrum usage, and security measures.

In the IEEE 1937.1 standard, swarm network architectural features and protocols specifically aimed at optimizing the efficiency of communications between UAVs and their collaboration capabilities are designed in order to support the dynamic implementation and management of cluster missions [30]. By using the standard protocol specification, it ensures high reliability and low-latency data exchange during the communications among the node UAVs within the cluster, and it adapts according to different mission scenarios' requirements on communications. Meanwhile, the collaboration mode is established for the coordination among different UAVs in dynamic complicated environments in order to accomplish tasks like distributed target detection, path planning, and real-time data processing. And dynamic topology adjustable protocols are designed for dynamic changes of topology for supporting the applicability demands on frequent topology updates of different missions' UAV swarms.

A short message mechanism is utilized in the IEEE 1937.3 standard in order to minimize the data transmission's communication resources significantly, in view of the unmanned clusters' high-frequency requirements for flight data update [31]. While optimizing the data transmission efficiency to ensure real-time and reliability, it reduces packet conflicts and improves the stability of the communication link to meet the needs of large-scale UAV swarm networks. At the same time, the standard provides efficient task scheduling and status reporting capabilities for UAV swarm networks through the short message mechanism, supporting distributed cooperative work of UAV swarms. It also enables real-time transmission of flight data, which helps quickly identify faulty nodes in the cluster and dynamically replace them. In the IEEE 802.11 standards, the wireless communication protocol has been improved. For the high dynamics in UAV swarms, the RTS/CTS (request-to-send/clear-to-send) mechanism is used to reduce the hidden terminal problem and ensure the reliability of communication [32]. Meanwhile, adaptive rate control is introduced to dynamically adjust the communication rate to adapt to the changes among network nodes. On the other hand, the standard improves the dedicated channel allocation mechanism for air-to-air communication. Ad-hoc mode is supported between UAVs, reducing the dependence on ground base stations and realizing efficient tasking and data sharing. This mechanism is particularly suitable for air-to-air data exchange and task collaboration in UAV swarms. Moreover, when a UAV node fails, the network can readjust the communication path through dynamic routing to ensure that the cluster's mission continues without interruption.

1.4 Organization of the Book

This book provides a comprehensive exploration of UAV swarm collaboration technologies from a networking perspective, with a focus on the key components of these network technologies. It covers topics such as network topology construction, efficient routing algorithms, and resource management strategies. Additionally, this book addresses the opportunities and challenges brought by emerging network technologies to UAV swarm networks, including distributed learning, semantic communication, and deterministic networking. The structure of the book is illustrated in Fig. 1.4. Each chapter revolves around the collaborative cooperation of UAV swarms, introducing state-of-the-art technologies and presenting algorithmic designs to analyze the challenges and solutions in practical applications.

This book begins by exploring the adaptive clustering and networking of UAV swarms. In dynamic environments, UAV swarms must adjust network topologies and swarm structures according to task requirements and environmental conditions to maintain efficient communication and collaboration. In Chap. 2, we delve into adaptive clustering and networking methods, focusing on how adaptive mechanisms enable UAV swarms to achieve dynamic clustering, improving communication efficiency and network stability in dynamic environments.

Next, Chap. 3 addresses the intelligent UAV swarm routing problem, designing intelligent routing algorithms for UAV swarms to enhance routing efficiency, network resilience, and adaptability in decentralized, dynamic environments. Unlike

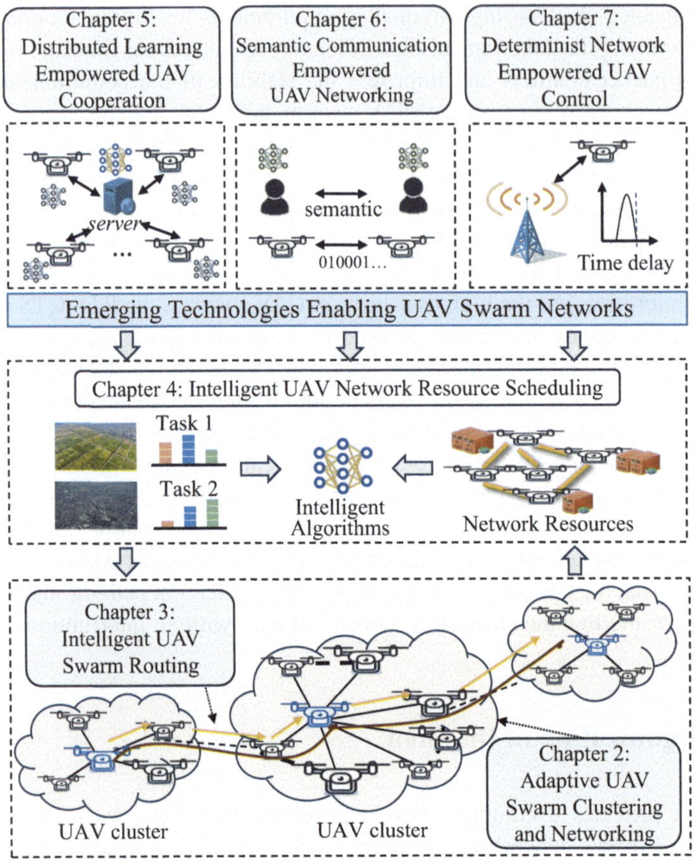

Fig. 1.4 Book organization

traditional static routing methods, intelligent routing enables UAVs to collaboratively learn optimal routing strategies, achieving joint optimization of network quality metrics such as lifetime, packet delivery rate, and throughput.

Following that, Chap. 4 discusses resource scheduling issues within UAV swarms. UAV swarms need to coordinate limited computational resources, communication bandwidth, and other factors to ensure efficient task execution. This chapter introduces intelligent resource scheduling algorithms, starting with virtualization techniques that abstract physical resources into virtual ones. This approach enables more flexible allocation of network resources to meet node demands and ensure optimal resource utilization.

Chapter 5 introduces semantic communication technology to enhance data transmission efficiency. UAVs can exchange critical semantic information more quickly, especially when facing large-scale data transfers and complex tasks, significantly improving communication efficiency and network response time. Given the dynamic changes in network topology and fluctuations in communication links, this chapter focuses on the challenge of training accurate semantic models in resource-constrained UAV devices.

Chapter 6 explores the application of distributed learning techniques in UAV swarm collaboration. Traditional UAV swarm coordination relies on centralized learning paradigms; however, by introducing techniques such as federated learning and distributed reinforcement learning, UAVs can collaborate efficiently without central control. This enables each UAV to independently learn and optimize decisions while sharing model parameters with other UAVs, thereby improving the overall performance of the UAV swarm network.

Finally, Chap. 7 discusses the application of deterministic networking in UAV swarm control. To ensure stability and real-time performance during high-precision tasks, deterministic networking technology can provide low-latency and highly reliable communication services, preventing task failure due to network delays or packet loss.

1.5 Summary

This chapter primarily provides an introduction to the background of the book. Firstly, we provide an overview of the fundamental concepts of UAV swarm networks and present research on UAV swarm networks facilitated by emerging technologies. Subsequently, we elaborate on the international standards and research developments of UAV swarm network technologies. Finally, the structure of the book is outlined.

References

1. C.-P. Wu, Y.-R. Li, J.-L. Wang, H.-P. Lin, L.-C. Wang, S.-S. Jeng, J.-Y. Chen, Learning-based downlink user selection algorithm for UAV-BS communication network, in *2020 IEEE 17th Annual Consumer Communications & Networking Conference (CCNC)* (IEEE, Piscataway, 2020), pp. 1–5
2. X. Dai, Y. Mao, T. Huang, N. Qin, D. Huang, Y. Li, Automatic obstacle avoidance of quadrotor UAV via CNN-based learning. Neurocomputing **402**, 346–358 (2020)
3. Y. Emami, H. Gao, K. Li, L. Almeida, E. Tovar, Z. Han, Age of information minimization using multi-agent UAVs based on AI-enhanced mean field resource allocation. IEEE Trans. Veh. Technol. **73**(9), 13368–13380 (2024)
4. Y. Hou, J. Zhao, R. Zhang, X. Cheng, L. Yang, UAV swarm cooperative target search: A multi-agent reinforcement learning approach. IEEE Trans. Intell. Veh. **9**, 568 (2023)
5. T. Li, S. Leng, Z. Wang, K. Zhang, L. Zhou, Intelligent resource allocation schemes for uav-swarm-based cooperative sensing. IEEE Internet Things J. **9**(21), 21570–21582 (2022)
6. E. Baskaya, M. Bronz, D. Delahaye, Fault detection & diagnosis for small UAVs via machine learning, in *2017 IEEE/AIAA 36th Digital Avionics Systems Conference (DASC)* (IEEE, Piscataway, 2017), pp. 1–6

7. H. Khalil, S.U. Rahman, I. Ullah, I. Khan, A.J. Alghadhban, M.H. Al-Adhaileh, G. Ali, M. ElAffendi, A UAV-swarm-communication model using a machine-learning approach for search-and-rescue applications. Drones **6**(12), 372 (2022)
8. L. Chen, R. Wang, Y. Cui, P. He, A. Duan, Joint client selection and model compression for efficient FL in UAV-assisted wireless networks. IEEE Trans. Veh. Technol. **73**(10), 15172–15184 (2024)
9. H. Yang, J. Zhao, Z. Xiong, K.-Y. Lam, S. Sun, L. Xiao, Privacy-preserving federated learning for UAV-enabled networks: Learning-based joint scheduling and resource management. IEEE J. Sel. Areas Commun. **39**(10), 3144–3159 (2021)
10. Z. Gao, J. Fu, Z. Jing, Y. Dai, L. Yang, MOIPC-MAAC: Communication-assisted multi-objective Marl for trajectory planning and task offloading in multi-UAV assisted MEC. IEEE Internet Things J. **11**(10), 18483–18502 (2024)
11. X. Liu, Y. Deng, T. Mahmoodi, Energy efficient user scheduling for hybrid split and federated learning in wireless UAV networks, in *ICC 2022-IEEE International Conference on Communications* (IEEE, Piscataway, 2022), pp. 1–6
12. H. Wang, H. Zhao, B. Ren et al., Cyber-physical framework for UAV intelligent communications. Sci. Sin. (Informationis) **52**(11), 2141–2154 (2022)
13. X. Luo, H.-H. Chen, Q. Guo, Semantic communications: Overview, open issues, and future research directions. IEEE Wirel. Commun. **29**(1), 210–219 (2022)
14. Y. Zhang, P. Zhang, J. Wei, H. Zhao, J. Xiong, Semantic communication for intelligent devices: Architectures and a paradigm. Sci. Sin. Inform. **52**(5), 907–921 (2022)
15. B. Varga, J. Farkas, F. Fejes, J. Ansari, I. Moldován, M. Máté, Robustness and reliability provided by deterministic packet networks (TSN and DetNet). IEEE Trans. Netw. Serv. Manag. **20**(3), 2309–2318 (2023)
16. J. Farkas, L.L. Bello, C. Gunther, Time-sensitive networking standards. IEEE Commun. Stand. Mag. **2**(2), 20–21 (2018)
17. A. Nasrallah, A.S. Thyagaturu, Z. Alharbi, C. Wang, X. Shao, M. Reisslein, H. ElBakoury, Ultra-low latency (ULL) networks: The IEEE TSN and IETF DetNet standards and related 5G ULL research. IEEE Commun. Surv. Tutorials **21**(1), 88–145 (2018)
18. J.H. Wang, Y. Zhou, Y. Xu, Deterministic network, in *Fundamentals of 6G Communications and Networking* (Springer, Cham, 2023), pp. 633–665
19. J. Fan, D. Xu, T. Zhang, R. Mumtaz, K. Yu, Deep reinforcement learning based dynamic time slot allocation in unmanned aerial vehicle, in *ICC 2024-IEEE International Conference on Communications* (IEEE, Piscataway, 2024), pp. 1310–1315
20. Y. Guan, L. Zhao, J. Hu, N. Lin, M.F. Alhamid, Softwarized industrial deterministic networking based on unmanned aerial vehicles. IEEE Trans. Ind. Inf. **17**(8), 5635–5644 (2020)
21. Z. Du, L. Tian, C. Xu, G. Zhao, Research on deterministic transmission services scheduling in UAV cluster network, in *2023 IEEE 23rd International Conference on Communication Technology (ICCT)* (IEEE, Piscataway, 2023), pp. 856–863
22. H. Shakhatreh, A.H. Sawalmeh, A. Al-Fuqaha, Z. Dou, E. Almaita, I. Khalil, N.S. Othman, A. Khreishah, M. Guizani, Unmanned aerial vehicles (UAVs): A survey on civil applications and key research challenges. IEEE Access **7**, 48572–48634 (2019)
23. G. Chmaj, H. Selvaraj, Distributed processing applications for UAV/drones: a survey, in *Progress in Systems Engineering: Proceedings of the Twenty-Third International Conference on Systems Engineering* (Springer, Cham, 2015), pp. 449–454
24. A.M. Vegni, V. Loscrí, C.T. Calafate, P. Manzoni, Communication technologies enabling effective UAV networks: a standards perspective. IEEE Commun. Stand. Mag. **5**(4), 33–40 (2021)
25. A.S. Abdalla, V. Marojevic, Communications standards for unmanned aircraft systems: The 3gpp perspective and research drivers. IEEE Commun. Stand. Mag. **5**(1), 70–77 (2021)
26. N.-B.-v. Le, H.-D. Thai, C.-W. Yoon, J.-H. Huh, Recent development of drone technology software engineering: A systematic survey. IEEE Access **12**, 128729–128751 (2024)
27. M. N. Bashir, S. Iqbal, Cooperative communication over UAVs-assisted networks: Review of techniques and challenges, in *2024 Advances in Science and Engineering Technology International Conferences (ASET)* (IEEE, Piscataway, 2024), pp. 1–7

28. R. Bajracharya, R. Shrestha, S. Kim, H. Jung, 6G NR-U based wireless infrastructure UAV: Standardization, opportunities, challenges and future scopes. IEEE Access **10**, 30536–30555 (2022)
29. D. Mishra, A.M. Vegni, V. Loscrí, E. Natalizio, Drone networking in the 6G era: A technology overview. IEEE Commun. Stand. Mag. **5**(4), 88–95 (2021)
30. M. K. Banafaa, Pepeoğlu, I. Shayea, A. Alhammadi, Z. A. Shamsan, M. A. Razaz, M. Alsagabi, S. Al-Sowayan, A comprehensive survey on 5g-and-beyond networks with uavs: Applications, emerging technologies, regulatory aspects, research trends and challenges. IEEE Access **12**, 7786–7826 (2024)
31. S. Javed, A. Hassan, R. Ahmad, W. Ahmed, R. Ahmed, A. Saadat, M. Guizani, State-of-the-art and future research challenges in uav swarms. IEEE Internet Things J. **11**(11), 19023–19045 (2024)
32. P. Cao, L. Lei, S. Cai, G. Shen, X. Liu, X. Wang, L. Zhang, L. Zhou, M. Guizani, Computational intelligence algorithms for uav swarm networking and collaboration: A comprehensive survey and future directions. IEEE Commun. Surveys Tutorials **26**(4), 2684–2728 (2024)

Chapter 2
Adaptive UAV Swarm Clustering and Networking

In contrast to terrestrial networks, UAV swarm networks are marked by dynamic topologies, unstable communication links, and inherent self-organizing behaviors, which present considerable challenges for their development and transmission. To address these issues, this chapter explores adaptive clustering and networking mechanisms for UAV swarms. By employing such strategically network clusters and organizational algorithms, both communication efficiency and network stability can be significantly enhanced in dynamic operational environments. To achieve this, we firstly introduce the Fission Spectral Clustering (FSC) strategy, which iteratively partitions the network through a fission process until predefined constraints are satisfied. This approach ensures adaptability to dynamic network conditions by dynamically adjusting clusters based on link effectiveness, ultimately improving communication performance [1]. Furthermore, considering the limited coverage range of individual UAVs, effective cooperative trajectory planning is essential for successful UAV swarm network deployment. Traditional trajectory planning methods, which depend on static user models, are insufficient for accommodating the fluctuating demands characteristic of dynamic environments. To address this shortfall, we propose a reinforcement learning architecture that incorporates centralized training with distributed execution. The experiment results demonstrate that the method significantly enhances the efficiency and adaptability of cooperative communication networks under dynamic conditions [2].

2.1 Fission Spectral Clustering Strategy for UAV Swarm Networks

The UAV swarm as the most powerful and exciting technology has quickly become a disruptive force reshaping how we live and work. UAV swarm refers to a coordinated unit of UAVs that work together to accomplish a specific task.

H. Yao, T. Mai, *UAV Swarm Cooperation*, Wireless Networks,
https://doi.org/10.1007/978-3-031-96444-2_2

Advantages to UAV swarm include easy deployable, time-savings, and reduction in operational expenses. Benefiting from the above advantages, the UAV swarm can be widely used in many fields ranging from disaster management to military operations [3]. However, compared to the terrestrial networks, UAV swarm networks possess the characteristic of dynamic topology, unstable links, self-organizing, and lack of central control. These characteristics result in a new network paradigm, termed flying ad-hoc networks (FANETs).

However, because of its unique nature, many network protocols of fixed networks will be no longer suitable in FANETs. Especially, the current flooding-based routing protocol extremely limits the scale of FANET. Once the network topology changes, flooding will be triggered to establish the new routing table, and therefore bring huge communication overhead in such dynamic networks.

Clustering, as a network organization method, can effectively prevent wide-range network flooding. The UAVs swarm is divided into several clusters, where each cluster is composed of one elected Cluster Head (CH) and several Cluster Members (CMs) [4]. In the UAV network, CH is responsible for cooperative information sharing between clusters and the management of all the CMs within the same cluster [5]. Based on this clustered structure, information transmission is hierarchical, which greatly reduces flooding overhead.

The typical UAV swarm clustering algorithms include the lowest ID algorithm (LOWID) [6], the highest node degree algorithm (HIGHD) [7], and the weight-based clustering algorithm (WCA) [8]. However, these algorithms neglect wireless link attributes (e.g. wireless link quality, wireless link connection duration, etc.). Therefore, the communication performance of the UAV swarm is always unsatisfactory after clustering. How to maintain high communication performance in the clustered UAV swarm is a critical challenge.

To address this challenge, spectral clustering (SC) becomes a viable option. SC is an unsupervised technology based on graph theory [9], which has been applied in the network field, such as load balancing [10], wireless communication [11], and so on. It treats UAVs as graph's vertexes and the wireless link between UAVs as graph's edges. The weight value of the edge reflects the communication quality of the link. SC transforms the UAV swarming problem into a graph cut problem. It aims to maximize the edge weights in the subgraph and minimize the edge weights between the subgraphs. Therefore, the high-performance communication link will be retained during clustering.

However, the SC algorithm is unstable in the clustering process, which performs in uncontrollable size and structure of each cluster. To achieve a more reasonable cluster result, in this part, we design a FSC strategy for UAV networks. FSC generates clusters by fission and stops the cluster fission process when it satisfies the constraints of size and structure. Moreover, the traditional SC algorithm adopts a static clustering scheme, and therefore cannot be applied to dynamic networks. Therefore, we propose that the effective value of the link represents the communication capability of the link within a specific time slot, enabling the capture of time-sequential wireless link attributes.

The major contributions of this part can be summarized as follows:

- We model the time-sequential attributes of nodes and edges as the link effective value, which can be further applied to the proposed FSC clustering algorithm.
- We propose the FSC algorithm to cluster the UAV swarm. FSC continuously clusters the UAV swarm by cutting off the edges with small weight values and keeping the edges with large weight values, where the cluster will fission until it satisfies the constraints of size and structure.
- We propose an on-demand cluster maintenance mechanism. It decides whether each cluster needs to fission by considering its size and structure in each time slot, therefore greatly reducing the computing overhead.

2.1.1 *System Model and Problem Formulation*

In this section, we first introduce the network system model. Then, we model the channel and relevant properties. Finally, we present the optimization problem of communication performance. For a clear exposition, the primary notations are summarized in Table 2.1.

2.1.1.1 System Model

As shown in Fig. 2.1, N UAVs are deployed in a 3D space. The set of UAVs are indicated as $\mathcal{N} = \{1, 2, \ldots, N\}$. N UAVs are divided into several clusters. In each UAV swarm, we elect one CH to gather the information of CMs in the network. We abstract the UAV network as a graph $G = (V, E)$, where V represents the N UAV nodes and E represents the set of edges. We define $(i, j) \in E^t$ if UAV i and UAV j are neighbors. Otherwise, $(i, j) \notin E^t$. The neighbors are conditioned on the minimum threshold value of the signal-to-noise ratio (SINR) between two UAVs.

In high dynamic UAV networks, the perception of features is discrete with a one-cycle interval [12]. Therefore, in this part, we analyze the UAV network based on the time slot. We set the maximum flight time of all UAVs as T and divide it into several equally spaced time slots. Let τ denote the length of each time slot and the s-th time slot is denoted by $(T_s, T_s + \tau)$. To further describe the connection relationship between UAVs, we model the mobility and channel as follows.

2.1.1.2 Mobility Model

UAVs are deployed in a three-dimensional space and the range is $(x_{\min}, y_{\min}, z_{\min}) \sim (x_{\max}, y_{\max}, z_{\max})$, where $x_{\min}, y_{\min}, z_{\min}, x_{\max}, y_{\max}, z_{\max}$ represent the minimum and maximum values for the x-axis, y-axis, and z-axis, respectively. The position $\mathbf{p}_{i,T_s}$ and speed $\mathbf{u}_{i,T_s}$ of UAV i at time T_s can be denoted by

Table 2.1 Related definitions

Notations	Descriptions
T	Total flight time of UAV
τ	Length of each time-slot
N	The set of N UAVs
$G = (V, E)$	UAV network graph, where V represents the N UAV nodes and E represents the set of edges
P	The transmission power of UAV
h	Channel gain
α	Mean path-loss exponent
σ^2	Noise power
d_{i,j,T_s}	Physical distance between UAV i and UAV j at time T_s
B	Channel bandwidth
γ	SINR threshold
ε	The maximum effective communication range of UAVs
$\mathbf{p}_{i,T_s}$	Three-dimensional position vector of UAV i at time T_s
$\mathbf{u}_{i,T_s}$	Three-dimensional speed vector of UAV i at time T_s
L_{i,j,T_s}	The link connection duration of UAV i and UAV j in time-slot τ from T_s
ε_{fs}	Energy consumption of the amplifier
E_{elec}	Energy consumed by the transmitter
E_{i,j,T_s}	Energy consumption of sending data packets in time T_s
β^s_{i,j,T_s}	Stability effectiveness factor
β^e_{i,j,T_s}	Cost effectiveness factor
θ	Effectiveness adjustable factor
β_{i,j,T_s}	Effectiveness factor
l_{i,j,T_s}	The number of bits sent between UAVs in time T_s
$n_{\max}$	The maximum UAV number in the cluster
Q_{i,j,T_s}	The link effective value of the channel in $(T_s, T_s + \tau)$

$$\begin{cases} \mathbf{p}_{i,T_s} = (p^x_{i,T_s}, p^y_{i,T_s}, p^z_{i,T_s}) \\ \mathbf{u}_{i,T_s} = (u^x_{i,T_s}, u^y_{i,T_s}, u^z_{i,T_s}) \end{cases}, \tag{2.1}$$

where p^x_{i,T_s}, p^y_{i,T_s}, p^z_{i,T_s}, u^x_{i,T_s}, u^y_{i,T_s}, u^z_{i,T_s} represent the position component and speed component of UAV i for the x-axis, y-axis, z-axis respectively. Thus, the position $\mathbf{p}_{i,T_{s+1}}$ of UAV i at time T_{s+1} can be denoted by

$$p^\kappa_{i,T_{s+1}} = p^\kappa_{i,T_s} + u^\kappa_{i,T_s}, \tag{2.2}$$

where $\kappa \in \{x, y, z\}$. In order to prevent UAVs from moving beyond the boundaries, a bouncing mechanism is designed. If a position component at time T_{s+1} exceeds the boundary, the speed component need to be reversed, which can be denoted by

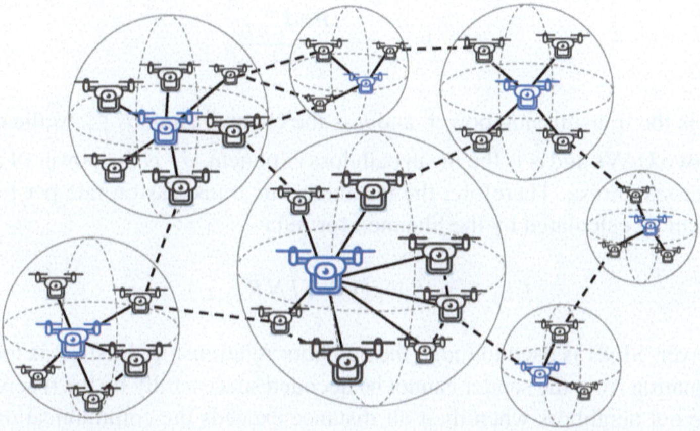

Fig. 2.1 UAV network clustering

$$p^{\kappa}_{i,T_{s+1}} = \begin{cases} p^{\kappa}_{i,T_s} + u^{\kappa}_{i,T_s} & p^{\kappa}_{i,T_s} + u^{\kappa}_{i,T_s} \in (\kappa_{\min}, \kappa_{\max}) \\ p^{\kappa}_{i,T_s} - u^{\kappa}_{i,T_s} & otherwise \end{cases} \tag{2.3}$$

In order to simulate the heterogeneity in the mobility of different UAVs, parameters with different distributions are employed to adjust the speed changes of UAVs. The speed $\mathbf{u}_{i,T_s} = (u^x_{i,T_s}, u^y_{i,T_s}, u^z_{i,T_s})$ can be transformed from a three-dimensional coordinate system to a spherical coordinate system, which can be denoted by

$$\begin{cases} u^x_{i,T_s} = |\mathbf{u}_{i,T_s}| \sin\theta_i \cos\varphi_i \\ u^y_{i,T_s} = |\mathbf{u}_{i,T_s}| \sin\theta_i \sin\varphi_i \\ u^z_{i,T_s} = |\mathbf{u}_{i,T_s}| \cos\varphi_i \end{cases} . \tag{2.4}$$

$|\mathbf{u}_{i,T_s}|$ is magnitude of the speed vector, which obeys the Gaussian distribution $|\mathbf{u}_{i,T_s}| \sim N(\mu_i, \sigma_i^2)$. θ_i is the angle between the speed of UAV and the z-axis, which obeys a uniform distribution $\theta_i \sim U(0, \pi)$. φ_i is the angle between the speed of UAV and the x-axis, which obeys a uniform distribution $\varphi_i \sim U(0, 2\pi)$. Compared to a uniform speed UAV model, this kind of modeling adheres more closely to physical principles, accounting for phenomena such as sharp turns, sudden braking, and body vibrations.

2.1.1.3 Channel Model

When UAV i transmits signals to UAV j at time T_s, the SINR of UAV j can be denoted by

$$SINR_{i,j,T_s} = \frac{Phd_{i,j,T_s}^{-\alpha}}{\sigma^2}, \tag{2.5}$$

where P is the transmission power, and h is the channel gain. $d_{i,j,T_s}^{-\alpha}$ is the distance between two UAVs and α is the mean path loss exponent. σ^2 is the power of additive white Gaussian noise. Therefore, the instantaneous transmission rate per bit of the channel can be calculated by the Shannon formula

$$C_{i,j,T_s} = B\log(1 + SINR_{i,j,T_s}). \tag{2.6}$$

Moreover, SINR is used to judge the neighbor relationship. If SINR is too small, the information from the sender cannot be decoded successfully by the receiver. Two UAVs are not neighbors when their air distance exceeds the communication range. Given a threshold γ, the maximum communication range ε can be calculated by

$$SINR_{i,j,T_s} = \frac{Ph\varepsilon^{-\alpha}}{\sigma^2} = \gamma \Longrightarrow \varepsilon = \sqrt[\alpha]{\frac{Ph}{\gamma\sigma^2}}. \tag{2.7}$$

2.1.1.4 Link Connection Duration Model

The channel capacity calculated by the Shannon formula gives the maximum speed of data transportation. However, it is an instantaneous value. Therefore, to calculate the throughput in the time slot, we need to consider the mobility of the UAV swarm and evaluate the link connection duration of UAVs. Meanwhile, the link connection duration among different UAVs can be considered as a reference indicator for the stability of the UAV network [12].

The link connection duration of UAV swarm is evaluated in $(T_s, T_s + \tau)$. As shown in Fig. 2.2, we assume that UAV j is a neighbor of UAV i at time T_s. We define $\Delta d_{i,j,T_s}$ as the relative distance difference between i and j from time T_{s-1} to T_s, which can be denoted by

$$\begin{aligned}\Delta d_{i,j,T_s} &= d_{i,j,T_s} - d_{i,j,T_{s-1}} \\ &= \left\|\mathbf{p}_{i,T_s} - \mathbf{p}_{j,T_s}\right\|_2 - \left\|\mathbf{p}_{i,T_{s-1}} - \mathbf{p}_{j,T_{s-1}}\right\|_2.\end{aligned} \tag{2.8}$$

When $\Delta d_{i,j,T_s} < 0$, the maximum relative distance that can stay connection is $\varepsilon + d_{i,j,T_s}$, where ε is the maximum communication range in Eq. (2.7). Otherwise, the maximum relative distance is $\varepsilon - d_{i,j,T_s}$.

Then, the connection time L_{i,j,T_s} between i and j can be denoted by

$$L_{i,j,T_s} = \begin{cases} \frac{\varepsilon + d_{i,j,T_s}}{u_{i,j,T_s}} & \Delta d_{i,j,T_s} < 0 \\ \infty & \Delta d_{i,j,T_s} = 0 \\ \frac{\varepsilon - d_{i,j,T_s}}{u_{i,j,T_s}} & \Delta d_{i,j,T_s} > 0 \end{cases}. \tag{2.9}$$

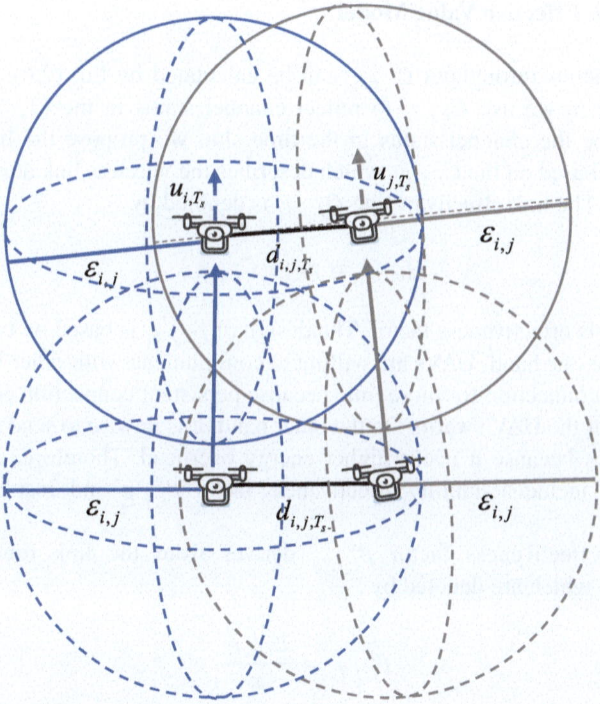

Fig. 2.2 The change of relative distance between two nodes

$\overline{u_{i,j,T_s}}$ is the relative speed, which is the absolute value of the projection of the UAV speed difference in the relative position direction

$$\overline{u_{i,j,T_s}} = \left| \frac{\left(\mathbf{p}_{i,T_s} - \mathbf{p}_{j,T_s}\right) \cdot \left(\mathbf{u}_{i,T_s} - \mathbf{u}_{j,T_s}\right)}{\left\|\mathbf{p}_{i,T_s} - \mathbf{p}_{j,T_s}\right\|_2} \right| . \tag{2.10}$$

Finally, we analyze the connection duration in a time slot, and Eq. (2.9) is revised to

$$L_{i,j,T_s} = \begin{cases} L_{i,j,T_s} & L_{i,j,T_s} < \tau \\ \tau & L_{i,j,T_s} \geq \tau \end{cases} . \tag{2.11}$$

Therefore, the throughput between UAV i and j can be denoted by

$$l_{i,j,T_s} = \int_{T_s}^{T_s + L_{i,j,T_s}} B \log(1 + SINR_{i,j,t}) dt. \tag{2.12}$$

2.1.1.5 Link Effective Value Model

The instantaneous throughput C_{i,j,T_s} can be calculated by Eq. (2.6). However, it is deviant when we use C_{i,j,T_s} to reflect channel status in the $(T_s, T_s + \tau)$. To better describe the channel status in the time slot, we propose the link effective value Q_{i,j,T_s} based on the C_{i,j,T_s}, which describes the wireless link attributes in the $(T_s, T_s + \tau)$. The link effective value Q_{i,j,T_s} is denoted by

$$Q_{i,j,T_s} = \beta_{i,j,T_s} C_{i,j,T_s}, \tag{2.13}$$

where β_{i,j,T_s} is effectiveness factor. The design of β_{i,j,T_s} is based on two considerations. On the one hand, UAVs are willing to communicate with other UAVs which can keep the connection for more time because persistent connection contributes to the stability of the UAV swarm. On the other hand, it is unwise to send packets over long distances because it needs higher energy overhead. Therefore, effectiveness factor β_{i,j,T_s} includes stability effectiveness factor β^s_{i,j,T_s} and cost effectiveness factor β^e_{i,j,T_s}.

Stability effectiveness factor β^s_{i,j,T_s} derives from the link time L_{i,j,T_s} in Sect. 2.1.1.4, which are denoted by

$$\beta^s_{i,j,T_s} = \frac{L_{i,j,T_s}}{\tau}. \tag{2.14}$$

Cost effectiveness factor β^e_{i,j,T_s} consider the energy consumption of transmitting packets. In [13], the sending energy E_{i,j,T_s} in the free space model can be formulated as

$$E_{i,j,T_s} = lE_{elec} + l\varepsilon_{fs} d^2_{i,j,T_s}, \tag{2.15}$$

where E_{elec} is the energy consumed by the transmitter, ε_{fs} is the energy consumption of the amplifier and l is the number of bits. Here, we will set l as 1 bit when calculating the cost effectiveness factor. In Sect. 2.1.1.3, we give the maximum effective communication range ε. The sending energy when the communication range is ε can be the measure of cost. The maximum energy consumes E_ε is

$$E_\varepsilon = lE_{elec} + l\varepsilon_{fs}\varepsilon^2. \tag{2.16}$$

Therefore, the cost effectiveness factor β^e_{i,j,T_s} is denoted by

$$\beta^e_{i,j,T_s} = 1 - \frac{E_{i,j,T_s}}{E_\varepsilon}. \tag{2.17}$$

The effectiveness factor β_{i,j,T_s} is a weighted value, which is denoted by

$$\beta_{i,j,T_s} = \theta\beta^s_{i,j,T_s} + (1-\theta)\beta^e_{i,j,T_s}, \tag{2.18}$$

where θ is an adjustable factor. The link effective value Q_{i,j,T_s} reflects wireless link attributes of UAV swarm in the $(T_s, T_s + \tau)$.

2.1.1.6 Problem Formulation

Our goal is to find an excellent UAV swarm structure to maximize the average cluster throughput in the UAV swarm network. The throughput of the cluster is the information amount that CH gathers from CMs in the time slot. Therefore our objective problem is mathematically formulated as follows

$$
\begin{aligned}
&\max \frac{1}{k} \sum_{k \in K} \sum_{i \in C_k/\{k\}} l_{k,i,T_s}, \\
&s.t. \quad C_1 : SINR_{k,i} \geq \gamma, \\
&\qquad C_2 : \bigcup_{k=1}^{K} C_k = N, \\
&\qquad C_3 : \bigcap_{k=1}^{K} C_k = \phi, \\
&\qquad C_4 : 1 \leq |C_k| \leq n_{\max}, \forall k \in \mathcal{K}.
\end{aligned}
\tag{2.19}
$$

where $\mathcal{K} = \{1, 2, \ldots, K\}$ is CH set and $C = \{C_1, \ldots, C_K\}$ is cluster set.

C_1 constraint shows that a large enough SINR is the premise of the communication between CH and CMs. C_2 constraint and C_3 constraint propose that all UAVs must belong to one cluster and can only belong to one cluster. C_4 constraint limits the number of UAVs in each cluster, where $n_{\max}$ is the maximum UAV number in the cluster. On the one hand, it can prevent a high collision rate in areas with very high node density when setting a limit UAV number value [14]. On the other hand, CH will cost a lot of energy when gathering information from too many CMs.

2.1.2 Fission Spectrum-Based Swarm Clustering Algorithm

In this part, we propose a clustering algorithm called fission spectrum clustering (FSC). The proposed clustering algorithm includes the initial clustering stage and cluster maintenance stage. The initial clustering stage includes cluster establishment and the CH selections. The cluster maintenance stage includes cluster changing, partial clustering, and re-clustering. When the network is first established, the initial clustering scheme is performed. Then, we implement maintaining clustering at the beginning of each time slot.

Algorithm 1 Fission spectral clustering

Input: Graph $G = (V, E)$, Maximum UAV number $n_{\max}$
Output: Clustering results C, The number of clusters K

```
Cluster result set results = ∅
Compute similarity matrix W of all UAVs by using Eq. (2.20)
Compute degree matrix D of all UAVs by using Eq. (2.22)
Compute Laplacian matrix L of all UAVs by using Eq. (2.21)
K = ⌈N / n_max⌉
Compute the smallest K eigenvalues of L and get eigenvector matrix F = {f_1, f_2, ..., f_K}
Decompose matrix F by row into {y_1, y_2, ..., y_N}
Cluster {y_1, y_2, ..., y_N} in K-Means and get the clustering result.
Put the result into the result and check the result in results
while result in results do
    while the size of result than n_max do
        K' = ⌈|result| / n_max⌉
        Split result into K' sub-cluster
        Update results
    end while
end while
while result in results do
    while there are no CH candidates in result do
        K' = 2
        Split result into K' sub-cluster
        Update results
    end while
end while
The number of clusters K is the size of results
C = {C_1, ..., C_K} = results
return Clustering results C and the number of clusters K
```

2.1.2.1 Initial Clustering Stage Based on Fission Spectrum Clustering

FSC is designed based on SC and SC is used to perform the first round of clustering. The SC involves selecting a cluster number, constructing a similarity matrix, creating a Laplace matrix, decomposing eigenvalues and clustering. To restrict the size of the cluster while ensuring throughput, the cluster number K is set to $\left\lceil \frac{N}{n_{\max}} \right\rceil$, where $\lceil . \rceil$ is rounding up. Similarity matrix $W \in R^{N \times N}$ is the important parameter, which quantitatively describes the link between nodes. In order to improve the network throughput after clustering, the link effective value Q_{i,j,T_s} is used to construct the similarity matrix W. Therefore, W is set to

$$w_{i,j} = \begin{cases} Q_{i,j,T_s} & SINR_{i,j,T_s} > \gamma \\ 0 & SINR_{i,j,T_s} \leq \gamma \end{cases}, \tag{2.20}$$

Then, the normalized Laplacian matrix L can be described as

$$L = I - D^{-\frac{1}{2}} W D^{-\frac{1}{2}}, \tag{2.21}$$

where D is called the degree matrix and it is a diagonal matrix formed by D_{ii}. D_{ii} is denoted as

$$D_{ii} = \sum_{j=1}^{N} w_{i,j}. \tag{2.22}$$

Compute the first K generalized eigenvalue problem $Lf = \lambda Df$, then take eigenvectors $f_1, f_2, \ldots, f_K$ as the columns of eigenvector matrix $F \in N \times K$. Each row in F is also considered a K-dimensional vector. Therefore F is decomposed into $y_1, y_2, \ldots, y_N$. Then cluster the vectors $y_i (i = 1, 2, \ldots, N)$ with the traditional K-Means clustering algorithm into clusters $A = \{A_1, \ldots, A_K\}$. Finally, the clustering results can be obtained according to the corresponding position

$$C_j = \{j | y_j \in A_i\}. \tag{2.23}$$

However, the initial clustering results were unsatisfactory, as they failed to ensure the validity of C_1 and C_4 constraint in Eq. (2.19) for each cluster. For constraint C_1, such a rough clustering method cannot ensure that a CH exists in the cluster which can communicate with all CMs. For constraint C_4, the number of UAVs in the cluster is likely to surpass the limit value.

Hence, FSC enhances traditional SC and conducts fission clustering on the results to satisfy the constraints of UAV swarm clustering. On the basis of the first clustering result, FSC algorithm will check whether each cluster satisfies our constraints. Supposed that the number of UAVs in cluster C_k is more than the limit value $n_{\max}$, this cluster will be split. The method of splitting this cluster C_k is executing SC algorithm again and this time K is $\left\lceil \frac{N_{C_k}}{n_{\max}} \right\rceil$. The clusters will undergo continuous fission and inspection iterations until the constraints are met C_4. Next, our proposed algorithm will check whether there are CH candidates in each cluster. CH candidates are the UAVs in the cluster that can communicate with all CMs. Each cluster should exist one or more CH candidates. If there are no CH candidates, this cluster will be split into two, i.e. K is 2. The choice of $K = 2$ is made to maximize the cluster size while ensuring satisfaction of the C_4 constraint, aiming to enhance the throughput of each cluster. The clusters will undergo continuous fission and inspection iterations until the constraints are met C_1. Figure 2.3 and Algorithm 1 describes this process.

Then it is the turn to elect CH from CH candidates. CH is responsible for gathering the information of CMs in the cluster and therefore the selection of CH is related to the throughput of the cluster. Therefore, CH should have the maximum link effective value in the CH candidates, and it is denoted by

$$k = \arg\max_{k \in cand} \sum_{i \in C_{k \setminus \{k\}}} Q_{k,i,T_t}, \tag{2.24}$$

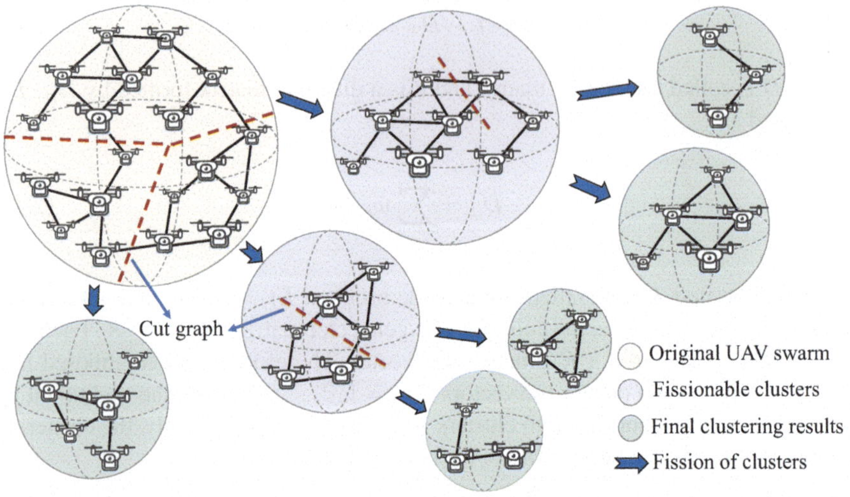

Fig. 2.3 The fission process of FSC. A cluster needs to fission when it doesn't satisfy the size and structure constraints

---

Algorithm 2 CH selection

Input: Cluster C_k
Output: The CH k of C_k

CH candidates set $cand = \phi$
for i in C_k **do**
  $count = 0$
  for j in C_k **do**
    if $i \neq j$ && $SINR_{i,j} \geq \gamma$ **then**
      $count = count + 1$
    end if
  end for
  if $count == len(C_k) - 1$ **then**
    put i into $cand$
  end if
end for
$k = \underset{k \in cand \cap i \in C_{k \setminus \{k\}}}{\arg\max} \; Q_{k,i,T_s}$
return CH k

---

where *cand* is CH candidates set. Algorithm 2 describes this process.

2.1.2.2 Cluster Maintenance Stage Based on Partial Fission Spectrum Clustering

The high dynamics of UAVs reduce the stability of the cluster structure. For this reason, we check our UAV network at intervals τ. We can choose to cluster the UAV swarm every τ by FSC. However, this method will cause a heavy computation

Algorithm 3 Cluster maintenance

Input: Cluster $C = \{C_1, \ldots, C_K\}$, Simulation duration T

Calculate the throughput $throughput$ of initial clustering
Set $threshold = 0.5 * throughput$
Set $new\_throughput = throughput$
$t = \tau$
while $t \leq T$ **do**
  if $new\_throughput < threshold$ **then**
    Execute Algorithm 1 for this cluster.
    $t = t + \tau$
    break
  end if
  Update the position of all UAVs.
  while a CM UAV cannot communicate with its CH **do**
    Get its new neighbor set $new\_neighbor$ by Eq. (2.5).
    Calculate link effective values of UAVs in $new\_neighbor$ and change itself own cluster number to the cluster number of the UAV which has the maximum effective value, i.e. $cluster\_number_j = \arg\max_{i \in new\_neighbor} Q_{i,j,t}$
  end while
  Check the statue of clusters
  if the size of the cluster than $n_{\max}$ or there are no CH candidates in cluster **then**
    Execute Algorithm 1 for this cluster.
  end if
  $t = t + \tau$
  Calculate the new average cluster throughput $new\_throughput$ of the current UAV swarm.
end while

overhead. Our proposed algorithm belongs to a centralized algorithm, which needs to build a complex UAV network and requires a great deal of calculation.

Although the UAV network is dynamic, many UAVs can still maintain communication with their clusters in the next time slot. These UAVs are unnecessary to be clustered again. Therefore, in the cluster maintenance stage, we only need to focus on the UAVs which are out of communication with their clusters. Note that the unconnected network is not considered in this article, i.e. there are no outlier UAVs in the network when UAVs move.

The cluster maintenance stage includes three steps: cluster changing, partial clustering, and re-clustering.

Cluster Changing The established association between a CM and its CH may be broken after the initial clustering stage because of the limited transmission range and random node mobility [14]. At the next time slot, if a CM UAV i finds that it moves out of its CH's coverage area, it will consider joining another cluster. It searches its new neighbors in its coverage scope ε and calculates the link effective values of them by Eq. (2.13). Suppose that UAV j has the maximal link effective value in the neighbors of UAV i. Then UAV i modifies its cluster number to UAV j's cluster number.

Partial Clustering After updating some UAVs' cluster numbers, our proposed algorithm will check the status of each cluster again. Each cluster can be regarded as a partial UAV network. If a cluster doesn't satisfy the constraints C_1 or C_4, it will execute the FSC algorithm. Compared with the initial clustering stage, the scale of the network is small and the computational complexity will be reduced a lot.

Re-clustering In our proposed algorithm, clusters will persistently fission. As a consequence, the size of each cluster will smaller and smaller, which will reduce the capability of each cluster. The algorithm takes the average cluster throughput in the initial clustering stage as the standard and stipulates a half of the standard is the threshold. If the average cluster throughput is under the threshold in a certain time slot, the capability of the current cluster structure is considered inefficient. Then, in the next time slot, all UAVs need to be clustered and the UAV swarm returns to the initial clustering stage. Algorithm 3 describe cluster maintenance stage.

2.1.3 Simulation Results and Discussions

In this section, we present the simulation results to evaluate the validity of the proposed algorithm. Our experiments simulate with Windows 10 operating system, which configuration is Intel Core i5 10400 central processing unit (CPU), 2.9 GHz.

For the proposed algorithm, we evaluate its performance, convergence, and stability. Then FSC is compared with other algorithms. Four UAV network clustering algorithms, termed Weighted Clustering Algorithm (WCA) [15], Gaussian mixture model (GMM) [16], Affinity Propagation (AP) [17], and Agglomerative Clustering (AC) [18] are set as the baseline.

2.1.3.1 Simulation Settings

In our experiment, the simulation environment is set to 2000×2000 m square area. The flight altitude of UAVs is between 100 and 150 m. The SINR threshold γ is defined by 0 dB which limits the communication range of UAVs. The noise power σ^2 is -100 dBm. The transmission power P of UAV is set to 20 dBm and the mean path-loss exponent α is set to 4. The channel bandwidth B is set to 10 MHz and the channel gain h is set to 0.5. The time slot τ is set to 10 s and we check the status of the network every 10 s. The simulation duration T is 300 s.

The detailed parameters setting can be found in Table 2.2. In the experiments, the defaults of UAVs' number, UAVs' speed and $n_{\max}$ are 100, 20 m/s, 20 unless otherwise noted.

Table 2.2 Simulation parameters

Parameter	Value
Simulation area	1000 m * 1000 m * (100–150) m
Number of UAVs	50–250
The Speed of UAVs	10–30 m/s
SINR threshold (γ)	0 dbm
Transmission power (P)	20 dbm
Mean path-loss exponent (α)	4
Channel bandwidth (B)	10 MHz
Channel gain (h)	0.5
Noise power (σ^2)	−100 dbm
Energy consumption of the amplifier (ε_{fs})	10×10^{-12} J/bit*m^2
Energy consumed by the transmitter (E_{elec})	50×10^{-9} J/bit
Adjustable factor (θ)	0.5
Time slot (τ)	10 s
Simulation duration (T)	300 s

2.1.3.2 Cluster Result

As shown in Fig. 2.4, we present cluster results in different maximum UAV numbers $n_{\max}$ in the initial clustering stage. Figure 2.4a shows the initial distribution of 100 UAVs at T_s. Then we set $n_{\max}$ as 15, 20, 25 and observe the change of cluster structure in (b), (c) and (d), where the $\star$ represents CH in each cluster. As the $n_{\max}$ increases and decreases, clusters can flexibly merge and fission, which is corresponding to tasks of different scales. What is more, we can notice that the ascription of UAVs in the high-density area is easier to change when $n_{\max}$ changes. It indicates that our clustering algorithm is an effective measure to reduce flooding interference in a high-density area.

2.1.3.3 Performance Analysis

In this part, we evaluate the algorithm's performance. The performance indicator is the average cluster throughput and we take the number of bits per cluster as the average cluster throughput by using Eq. (2.19).

Figure 2.5 shows the average cluster throughput of the different $n_{\max}$ when we execute the FSC algorithm in the initial clustering stage. Generally, a lower $n_{\max}$ will lead to more clusters when the number of UAVs is the same, which is an effective measure to reduce flooding interference. However, it cost the communication performance of the UAV swarm. Besides, with the number of UAVs increasing, the average cluster throughput increases. However, this increasing trend is converging because the number of clusters also increases with the average cluster throughput increasing.

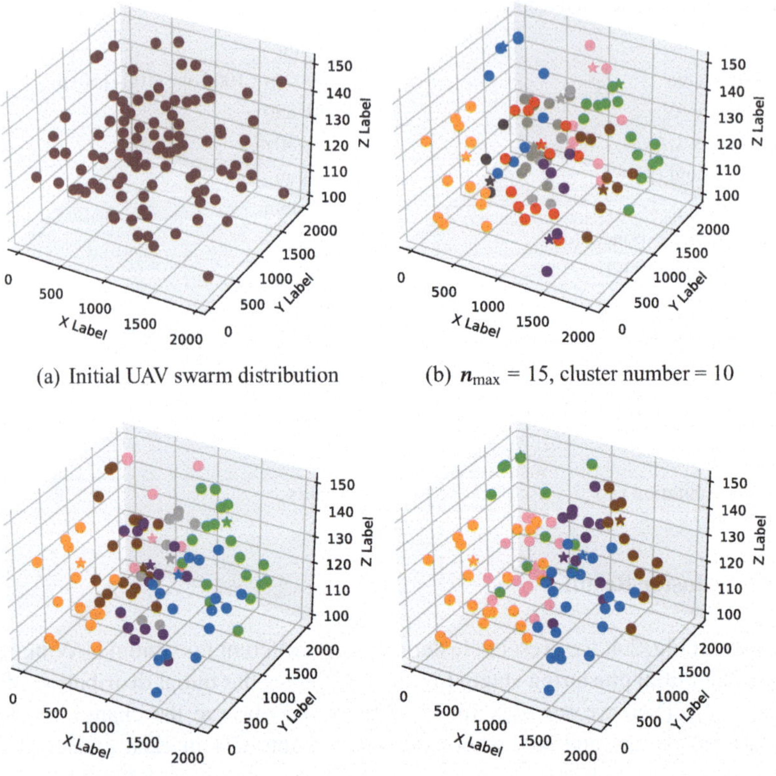

(a) Initial UAV swarm distribution

(b) n_{max} = 15, cluster number = 10

(c) n_{max} = 20, cluster number = 7

(d) n_{max} = 25, cluster number = 6

Fig. 2.4 The clustering results of UAV swarm under different n_{max}. **(a)** Initial UAV swarm distribution. **(b)** $n_{max} = 15$, cluster number = 10. **(c)** $n_{max} = 20$, cluster number = 7. **(d)** $n_{max} = 25$, cluster number = 6

Figure 2.6 shows the average cluster throughput under different n_{max} varies with time. With the n_{max} increasing, the average cluster throughput increases. This is mainly caused by the decrease in the number of clusters with the n_{max} increasing. At the beginning of simulation time, the UAV network maintains its cluster structure by fissioning the cluster. However, the average cluster throughput decreases as the increase of clusters. We have to re-cluster to improve the average cluster throughput when the average cluster throughput is lower than the threshold, which is a cycle process. The dotted line represents the re-clustering of the UAV swarm in this time slot.

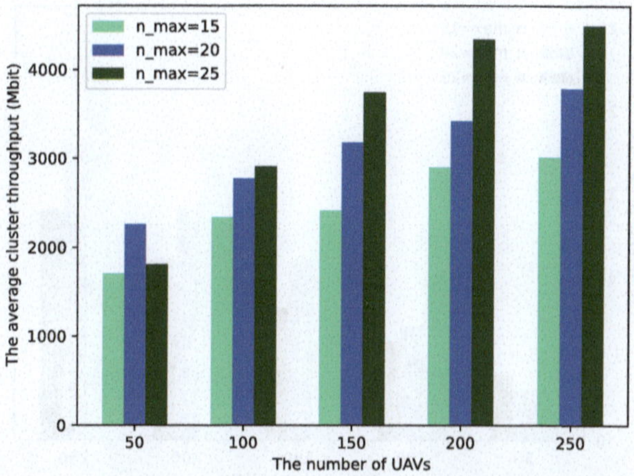

Fig. 2.5 The average cluster throughput under different n_{max} when using FSC in the initial clustering stage

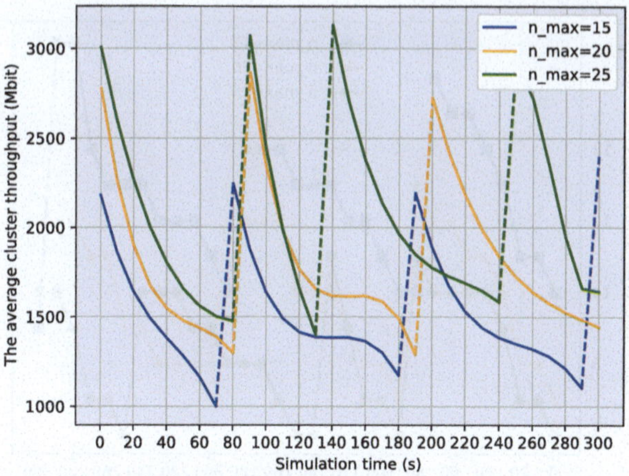

Fig. 2.6 The average cluster throughput under different n_{max} when using FSC in the cluster maintenance stage

2.1.3.4 Convergence Analysis

Our proposed algorithm is an adaptive fission algorithm. Clusters will stop fission when the UAV network structure satisfies the constraints. We don't expect unrestricted fission because too many clusters will decrease the amount of UAVs in each cluster, which weakens the communication performance of each cluster. Therefore, our algorithm is convergent.

Figure 2.7 shows the number of clusters in the different numbers of UAVs and n_{max}. The number of clusters increases with the number of UAVs increasing. We

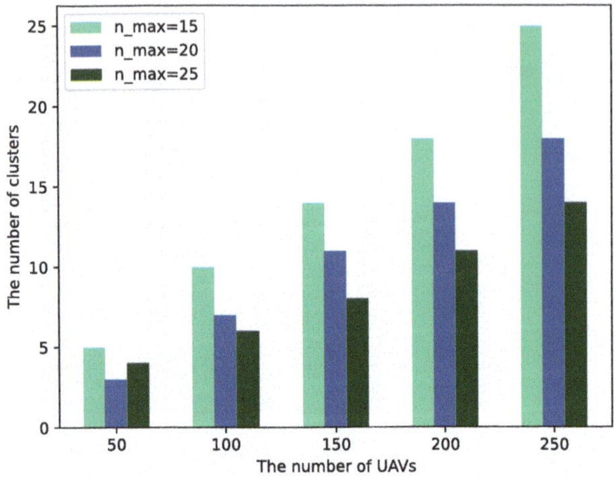

Fig. 2.7 The number of clusters under different $n_{\max}$ when using FSC in the initial clustering stage

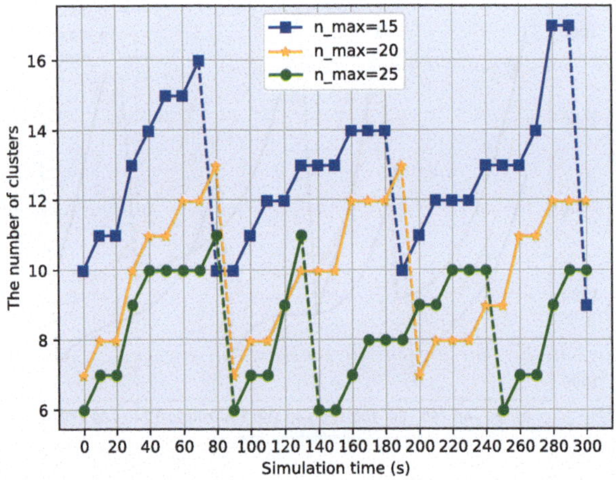

Fig. 2.8 The number of clusters under different $n_{\max}$ when using FSC in the cluster maintenance stage

can notice that usually the actual number of clusters is more than $\left\lceil \frac{|UAVs|}{n_{\max}} \right\rceil$. This is because UAVs have high dynamics and the UAV network has irregularity. If we cannot satisfy the constraints when we get a clustering result, we must continue to split this cluster.

Figure 2.8 shows the number of clusters under different $n_{\max}$ varies with time, which is corresponding to Fig. 2.6. From Fig. 2.8, We find that the UAV network in our algorithm tries its best to maintain the current cluster structure and Only a small number of clusters that don't meet the constraints will undergo fission.

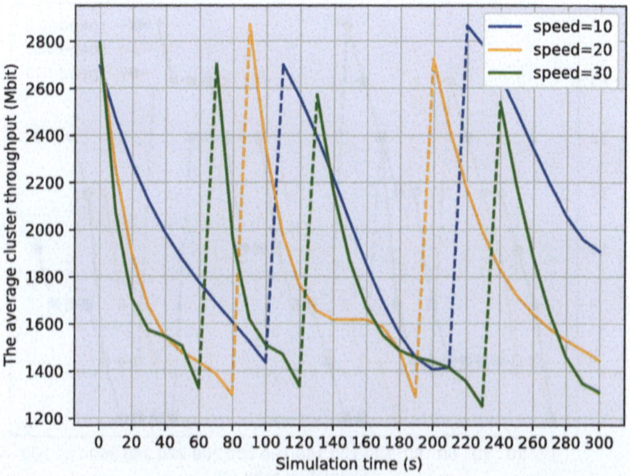

Fig. 2.9 The average cluster throughput under different speeds when using FSC in the cluster maintenance stage

2.1.3.5 Stability Analysis

In this part, we discuss the stability of the UAV network and mainly analyze how mobility affects the clustering results.

Figure 2.9 shows the average cluster throughput under different speed varies with time. When time is $0s$, the UAV network is in the initial clustering stage. In the three experiments, their $n_{\max}$ are the same. Therefore their initial clustering results are the same at the beginning when executing the FSC algorithm. Then the UAVs begin to move, which leads to different graphs. From Fig. 2.9, we can conclude that the faster the speed is, the shorter the cluster maintenance stage will be. In a high-speed environment, cluster structure is unstable and we have to fission the cluster frequently to maintain cluster structure, which leads to the falling average cluster throughput rapidly.

Figure 2.10 shows the number of clusters under different speed varies with time, which is corresponding to Fig. 2.9. We can find from the figure that cluster fissions frequently in a highly dynamic environment. Contrarily, a cluster structure can keep for several time slots.

2.1.3.6 Comparison

In this part, we compare our proposed algorithm FSC with WCA, GMM, AP, and AC. When the number of UAVs is N and the number of clusters is k, the complexity analysis of these algorithms is as follows. WCA is essentially a K-Means algorithm with a time complexity of $O(tNk)$, where t is the number of iterations. The GMM process is similar to K-Means, but it falls under soft clustering and the output consists of probabilities of belonging. The computational

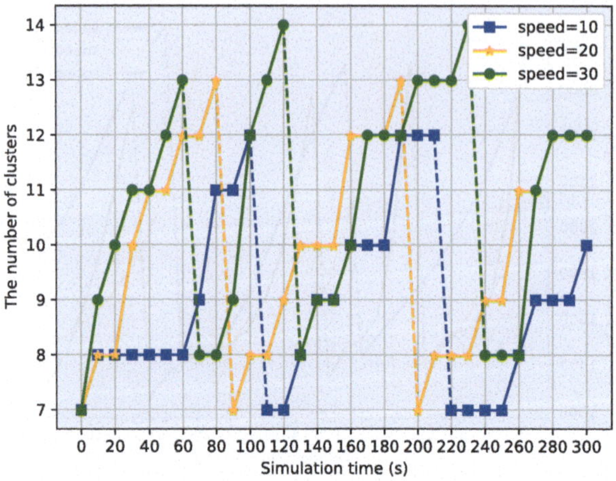

Fig. 2.10 The number of clusters under different speeds when using FSC in the cluster maintenance stage

complexity of GMM is high, with each iteration being greater than K-Means. The AP algorithm can automatically cluster based on sample data. However, it is necessary to calculate the similarity between each pair of data objects in advance, which requires a large amount of computation. Its time complexity is $O(N^2t)$, where t is the number of iterations. AC is a method of clustering analysis by establishing a hierarchical structure of clusters. AC requires N iterations, each of which requires updating and storing the similarity matrix. Therefore, the time complexity is the cubic order of the number of samples N, which is $O(N^3)$. The core of FSC is SC. The time complexity of SC comes from constructing similarity matrix $O(N^2)$ and decomposing eigenvalues $O(N^3)$. Thus, the time complexity of FSC is $O(N^3t)$, where t is the number of fission. The time complexity of FSC is the highest during the initial clustering stage. However, during the cluster maintenance stage, FSC pre-detects the state of the cluster and only performs partial fission on clusters which don't meet the constraints. Therefore, the number of samples N has been significantly reduced, greatly saving computational costs. Additionally, FSC considers the dynamics of the UAV swarm, enhancing the stability of the cluster and reducing the number of re-clustering.

As shown in Fig. 2.11, we present the average cluster throughput in the network of the different algorithms. With the number of UAVs increasing, the average cluster throughput increases. This is mainly caused by the increase in communication between UAVs with the number of clusters increasing. FSC algorithm considers the network status in the time slot and has a time effect, which is an advantage over other algorithms. Figure 2.11 demonstrates the validity of our proposed algorithm.

As shown in Fig. 2.12, we present the number of UAVs in each cluster when using different algorithms. In the box boxplot, we consider two aspects: one is the difference between the maximum value and minimum value, and the other is

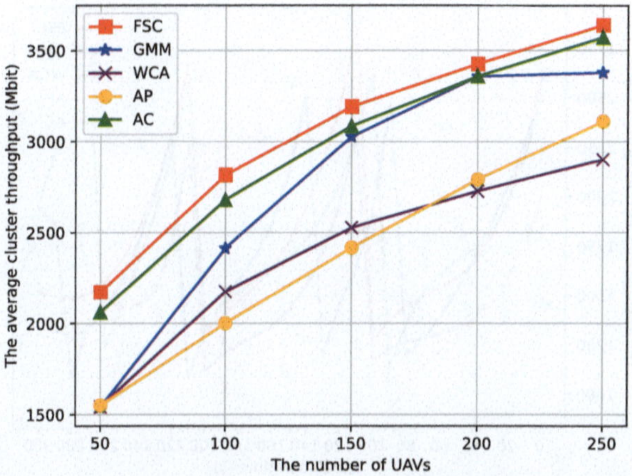

Fig. 2.11 The average cluster throughput when using different algorithms in the initial clustering stage

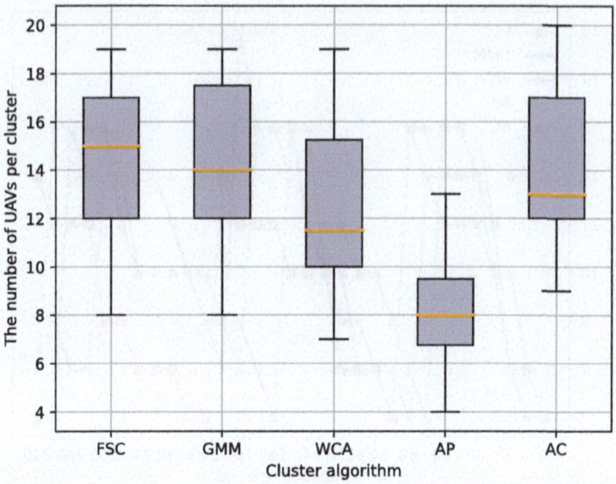

Fig. 2.12 The number of UAVs in each cluster when using different algorithms in the initial clustering stage

the median number. From the perspective of maximum value and minimum value, the effect of AP is the best. However, the median number of AP is significantly lower than the other four algorithms. It demonstrates the size of clusters in the AP algorithm. This clustering strategy will affect the average cluster throughput. Excluding AP, FSC, GMM, and AC all have the second small difference. Although the maximum value of AC reaches the $m_{\max}$, FSC has the largest median number, which demonstrates the clustering results of FSC distributes more uniformly.

Figure 2.13 shows the average cluster throughput under different algorithms varies with time. When the simulation time is 100 s, 80 s, 100 s, 90 s, 90 s, FSC,

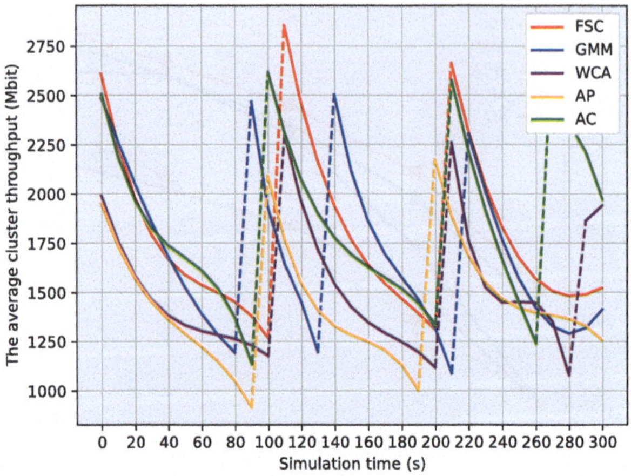

Fig. 2.13 The average cluster throughput when using different algorithms in the cluster maintenance clustering stage

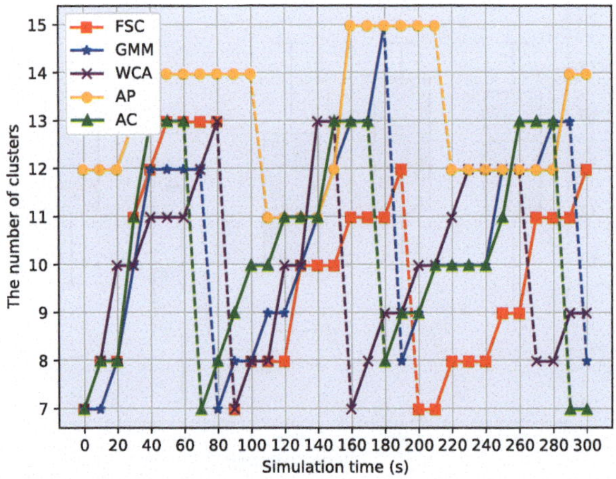

Fig. 2.14 The number of clusters when using different algorithms in the cluster maintenance clustering stage

GMM, WCA, AP, and AC respectively reach the threshold for the first time. Then, in the next time slot, they return to the initial clustering stage and re-clustering. A longer cluster maintenance stage will save computing overhead. In addition, although FSC and WCA both keep the cluster structure for 100*s*, the average cluster throughput of FSC is significantly higher than WCA.

Figure 2.14 shows the number of clusters under different algorithms varies with time, which is corresponding to Fig. 2.13. Within 300 s of simulation time, the re-clustering number of FSC, GMM, WCA, AP, AC are respectively 3, 4, 4, 3 and 4. In addition, compared with other algorithms, FSC has less clusters, which means that the size of the cluster in the FSC algorithm is larger and therefore each cluster has a stronger capability.

2.2 Cooperative Intelligence-Based Network Dynamic Planning for UAV Swarm Networks

The UAV swarm is one of the most powerful and exciting technologies, quickly becoming a disruptive force that reshapes how we live and work. In particular, the UAV swarm is crucial in emergency communication fields, thanks to its ability for fast and flexible deployment [19]. When ground communication base stations are damaged, UAV swarms can be used to reconstruct emergency communication networks. However, due to the limited coverage power of single UAV devices, coverage trajectory planning for cooperating UAV swarms is the key problem in the emergency communication system.

Traditional trajectory planning methods are typically based on a static user model, which divides the time into time slices and assumes that the user location remains fixed during each time slice [20–22]. Optimization algorithms, such as ant colony and particle swarm optimization, are then applied to solve the optimized trajectory of a UAV swarm. However, such methods are hardly adaptable to user dynamics in real-world scenarios. When time slices are divided too large, it cannot represent the mobility of users accurately. Conversely, if they are divided too small, iterative algorithms struggle to converge effectively [23, 24]. As a result, designing a continuous trajectory control scheme has become a popular research issue and a challenge. The development of such a scheme is needed to overcome limitations posed by traditional methods and to enable effective coordination of UAV swarms with dynamic user movements.

Recently, reinforcement learning has emerged as a promising approach for UAV trajectory control, gaining significant attention from academia. The continuous control optimization capability of reinforcement learning can be effectively applied to the trajectory planning of UAV base stations (UAV BSs) problem with a dynamic user movements model. Current reinforcement learning based methods can be categorized into centralized and distributed approaches. In [25], Bouhamed et al. proposed a deep deterministic policy gradient (DDPG) algorithm for task division into time slots and making decisions for UAV trajectory planning within those slots. Another study by Liu et al. Liu et al. [26] presented a distributed dual Q-learning-based trajectory planning algorithm for maximizing the number of users in UAV-assisted cellular networks. However, distributed reinforcement learning exists in the "credit assignment" and "spurious reward signals" problem, which poses a difficulty in achieving mutual cooperation among UAV BSs. The independence of UAV BSs hinders the attainment of global optimization, making it challenging to coordinate their actions.

Centralized reinforcement learning methods have also been explored. In [27], Saxena et al. proposed a deep reinforcement learning (DRL) method that optimized the UAV BS flight trajectory through traffic characteristics and load balancing. In [28], Zhang et al. proposed the use of deep Q-network (DQN) to control the flight of UAV BSs, helping to avoid data overflow and device battery drain. However, a key drawback of centralized reinforcement learning is the need to collect all user and global network state information, which greatly reduces the reliability and scalability of the system. At the same time, the increased number of UAV BSs will rapidly increase the complexity of the centralized algorithm.

Therefore, we propose a centralized training and distributed execution multi-agent reinforcement learning deployment architecture. And we propose a trajectory planning algorithm based on the Q-function mixing network. The contributions of this work are summarized as follows:

1. We first propose to model the UAV-assisted emergency communication system as a classic "Predator-Prey" model for reinforcement learning and express it mathematically through the decentralized partially observable Markov decision process. Among them, UAV BSs chase users and provide communication services.
2. We propose a UAV-assisted emergency communication method without global information. This method uses a joint state composed of local information to replace the global information. Then, we propose the Q-function Mixing network to solve the "credit assignment" and "spurious reward signals" problem. Experimental results show that the Q-function mixing network approach effectively improves the throughput of multi-UAV BSs-assisted communication networks.

2.2.1 System Model and Problem Formulation

In this section, we present our system model, mobile model, transmission model, coverage model and problem formulation.

2.2.1.1 System Model

As shown in Fig. 2.15, we consider a UAV-assisted emergency communication network consisting of F UAV BSs, N users, temporary base stations and edge computing nodes. UAV BSs provide users with communication services to establish wireless links. The network traffic is transmitted through the wireless link to the temporary base station. The temporary base station will establish communication with the convergence data center to realize communication with the outside world.

Each UAV BS collects the location information of users within the coverage area and the positions of adjacent UAV BSs. The UAV BS's state-action value function (Q-function) is transmitted to the edge computing node for training the joint state-action value function. We consider that both the UAV BS and the user have mobility.

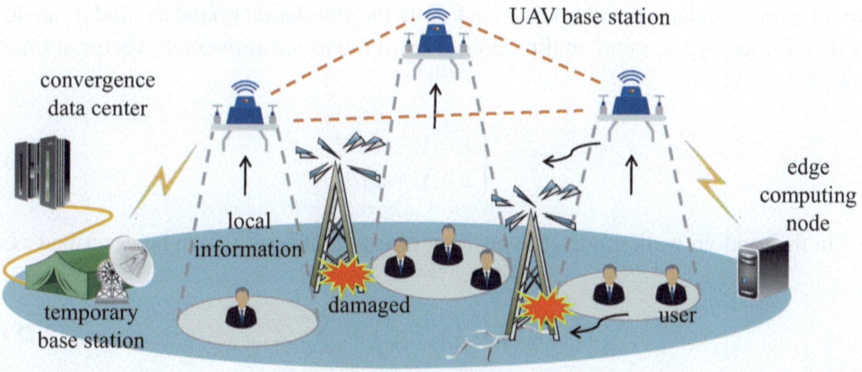

Fig. 2.15 UAV base station-assisted emergency communication scene

2.2.1.2 Mobile Model

The task of the UAV BS is to provide communication services to ground users. The task duration is T time slots. The time interval for each time slot is the same. At the initial moment of the task, each UAV BS takes off at a random position. The flight altitude of the UAV BS is fixed at H. Then, the UAV BS uses the trajectory planning algorithm proposed by our model to adjust its movement continuously. The UAV BS's movement can be represented by a binary (e, φ), where e represents the moving speed of the UAV BS. Let φ represent the moving direction of the UAV BS. The position update of the UAV BS can be described as:

$$\begin{cases} x_f^t = x_f^{t-1} + e_f^{t-1} \cdot \Delta t \cdot \cos \varphi_f^{t-1} \\ y_f^t = y_f^{t-1} + e_f^{t-1} \cdot \Delta t \cdot \sin \varphi_f^{t-1} \\ z_f^t = H \end{cases}, \tag{2.25}$$

where t represents the t-th time slot; Δt is the size of the time slot. Let e_f^{t-1} represent the speed of the UAV BS f at time slot $t-1$, where f represents the f-th UAV BS. Equation (2.25) describes the movement of the UAV BS from time slot $t-1$ to time slot t. We express the user's mobility model through a binary group (v, θ), where v represents the user's moving speed. Let θ represent the moving direction of the user. Let N be the number of users. The user's location update equation can be described as:

$$\begin{cases} x_n^t = x_n^{t-1} + v \cdot \Delta t \cdot \cos \theta_n \\ y_n^t = y_n^{t-1} + v \cdot \Delta t \cdot \sin \theta_n \end{cases}. \tag{2.26}$$

The user mobility model adopts a random walk model. We assume that users move at the same speed v. Speed v is a fixed constant value. At the same time, user movement direction θ follows the normal distribution. If the user reaches the simulation boundary, it will bounce back from the simulation boundary, and its angle is determined by the incident direction. Record the user n movement vector at time t as Z_n^t:

$$Z_n^t = \begin{cases} v \cdot \Delta t \cdot \cos\theta_n^t \\ v \cdot \Delta t \cdot \sin\theta_n^t \end{cases}. \tag{2.27}$$

In the random walk model, the position of user n at time T_u can be described as:

$$\left(x_n^{T_u}, y_n^{T_u}\right) = \left(x_n^0, y_n^0\right) + \sum_{t=1}^{T_u-1} Z_n^t, \tag{2.28}$$

where $\left(x_n^0, y_n^0\right)$ represents the user's initial position. Combining Eqs. (2.25) and (2.26), the distance between user n and UAV BS f at time slot t can be described as:

$$d_{n,f}^t = \sqrt{\left(x_f^t - x_n^t\right)^2 + \left(y_f^t - y_n^t\right)^2 + H^2}. \tag{2.29}$$

2.2.1.3 Transmission Model

We set the free space propagation model as the transmission model. There are two methods for calculating link loss in the free space propagation model: line-of-sight and non-line-of-sight. The line-of-sight (LoS) link loss of the UAV BS f to the ground user n is as follows:

$$L_{n,f}^{LoS}(t) = 20\log\left(\frac{4\pi f_c d_{n,f}^t}{c}\right) + \xi_{LoS}, \tag{2.30}$$

where ξ_{LoS} is a constant. $d_{n,f}^t$ is the distance between the UAV BS f and user n; f_c is the carrier frequency; c is the speed of light. The non-line-of-sight (NLoS) link loss of UAV BS f to ground user n as follow:

$$L_{n,f}^{NLoS}(t) = 20\log\left(\frac{4\pi f_c d_{n,f}^t}{c}\right) + \xi_{NLoS}. \tag{2.31}$$

Because the user's movement will change the signal state, the non-line-of-sight link transmission model is used when there are obstacles. Otherwise, the line-of-sight link transmission model is selected. The probability $P_{n,f}^{LoS}$ of a line-of-sight link between user n and UAV BS f is set as [29]:

$$P_{n,f}^{LoS}(t) = \frac{1}{1 + ae^{-b\left(\arcsin\left(\frac{H}{d_{n,f}^{t}}\right) - a\right)}}, \tag{2.32}$$

where a and b are two parameters, determined according to the scene (such as urban, rural). Combining Eqs. (2.30)–(2.32), the path loss of the UAV BS f to user n can be described as:

$$L_{n,f}(t) = P_{n,f}^{LoS}(t)L_{n,f}^{LoS}(t) + \left(1 - P_{n,f}^{LoS}(t)\right) L_{n,f}^{NLoS}(t). \tag{2.33}$$

When the transmit signal strength of the UAV BS f is P_f, the actual signal strength received by the user is $P_{n,f}$:

$$P_{n,f} = P_f + L_{n,f}(t). \tag{2.34}$$

2.2.1.4 Coverage Model

When the actual receiving strength $P_{n,f}$ provided by the UAV BS is greater than the threshold P_q. The UAV BS can establish a wireless link with the user. Then, the UAV BS provides communication services for ground users [30]. The binary variable $c_{n,f}^{t}$ represents whether the ground user n establishes a wireless link with the UAV BS f in the t time slot.

$$\begin{cases} c_{n,f}^{t} = 1, \; \exists f \in F, \; P_{n,f} \geq P_q \\ c_{n,f}^{t} = 0, \; \forall f \in F, \; P_{n,f} < P_q \end{cases}, \tag{2.35}$$

where P_q represents the minimum signal strength required for the UAV BS to establish a wireless link with the user. In fact, one UAV BS can establish wireless links with multiple users, but one user can only establish a wireless link with one UAV BS. It is assumed the user selects the UAV BS with the highest signal strength for establishing a wireless link:

$$\begin{aligned} & f^{*} = \operatorname{argmax} \left\{P_{n,f}\right\}_{f \in F, P_{n,f} \geq P_q} \\ & \begin{cases} c_{n,f}^{t} = 1, \; f = f^{*} \\ c_{n,f}^{t} = 0, \; f \neq f^{*} \end{cases} \end{aligned} . \tag{2.36}$$

2.2.1.5 Problem Formulation

Our goal is to improve the network throughput of the UAV BSs service area. The network throughput in the overall area is provided by the UAV BSs, where the throughput of the UAV BSs can be described as:

$$C_{sum}^{t} = \sum_{f=1}^{F} \sum_{n=1}^{N} R_{n,f}^{t} c_{n,f}^{t}, \tag{2.37}$$

where $c_{n,f}^{t}$ is shown in Eqs. (2.35). Let $R_{n,f}^{t}$ represent the wireless transmission rate between the $n-th$ user and the UAV BS f. According to the Shannon formula, $R_{n,f}^{t}$ can be described as:

$$R_{n,f} = W_{n,f} \log_2 \left(1 + 10^{SNR_{n,f}/10}\right), \tag{2.38}$$

where the unit of $R_{n,f}$ is bit/s. $W_{n,f}$ represents the communication channel bandwidth between the UAV BS f and user n. $SNR_{n,f}$ is the signal-to-noise ratio. Our goal is to maximize the throughput of the network. The objective is solved in a multi-constraint environment, and its constraint expression is as follows:

$$\text{MAX}\left(C_{sum}^{t}\right), \tag{2.39}$$

$$\text{s.t.} \left|\left(x_i^t, y_i^t\right) - \left(x_j^t, y_j^t\right)\right|_{i \neq j} > d, \forall i, j \in F \tag{2.39a}$$

$$c_{n,f}^{t} = \{0, 1\}, \forall f \in F, n = 1, 2, \cdots, N \tag{2.39b}$$

$$\left(x_i^t, y_i^t\right) \in \Delta G, i \in F \tag{2.39c}$$

$$e_f \leq e_{max}, f \in F \tag{2.39d}$$

where ΔG represents the area to be used for ground communication assisted by the UAV BS. Constrant (a) limits the distance between UAV BSs. Due to the UAV BSs working at the same height H, we add the spacing range d between UAVs to prevent collisions. Constrant (b) represents the link relationship between the UAV BS and the user. Constrant (c) limits the range of UAV BS movement. Constrant (d) limits the movement speed of the UAV and the movement range of the UAV in a time slot.

2.2.2 *Q-Mixing Learning-Based Nodes Trajectory Planning for UAV Swarm Networks*

In this section, we model the UAV-assisted emergency communication system as the "Predator-Prey" model. And we express it mathematically through the Markov team game. Then, we introduce the training and deployment of the Q-function mixing network model.

2.2.2.1 Multi UAV Base Station Assisted Ground Communication System Based on Markov Team Game

A UAV-assisted emergency communication system can be abstracted as the task of multiple UAVs cooperating to chase users. The fully cooperative multi-agent task can be represented by a decentralized partially observable Markov decision process (Dec-POMDP) model. The model consists of a tuple $G = (S, A, P, r, f, \gamma)$. Let agent f represent the UAV BSs, where agent $f \in \{1, 2, \ldots, F\}$. Let s_f^t represent the state of the environment of the agent f; Let $s^t = (s_1^t, s_2^t, \cdots, s_f^t)$ represent the joint status information; Let r represent the sum of the reward values obtained by all agents. Each agent will choose its action $a_f \in A$. The actions of all agents form a joint action $a^t = (a_1^t, a_2^t, \cdots, a_f^t)$. The state will change according to the state transition function $P\left(s^{t+1} \mid s^t, a^t\right)$. All agents use the same reward function $r\left(s^t, a^t\right)$. $\gamma \in [0, 1)$, where γ is a discount factor.

In this model, the agent periodically collects the local information. The UAV BS maintains its own Q-function. Based on the design of the reinforcement learning algorithm, we define the state, action and reward as follows:

(1) State: In each time slot t in the task cycle, we define state $s_f^t = \left\{c_f^t, \text{pos}^{user}, \text{pos}_f^{uav}\right\}$, which includes the following three parts: The first part is c_f^t. c_f^t represents the connection status between the UAV BS f and the users, where $c_f^t = \left\{c_{1,f}^t, c_{2,f}^t, \ldots, c_{n,f}^t\right\}$. The second part is pos^{user}. pos^{user} is a coordinate set that contains multiple user coordinates within the coverage range of UAV BS. The third part is pos_f^{uav}. pos_f^{uav} is the position of the UAV BS f.

(2) Reward: The reward value obtained after taking action a^t at time slot t is defined as:

$$r^t = C_{sum}^t - \sum_{f=1}^{F} \eta e_f, \tag{2.40}$$

where C_{sum}^t represents the throughput of the UAV-assisted communication network in time slot t. The latter term is the penalty term, where η is the penalty coefficient.

(3) Action: Action controls the moving direction φ and speed e of the UAV BS. (a) $\varphi_f^t \in [0, 2\pi)$: The moving direction angle of the UAV BS f at time slot t. (b) $e_f^t \in [0, e_{max}]$: The flying speed e_f^t of the UAV BS f at time slot t, where e_{max} represents the maximum speed that the UAV BS can reach in each time slot. It is worth noting that if $e_f^t = 0$, then it means that the UAV BS f remains to hover at time slot t. Action selection can be described as:

$$a^* \leftarrow \arg\max_a Q\left(\tau^f, a\right), \tag{2.41}$$

where a^* is the action with the largest expected value in the current state. τ^f represents the joint action-observation state history of the UAV BS f, where $\tau^f = (a_f^0, s_f^1, \cdots, a_f^{t-1}, s_f^t)$. We use greedy algorithms to select actions as follows:

$$\begin{cases} 1 - \epsilon + \epsilon / |A|, \text{ if } a = a^* \\ \epsilon / |A|, \text{ if } a \neq a^* \end{cases}. \tag{2.42}$$

For action a^*, it has the highest probability of being chosen. a^* is chosen with probability $1 - \epsilon + \epsilon / |A|$. The remaining actions are chosen with probability $\epsilon / |A|$, where $\epsilon \in (0, 1)$.

The algorithm framework design is shown in Fig. 2.16. The Q-function mixing network model consists of two parts: the MIX network and the agent network. The input of the agent network is the observation state s_f^t of the agent and action a_f^{t-1} at the previous moment. The output of the agent network is the state-action value $Q(\tau^f, a_f^t, \varepsilon_f)$ of a single agent. The Gate Recurrent Unit (GRU) network used by the agent network is a Recurrent Neural Network (RNN). Using the RNN, agents can use historical information to represent the current state. The input of the mixing network is $Q(\tau^f, a_f^t, \varepsilon_f)$, and the output is the joint value function $Q_{tot}\left(\tau, a^t; \varepsilon\right)$. The MLP represents Multilayer Perceptron.

To solve a completely decentralized policy, make the centralized policy and the distributed policy consistent. This requires that the executive action of global $Q_{tot}\left(\tau, a^t; \varepsilon\right)$ is the same as that of single-agent $Q(\tau^f, a_f^t, \varepsilon_f)$:

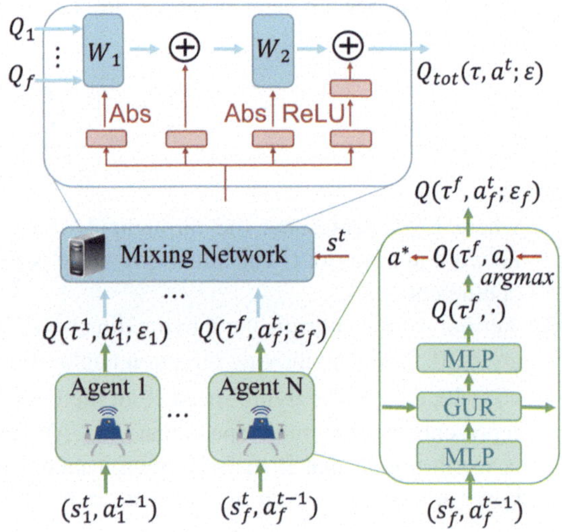

Fig. 2.16 Q-function mixing network algorithm architecture

$$\underset{a}{\operatorname{argmax}} Q_{tot}\left(\tau, a^t\right) = \begin{pmatrix} \underset{a_1^t}{\operatorname{argmax}} Q_1(\tau^1, a_1^t) \\ \vdots \\ \underset{a_f^t}{\operatorname{argmax}} Q_f(\tau^f, a_f^t) \end{pmatrix}. \tag{2.43}$$

Equation (2.43) indicates that the joint actions of all agents are equivalent to the set of each agent's actions. For the relationship of Eq. (2.19), it is only necessary to satisfy that Q_{tot} is monotonically increasing for any Q_f:

$$\frac{\partial Q_{tot}}{\partial Q_f} \geq 0, \forall f \in F. \tag{2.44}$$

To realize Eq. (2.44), the MIX network weight needs to be greater than 0 or equal to 0. The adopted activation function is the absolute function. The deviation doesn't require a value range, so the ReLU activation function is used.

2.2.2.2 Deployment and Training of Q-Function Mixing Network

The edge computing node deploys a Mixing network. The UAV BS deploys the agent network. The UAV BS collects user information within the coverage area to maintain the current agent network and uploads the single-agent network to the edge computing node for mixing network training. The edge computing node delivers training parameter updates. The UAV BS perceives local information and performs optimal trajectory planning through a single-agent network. The Q-function of each agent can be described as $Q(\tau^f, a_f^t, \varepsilon_f)$. ε_f represents the Q-function network parameter of the agent f; a_f^t is the action of the agent f at time slot t; τ^f represents the joint action-observation state history of the UAV BS f. The agent generates a Q-function through the neural network and executes action a_f^t in the state of s_f^t. The global Q-value function can be described as $Q_{tot}\left(\tau, a^t; \varepsilon\right)$. Where $\varepsilon = \left\{\varepsilon_1, \varepsilon_2, \cdots, \varepsilon_f; \delta\right\}$, δ is the Mixing network parameter.

By minimizing the square term of the Temporal-Difference (TD) error, the Q-function mixing network model uses the following loss function to learn the network parameter ε:

$$L(\varepsilon) = \sum_{i=1}^{m} \left(y_m^{tot} - Q_{tot}\left(\tau, a, s; \varepsilon\right)\right)^2, \tag{2.45}$$

where m is the number of transition groups. The target y^{tot} can be described as:

$$y^{tot} = r + \gamma \max_a \left\{Q_{tot}\left(\tau, a, s; \varepsilon\right)\right\}. \tag{2.46}$$

Table 2.3 Simulation parameters

Parameter	Value	Parameter	Value
f_c	2.6 GHz	N	50
$N\eta_{LoS}$	1 dB	F	6
$N\eta_{NLoS}$	20 dB	a	75
H	120 m	b	8
e_{max}	1.5 m/s	γ	0.9
P_f	[150,170] dBm	v	0.6 m/s
P_q	85 dBm	Δt	5 s

2.2.3 Simulation Results and Discussions

In the simulation experiment, we set a rectangular target area with a size of 1000 × 1000 m, the number of ground users is 50, and 6 UAV BSs are deployed in the network. The experimental parameters are listed in Table 2.3. All three models are trained in 500 steps; each step includes 300-time slots. For the user mobility model, we consider using the Random Walk Mobility Model. The Random Walk Mobility Model refers to [31].

2.2.3.1 Convergence Analysis

Deep Q-Network (DQN) and Deep Deterministic Policy Gradient (DDPG) are currently two reinforcement learning methods that are better applied to UAV BS trajectory planning [28, 32]. We conduct a comparison experiment on the algorithm convergence characteristics of DQN, DDPG, and the Q-function mixing network (QMIX). The experimental results are shown in Fig. 2.17:

According to Fig. 2.17, we can see that DQN is difficult to converge in high-dynamic UAV BS-assisted communication scenarios. Compared to the DQN algorithm, the DDPG algorithm improves the convergence characteristics of the algorithm through the Actor-Critic (AC) framework. DDPG can be adapted to large-scale UAV BS-assisted communication scenarios. However, for high dynamic network scenarios, the DDPG algorithm fluctuates greatly. The reason is the problem of overestimation, which is easy to cause difficulty in policy convergence. The joint value function method adopted by the Q-function mixing network pays more attention to the interaction between agents, so it can better solve the dynamic impact problem between multiple UAV BSs. Compared with DDPG and DQN, the Q-function mixing network algorithm has the characteristics of easy convergence and high stability for large-scale and highly dynamic UAV BS-assisted communication scenario. In addition, we also demonstrate the coverage effect of the Q-function mixing network model for UAV BS-assisted ground communication in Fig. 2.18. The experiment included 6 UAV BSs and 50 users. We have drawn the signal strength distribution equipotential diagram. When the signal strength exceeds 85

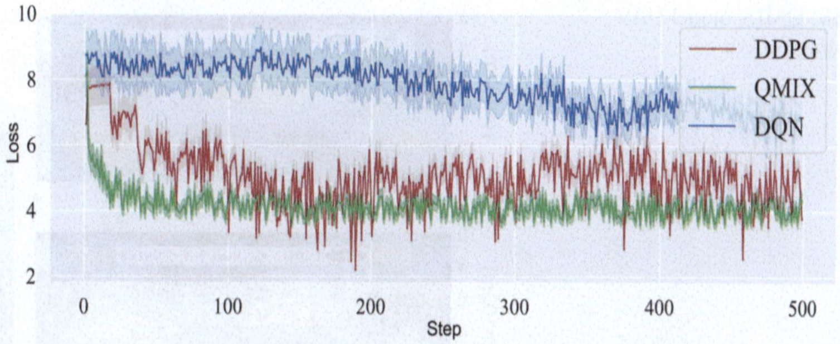

(a) DQN, DDPG, QMIX algorithm convergence comparison

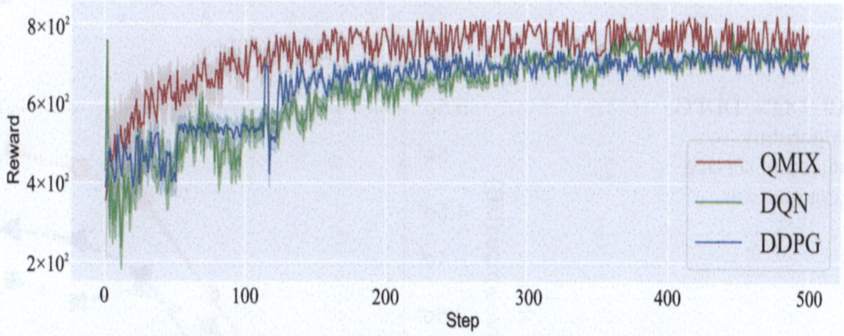

(b) DQN, DDPG, QMIX algorithm reward value comparison

Fig. 2.17 Performance comparison of DQN, DDPG, and QMIX models. (**a**) DQN, DDPG, QMIX algorithm convergence comparison. (**b**) DQN, DDPG, QMIX algorithm reward value comparison

dBm, users can establish a connection with the UAV base station through the wireless link. The experimental results show that the Q-function mixing network model can be deployed appropriately based on user distribution and achieve 90% user communication coverage.

2.2.3.2 Throughput Analysis

We compare the cumulative throughput of DQN, DDPG and Q-function mixing network model, as shown in Fig. 2.19:

All three models make the network throughput larger as the number of UAV BSs increases. However, the DQN model has a significantly lower growth trend than the other two models. The Q-function mixing network model presents approximately linear experimental results because the Q-function is deployed in a distributed manner on the UAV BS. For each UAV BS, its trajectory optimization action is independent, thus avoiding the explosion of state space and action space dimensions. Experiments show that DQN is unsuitable for large-scale UAV BS cooperative assisted communication. The DDPG model curve is relatively jittery, and its stability is weaker than that of the Q-function mixing network model. As the number of UAV

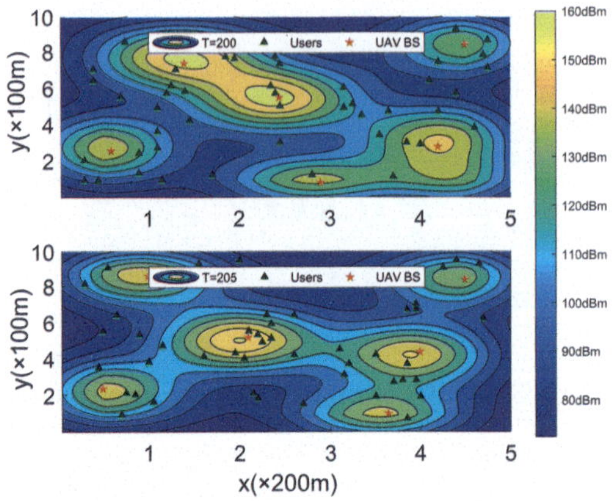

Fig. 2.18 QMIX model UAV BS coverage effect

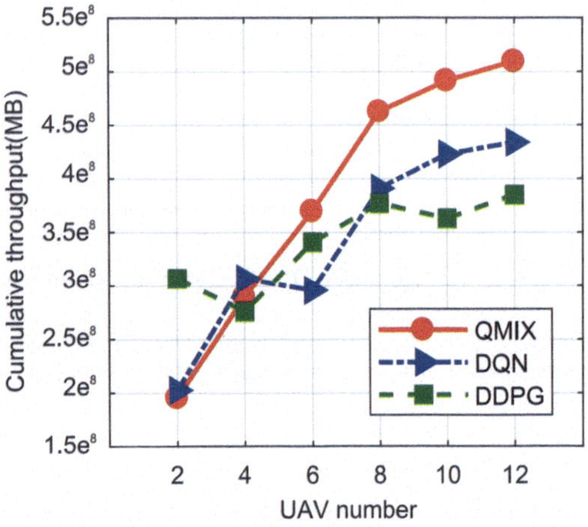

Fig. 2.19 DQN, DDPG, QMIX algorithm communication service quality comparison

BSs increases, the environment becomes more dynamic. In our experiment, the Q-function mixing network model performs better under the ground communication condition assisted by 12 UAV BSs. Compared with DDPG and DQN, the network throughput is increased by 19.82% and 31.39%.

2.2.3.3 Collaborative Analysis

Traditional methods are usually difficult to realize the cooperative auxiliary communication of UAV BSs. To this end, we adopt the method of joint value function to strengthen the cooperation between agents. The experimental results are shown in Fig. 2.20.

We perform an analysis of variance on the throughput of the three models of UAV BSs. SS represents the sum of squares; df represents degrees of freedom; MS represents mean square error; F represents the F-statistic. It can be seen from the

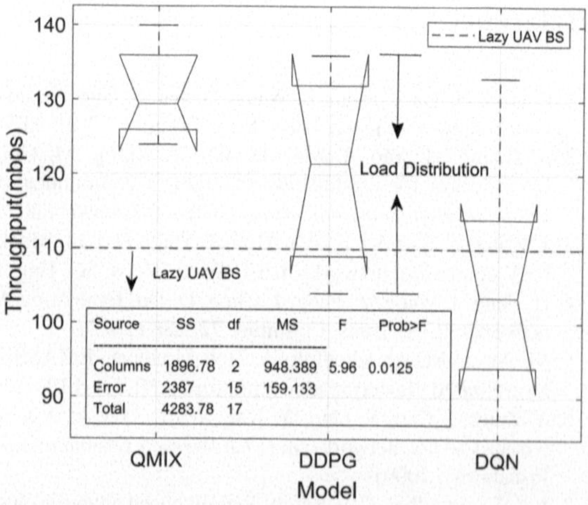

Fig. 2.20 Collaborative analysis of QMIX algorithm

experimental results that the DQN model has serious credit assignment and spurious reward signals. The throughput variance of DQN UAV BS is very large, and there is a large number of lazy agents. These agents do not work by relying on the reward value brought by other agents. Comparing the three models, the Q-function mixing network model's agent is more cooperative and more balanced load distribution.

2.3 Summary

This chapter explores adaptive UAV swarm clustering and networking optimization methods. Due to the high mobility and limited resources of UAV swarms, effective clustering and networking optimization are often required during communication. This chapter proposes an FSC-based UAV network clustering method, where the "link effective value" is used to describe the time-sequential attributes of UAV nodes and edges. The method employs Laplacian matrix decomposition to achieve UAV swarming. Once the cluster size and structural requirements are satisfied, clustering splitting is applied for continuous cluster maintenance. This approach effectively enhances communication performance and extends cluster maintenance time. Furthermore, in scenarios where ground communication base stations are damaged, and UAV base stations need to provide wireless links to ground users, this chapter introduces a Q-function mixing algorithm. This method addresses the challenges of information acquisition and the lack of cooperation in traditional trajectory planning approaches. Compared to DQN and DDPG algorithms, the Q-function mixing network model demonstrates superior performance in terms of throughput and stability.

References

1. G. Zhu, H. Yao, T. Mai, Z. Wang, D. Wu, S. Guo, Fission spectral clustering strategy for UAV swarm networks. IEEE Trans. Serv. Comput. **17**(2), 537–548 (2024)
2. S. Huang, H. Yao, T. Mai, D. Wu, Z. Xiong, M. Guizani, Cooperative intelligence-based UAV swarm for establishing emergency communication, in *ICC 2024-IEEE International Conference on Communications* (IEEE, Piscataway, 2024), pp. 3919–3924
3. J. Wang, Z. Jiao, J. Chen, X. Hou, T. Yang, D. Lan, Blockchain-aided secure access control for UAV computing networks. IEEE Trans. Netw. Sci. Eng. **11**(6), 5267–5279 (2024)
4. H. Feng, J. Wang, Z. Fang, J. Chen, D. Do, Evaluating AoI-centric HARQ protocols for UAV networks. IEEE Trans. Commun. **72**, 288 (2023)
5. M. Aissa, M. Abdelhafidh, A. Ben Mnaouer, EMASS: A novel energy, safety and mobility aware-based clustering algorithm for FANETs. IEEE Access **9**, 105506–105520 (2021)
6. Y. Zeng, J. Cao, S. Guo, Y. Kai, X. Li, SWCA: A secure weighted clustering algorithm in wireless ad hoc networks, in *IEEE Wireless Communications & Networking Conference* (IEEE, Piscataway, 2009)
7. S.K. Dhurandher, G.V. Singh, Weight based adaptive clustering in wireless ad hoc networks, in *Personal Wireless Communications (2005) ICPWC 2005. 2005 IEEE International Conference on* (IEEE, Piscataway, 2005)
8. M. Chatterjee, An on-demand weighted clustering algorithm (WCA) for ad hoc networks, in *IEEE Global Telecommunications Conference* (IEEE, Piscataway, 2000)
9. Z. JingMao, S. YanXia, Review on spectral methods for clustering, in *2015 34th Chinese Control Conference (CCC)* (IEEE, Piscataway, 2015), pp. 3791–3796
10. R. Van Driessche, D. Roose, An improved spectral bisection algorithm and its application to dynamic load balancing. Parallel Comput. **21**(1), 29–48 (1995)
11. K. Hou, Q. Xu, X. Zhang, Y. Huang, L. Yang, User association and power allocation based on unsupervised graph model in ultra-dense network, in *2021 IEEE Wireless Communications and Networking Conference (WCNC)*, pp. 1–6 (IEEE, Piscataway, 2021)
12. L. Hong, H. Guo, J. Liu, Y. Zhang, Toward swarm coordination: Topology-aware inter-UAV routing optimization. IEEE Trans. Veh. Technol. **PP**(99), 1 (2020)
13. W.B. Wendi, An application-specific protocol architecture for wireless microsensor networks. IEEE Trans. Wirel. Commun. **1**, 660 (2002)
14. M. Ni, Z. Zhong, D. Zhao, MPBC: A mobility prediction-based clustering scheme for ad hoc networks. IEEE Trans. Veh. Technol. **60**(9), 4549–4559 (2011)
15. M. Chatterjee, S.K. Das, D. Turgut, WCA: A weighted clustering algorithm for mobile ad hoc networks. Cluster Comput. **5**(2), 193–204 (2002)
16. C. Stauffer, W.E.L. Grimson, Adaptive background mixture models for real-time tracking, in *Proceedings. 1999 IEEE Computer Society Conference on Computer Vision and Pattern Recognition (Cat. No PR00149)*, vol. 2 (1999), pp. 246–252
17. B.J. Frey, D. Dueck, Clustering by passing messages between data points. Science **315**(5814), 972–976 (2007)
18. D. Müllner, Modern hierarchical, agglomerative clustering algorithms. Preprint. arXiv:1109.2378 (2011)
19. B. Sharma, G. Srivastava, J.C.-W. Lin, A bidirectional congestion control transport protocol for the internet of drones. Comput. Commun. **153**, 102–116 (2020)
20. A.L. Jennings, R. Ordonez, N. Ceccarelli, Dynamic programming applied to UAV way point path planning in wind, in *2008 IEEE International Conference on Computer-Aided Control Systems* (2008), pp. 215–220. https://doi.org/10.1109/CACSD.2008.4627357
21. E.J. Forsmo, E.I. Grtli, T.I. Fossen, T.A. Johansen, Optimal search mission with unmanned aerial vehicles using mixed integer linear programming, in *2013 International Conference on Unmanned Aircraft Systems (ICUAS)* (IEEE, Piscataway, 2013), pp. 253–259. https://doi.org/10.1109/ICUAS.2013.6564697

22. C. Zhang, Z. Zhen, D. Wang, M. Li, UAV path planning method based on ant colony optimization, in *2010 Chinese Control and Decision Conference* (IEEE, Piscataway, 2010), pp. 3790–3792. https://doi.org/10.1109/CCDC.2010.5498477
23. H. Shiri, J. Park, M. Bennis, Remote UAV online path planning via neural network-based opportunistic control. IEEE Wirel. Commun. Lett. **9**(6), 861–865 (2020)
24. X. Yue, W. Zhang, UAV path planning based on K-means algorithm and simulated annealing algorithm, in *2018 37th Chinese Control Conference (CCC)* (IEEE, Piscataway, 2018), pp. 2290–2295. https://doi.org/10.23919/ChiCC.2018.8483993
25. O. Bouhamed et al., Autonomous UAV navigation: A DDPG-based deep reinforcement learning approach, in *2020 IEEE International Symposium on Circuits and Systems (ISCAS)* (IEEE, Piscataway, 2020)
26. X. Liu, M. Chen, C. Yin, Optimized trajectory design in UAV based cellular networks for 3D users: A double Q-learning approach. J. Commun. Inf. Netw. **4**(1), 24–32 (2019). https://doi.org/10.23919/JCIN.2019.8916643
27. V. Saxena, J. Jaldn, H. Klessig, Optimal UAV base station trajectories using flow-level models for reinforcement learning. IEEE Trans. Cogn. Commun. Netw. **5**(4), 1101–1112 (2019)
28. J. Zhang et al., Trajectory planning of UAV in wireless powered IoT system based on deep reinforcement learning, in *2020 IEEE/CIC International Conference on Communications in China (ICCC)* (IEEE, Piscataway, 2020), pp. 645–650. https://doi.org/10.1109/ICCC49849.2020.9238842
29. A. Al-Hourani, S. Kandeepan, S. Lardner, Optimal LAP altitude for maximum coverage. IEEE Wirel. Commun. Lett. **3**(6), 569–572 (2014)
30. A. Jee, K. Janghel, S. Prakriya, Performance of adaptive multi-user underlay NOMA transmission with simple user selection. IEEE Trans. Cogn. Commun. Netw. **8**(2), 871–887 (2022)
31. K.H. Chiang, N. Shenoy, A 2-D random-walk mobility model for location-management studies in wireless networks. IEEE Trans. Veh. Technol. **53**(2), 413–424 (2004)
32. R. Wu, F. Gu, H. Liu et al., UAV path planning based on multicritic-delayed deep deterministic policy gradient. Wirel. Commun. Mob. Comput. **2022**, 1–12 (2022)

Chapter 3
Intelligent UAV Swarm Routing

The design of routing algorithms for UAV swarm networks presents significant challenges due to their dynamic nature and constrained resources. Traditional routing approaches, such as flooding-based algorithms (e.g., OLSR, AODV), often incur substantial communication overhead, while greedy algorithms suffer from inefficiencies and the risk of converging to local optima. To address these limitations, this chapter explores intelligent UAV swarm routing algorithms. By leveraging replicator dynamics, evolutionary game theory facilitates adaptation to fluctuating network conditions, encouraging UAV nodes to share information and forward data packets. This approach effectively extends network longevity while maximizing packet delivery rates and overall network performance [1]. To further optimize network quality—in terms of survival time, packet delivery rate, and throughput—we propose a Multi-Agent Policy Gradient-based dynamic routing algorithm. This framework adopts centralized training with distributed execution, enabling UAVs to collaboratively learn and refine optimal routing strategies in dynamic network environments [2]. Additionally, a Multilayer Perceptron (MLP) algorithm is introduced to streamline UAV selection for routing information sharing, thereby reducing communication overhead. To optimize data transmission, a Gated Recurrent Unit (GRU) algorithm is integrated to compress communication data. Furthermore, a Multi-Agent Actor-Critic algorithm is designed to enhance routing decision-making, thereby improving global optimization and the stability of routing strategies [3].

3.1 Evolutionary Game-Based Routing Strategy for UAV Swarm Networks

UAVs are aerial devices operated either by wireless remote control or autonomous programs [4]. They are known for their cost-effectiveness and high flexibility.

H. Yao, T. Mai, *UAV Swarm Cooperation*, Wireless Networks,
https://doi.org/10.1007/978-3-031-96444-2_3

With the continuous advancements in Information and Communication Technology (ICT), communication technologies for UAV swarms have rapidly evolved, leading to the emergence of large-scale swarm networks [5]. The applications of UAVs have expanded beyond their traditional military uses and found relevance in civilian domains as well.

UAV swarms exhibit unique characteristics, including large scale, flexibility, and high dynamism. As a key component of intelligent warfare on future battlefields, the autonomous collaborative capability of UAV swarms significantly influences combat effectiveness. High-dynamic networks formed by multiple UAV swarms possess decentralized, self-organizing, multi-hop routing, and dynamic topologies, categorizing them as wireless self-organizing network architectures [6].

As a type of wireless self-organizing networks, UAV swarm networks exhibit dynamic and adaptable network topologies. Establishing an effective communication mechanism between UAV swarms requires research on routing methods specifically designed for UAV swarm networks. Routing algorithms play a crucial role in determining the most efficient route for packet transmission from source nodes to destination nodes by computing the forwarding table. Common routing algorithms include topology-based routing algorithms and geographic-based routing algorithms [7, 8]. However, compared to traditional fixed networks, UAV-based mobile self-organizing networks face challenges such as greater node mobility, rapidly fluctuating network topologies, and reduced link stability. In such highly dynamic scenarios, conventional routing algorithms struggle with issues like elevated computational costs and slow convergence rates. Additionally, their often high computational complexity makes it difficult to meet the demands of energy efficiency and resource optimization, which are crucial for the sustainable operation of UAV networks. Improving energy consumption not only prolongs UAV operational time but also mitigates environmental impact, adhering to sustainable computing principles. To sum up, it is evident that existing traditional routing algorithms are inadequate to meet the communication needs of UAV swarms, necessitating the development of new routing algorithms to support their efficient operation [9].

Game theory is a branch of applied mathematics that has been applied mostly in the field of economics but has also successfully penetrated into many other disciplines such as wireless and communication networks [10, 11]. In recent years, many scholars have applied game theory to the FANET [12, 13], successfully enhancing the performance of mobile ad-hoc networks. Facing with the rational behavior, environmental dynamics, and various preferences of nodes in wireless networks, game theory, as a distributed algorithm, performs well in the modeling, analysis, and design of wireless networks, especially in routing strategy. However, with the increase in the number of UAVs in ad-hoc networks, traditional game theoretic algorithms may become intractable. In highly dynamic FANET, the increasing number of UAVs can lead to rapid decision-making based on traditional game theory models. Therefore, the pace of changes in routing strategies may become too fast, making it challenging to achieve Nash equilibrium.

Evolutionary game theory represents a sophisticated approach within game theory, focused on modeling and analyzing the evolution of populations and the interactions among rational agents. It emphasizes the dynamics of strategic change, wherein optimal strategies evolve over time through the process of adaptation. This framework provides a powerful tool for understanding how strategies develop and adjust within competitive environments. In recent years, the application of evolutionary game theory [14] in routing optimization for wireless self-organizing networks has gained momentum. The evolutionary game theory focuses on mixed strategies and the decision-making dynamics of groups employing these strategies. It explores how individual decision-makers imitate and learn from each other over time, leading to a state where the entire network community benefits. Additionally, it provides more general convergence theories and stable results. Given the widespread application of evolutionary game theory in addressing cooperative problems, utilizing this theory to resolve collaborative issues within UAV swarms holds significant theoretical and practical significance [15]. Evolutionary game theory is a theoretical framework that models group interactions and encompasses multiple disciplines. While traditional game theory emphasizes static equilibria and relatively stable states, evolutionary game theory focuses on dynamic equilibria. It explores how groups, under conditions of bounded rationality, information asymmetry, and deviations from environmental expectations, continuously adapt through learning, imitation, and trial-and-error to reach a stable state known as the Evolutionary Stable Strategy (ESS) [16]. Evolutionary game theory is based on the fundamental assumption of bounded rationality and does not require participants to be fully rational or possess complete information. Additionally, we introduce replicator dynamics in the context of ad-hoc network games, examining how UAV nodes progressively improve their strategy selection over time until convergence to an ESS is achieved. Once the strategies reach ESS, the population becomes resistant to invasion by alternative node populations, resulting in a stable system. From the perspective of evolutionary game theory, it offers meaningful exploration and inspiration for effective collaborative control of UAV swarms and the development of new design methodologies.

The main contributions of this part are listed as follows:

1. We present a novel UAV swarm routing optimization method using evolutionary game theory with low computation complexity to extend the survival time of UAV swarm networks.
2. We study the process of updating the routing strategy by employing replicator dynamics, which provides a framework for modeling the evolutionary adjustment of strategies, confirming the stability and resilience of certain routing strategies in highly dynamic environments.
3. We incorporate cooperative incentives to incentivize cooperation among self-interested UAV nodes. By encouraging nodes to forward data packets from other nodes, our proposed method aims to improve network longevity, maximize the number of successfully forwarded packets, and enhance overall network performance.

3.1.1 System Model and Problem Formulation

In this section, we initially present the system model, routing method, and queue model. Subsequently, we formulate the optimization problem for UAV swarm networks. The key notations used in this paper are summarized in Table 3.1.

3.1.1.1 System Model

As shown in Fig. 3.1, we consider a UAV swarm with N UAV nodes in the environment, which can be denoted as $K = (k_0, k_1, \cdots, k_N)$. This UAV swarm resides within a confined 3D space, delineated by the boundary coordinates $(x_{max}, 0, 0)$, $(0, y_{max}, 0)$ and $(0, 0, z_{max})$. At a discrete time step n, the position

Table 3.1 Summary of key notations

Notation	Definition
K	UAV swarm
$\overrightarrow{p_i(n)}$	The position coordinates of node k_i
PKT_NUM	Number of data packets
PKT_SPD	The rate of generating new data packets
L_{max}	Maximum length of UAV node's has buffer queue
$full\_percent$	Congestion level percentage
$Q_{k_i}(n)$	The queue length of node k_i at time n
$T(n)$	The time that has elapsed up to time n
Suc_{pkt}	The number of data packets transmitted at time i

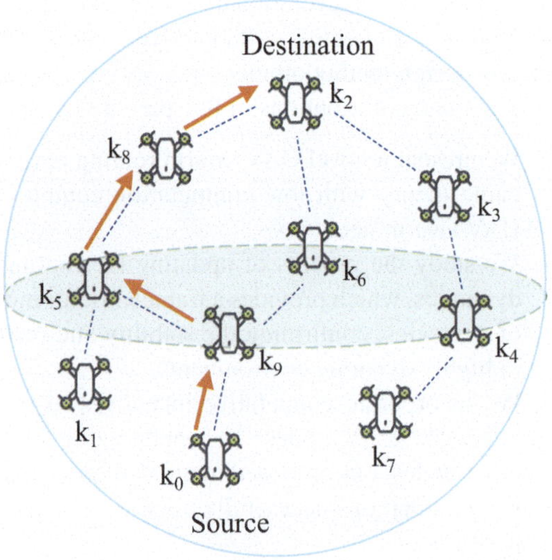

Fig. 3.1 UAV swarm flight environment model

coordinates of node k_i in the environment can be represented as a vector $\overrightarrow{p_i(n)} = [p_i^x(n), p_i^y(n), p_i^z(n)]$, and the velocity can be represented as a vector $\overrightarrow{v_i(n)} = [v_i^x(n), v_i^y(n), v_i^z(n)]$ [17].

Based on previous research [18], we can define the generation and motion rules for UAV nodes in the network. In the initial state, UAV nodes uniformly and randomly select their initial positions in the coordinate range, obtaining initial position coordinates $\overrightarrow{p_i(0)} = [p_i^x(0), p_i^y(0), p_i^z(0)]$. They also uniformly and randomly select their initial velocities within a specified speed range $[v_{min}, v_{max}]$ for all three coordinate dimensions, resulting in the initial velocity of the UAV node $\overrightarrow{v_i(0)} = [v_i^x(0), v_i^y(0), v_i^z(0)]$.

Due to the high dynamic nature of the UAV swarm network, each UAV node continuously moves and updates its position in the environment. Assuming that the three-dimensional movement process from position $p_i(n)$ to $p_i(n+1)$ is updated as follows:

$$p_i^\alpha(n+1) = \begin{cases} p_i^\alpha(n) + v_i^\alpha(n), \text{if } p_i^\alpha(n) + v_i^\alpha(n) \in [0, p_{max}^\alpha] \\ p_i^\alpha(n) - v_i^\alpha(n), \text{otherwise} \end{cases}, \tag{3.1}$$

where $\alpha \in \{x, y, z\}$ represents each dimension in the three-dimensional space. If the movement results in the node's position exceeding the spatial boundaries, the velocity in the corresponding dimension must be reversed, and the movement and position update should occur with that adjusted velocity. Once a node completes its movement and position update, it remains at the new position $p_i(n+1)$ for a specified duration, during which it participates in interaction processes with the environment and the algorithms. After completing the interaction task, the node will update its velocity within a certain range, namely $[v_{min}, v_{max}]$, with a certain probability p. At this point, the UAV node completes a single period of motion. In the context of this part, as the number of game periods continues to increase, the motion of the UAV nodes will repeat the above process until the game stage concludes.

3.1.1.2 Routing Model and Queue Model

In the routing model presented in this part, each UAV node can only communicate with other UAV nodes within its communication range. The data packet delivery and forwarding process follows a hop-by-hop routing scheme [19]. In the UAV swarm network, it is evident that a typical path between the source and destination nodes consists of multiple hops, with intermediate nodes serving as relays. In other words, each UAV node's data packet, depending on its destination, will successively choose neighboring UAV nodes as the next hop until it reaches the intended destination.We assume that there are $N = 10$ UAVs in the environment (corresponding to $K = (k_0, k_1, \cdots, k_8, k_9)$), as shown in Fig. 3.1. Each UAV node is randomly distributed in the 3D space. In the example shown in Fig. 3.1, the transmission path of a data packet from the source node S to the destination node D is as follows. It starts from node k_0 (source node) and passes through the following nodes in order: k_9, k_5 and k_8, and finally reaches k_2 (destination node).

In the initial state of the system, each UAV node has a certain number of data packets, denoted as PKT_NUM. As time progresses, nodes generate new data packets at a rate of PKT_SPD. Each UAV node has an internal buffer queue with a maximum length, denoted as L_{max}, and follows a queuing model for data packet insertion and removal. When a new data packet arrives, it is placed at the end of the queue, and when a data packet is sent, it is removed from the front of the queue. The rate at which the queue length of a UAV node grows is equal to the rate at which data packets enter the buffer queue minus the rate at which data packets leave the buffer queue. By setting the buffer queue, you can prevent significant traffic fluctuations due to data packet transmissions. When the queue length of a node reaches its maximum value, it is considered a network congestion, and any incoming data packets beyond that point will be discarded. Therefore, we define a congestion level percentage, $full\_percent$, which is calculated by dividing the current queue length of the node by the maximum queue length of the node. A higher percentage indicates a higher likelihood of congestion.

3.1.1.3 Problem Formulation

The performance evaluation of routing algorithms for UAV swarm networks relies on two critical criteria: network load and data packet transmission. In the described routing model, congestion occurs when the buffer queue of any UAV node reaches its maximum capacity. The primary goal is to proactively prevent congestion, thereby extending the network's operational duration from its initial state to the onset of congestion, referred to as the network survival time. To monitor the network load, we use the maximum buffer queue length across all UAV nodes. The aim is to minimize the growth rate of this maximum value, thereby continuously extending the network's survival time. Additionally, the average number of successfully transmitted data packets during the network survival time serves as a key metric to assess the efficiency and effectiveness of data transmission. These two factors—network load and transmission efficiency—jointly determine the overall performance of the routing algorithm. Accordingly, the optimization objective can be mathematically defined as follows:

$$\lim_{n\to\infty}(min[max(Q_{k_i}(n))] + max[\frac{1}{T(n)} \times \sum_{i=0}^{T(n)-1} Suc_{pkt_i}]). \tag{3.2}$$

In Eq. (3.34), the first part is dedicated to the minimization of the maximum buffer queue length among UAV nodes. Here, $Q_{k_i}(n)$ represents the queue length of node k_i at time n. The second part aims to maximize the number of successfully transmitted data packets within the network's survival time. $T(n)$ signifies the time that has elapsed up to time n, and Suc_{pkt_i} represents the number of data packets successfully transmitted at time i. It is important to emphasize that data packets are assumed to be transmitted from one node to another, with a preference for neighboring nodes closer to the destination, thereby showcasing the algorithm's ability to minimize the transmission distance. The effectiveness of the routing algorithm is ultimately evaluated through the use of curve plotting, which provides insights into the algorithm's ability to mitigate congestion and increase the number of successfully transmitted data packets.

3.1.2 Evolutionary Game Theory-Based Routing Algorithm

In this section, we elaborate on the algorithm design and implementation process based on evolutionary game theory. We provide a detailed explanation about the evolution process of strategies for UAV swarms.

3.1.2.1 Selection and Mutation

In general, the construction of evolutionary game models involves two fundamental aspects: selection and mutation. The former signifies that strategies that yield higher payoffs after a game will be adopted by participants in subsequent games. The latter refers to the random choice of strategies different from those chosen by the majority of the population by some participants. This strategy may lead to either higher or lower payoffs for the individual. Essentially, mutation is an ongoing trial-and-error process, and it is also an adaptive and continuously improving selection process based on imitation and learning. Therefore, in evolutionary game models, both of these processes are indispensable.

3.1.2.2 Evolutionary Game Formulation

Evolutionary game theory can be represented as a triplet $G =< P, S, F >$, where P represents the players in the game, S is the set of strategies chosen by the players, and F denotes the payoff matrix of the game. In the routing method we design, the objective is to maintain the network in a good state by allowing UAVs to autonomously decide whether to participate in forwarding. Therefore, we define the game triplet for the UAV swarm routing method as follows:

Players The N UAV nodes in the network, denoted as $P = \{A, B\}$, where A represents all individuals choosing the cooperative strategy, and B represents all individuals choosing the non-cooperative strategy.

Strategies The set of strategies available to the players in the game includes two options, cooperation and defection, denoted as $S = \{Cooperate, Defect\}$, where

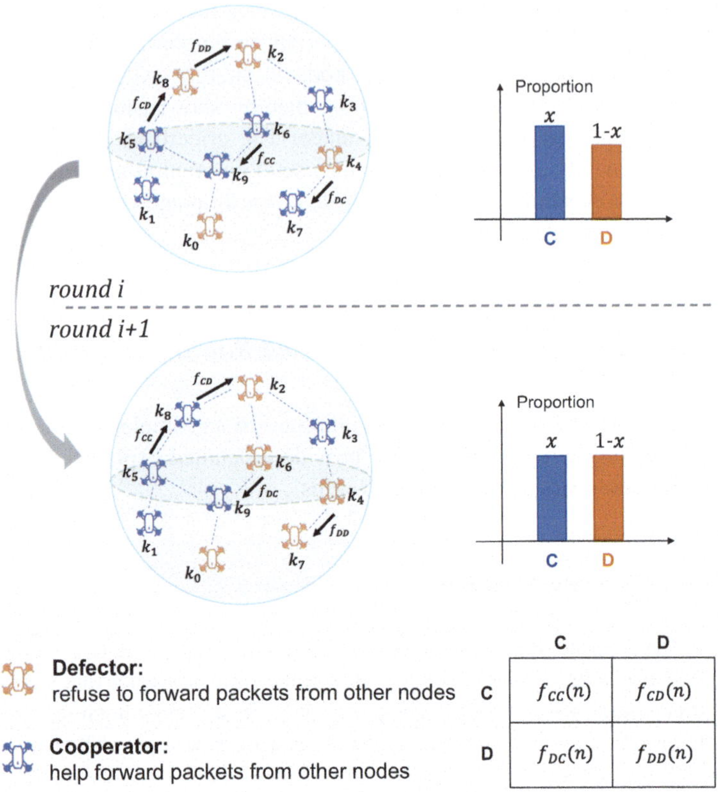

Fig. 3.2 Evolutionary game theory-based routing optimization in UAV networks

Cooperate represents the cooperative strategy, and *Defect* represents the non-cooperative strategy. When a node opts for the *Cooperate* strategy, it not only transmits the data packet at the head of its own buffer queue to neighboring nodes that have also chosen the *Cooperate* strategy but also accepts data packets transmitted to it from other nodes. On the other hand, when a node selects the *Defect* strategy, it only forwards its own data packets to neighboring nodes that have chosen the *Cooperate* strategy and will not accept data packets from other nodes.

Payoff We design a dynamic adjustment mechanism for the payoff matrix, which is based on the network's congestion conditions. This mechanism enables nodes to dynamically adapt their strategy choices in response to changing network states. The details of the payoff function are discussed in the next subsection.

In the continuous process of game theory and data packet forwarding between nodes, the UAV nodes interact continuously with each other and adjust the various parameter values in the payoff matrix based on the provided queue status, influencing the direction of the game's evolution. As shown in Fig. 3.2, after a round of gaming, the algorithm, following the principle of imitating the best performer [20], guides the strategy selection of each node for the next round.

During the game, when the lead data packet leaves the original UAV node, the length of the node's buffer queue will decrease by 1. At the same time, another neighbor node that is closer to the destination node of the data packet and selects a cooperative strategy will put the data packet at the end of its own queue and increase its own buffer queue length by 1, thereby completing the forwarding operation of the data packet.

In the routing scenario built in this part, we hope to maximize the network's survival time and avoid congestion as much as possible. Therefore, when the buffer queue of UAV nodes is about to reach its maximum length L_{max}, we hope that nodes can avoid cooperation and have a higher probability of choosing to refuse cooperation, i.e., not forwarding data packets from other nodes, to reduce their buffer queue length. Therefore, in the designed routing scenario, when the length of the node's buffer queue first reaches $(L_{max} - 5)$, regardless of the strategy chosen by the node in the environment at this time, the algorithm will force it to choose a non-cooperation strategy. In other cases, the algorithm automatically adjusts the benefit function of the environment based on the congestion percentage $full\_percent$, constantly changing the direction of game evolution and improving network survival time.

3.1.2.3 Payoff Function

In the game and packet forwarding process, we need to define the cost of packet forwarding in the network and the payoff of whether nodes choose to forward packets.

We assume that the cost of each node to transmit packets to other UAV nodes within its communication range is a constant value C [21], which does not change with the distance between nodes. In addition, we assume that each UAV node can choose to forward or refuse to forward packets from other nodes according to its current payoff status, i.e., corresponding to the two strategies of cooperation and refusal to cooperate [22]. The payoff values corresponding to different strategies chosen by nodes are also different, as shown in Table 3.2.

Then we can obtain the payoff matrix in the process of data packet forwarding, as shown in Table 3.3.

According to the length of the node packet queue in the environment, the four parameter values in Table 3.3 will be adjusted in time to prevent congestion to the greatest extent. In the initial situation, the buffer queue length of each node is PKT_NUM, so congestion will not occur in a short time. Considering our second optimization objective, which is to maximize the number of successfully transmitted packets, we hope that in the early stage of the game, each UAV node will choose the cooperation strategy as much as possible. Therefore, we set a large cooperation incentive value to encourage cooperation among UAV nodes.

Table 3.2 The parameter meaning of packet forwarding in system model

Parameter	Meaning
R_T	Payoff when a node chooses the cooperation strategy
R_D	Payoff when a node chooses the defection strategy
R_C	Additional payoff when the opponent chooses the cooperation strategy
C	Cost paid by a node to forward a packet
W	Additional payoff when a node chooses the cooperation strategy (i.e. cooperation incentive)
L	Loss when a node chooses the cooperation strategy but the opponent refuses to cooperate

Table 3.3 The payoff matrix of packet forwarding process in system model

Payoff \ Strategy Strategy	Cooperate	Defect
Cooperate	$R_T + R_C + W - C$, $R_T + R_C + W - C$	$R_T + W - C - L$, $R_C + R_D$
Defect	$R_C + R_D$, $R_T + W - C - L$	R_D, R_D

Furthermore, we note that as the data packet forwarding and game process continue, the buffer queue length of UAV nodes is gradually increasing, which also means that congestion is more likely to occur. Considering our first optimization objective, which is to maximize the network survival time, we should prevent nodes from choosing the cooperation strategy and forwarding data packets from other nodes. Therefore, we set the $full\_percent$ to dynamically adjust the cooperation incentive parameter W in the payoff matrix, thereby influencing the node's strategy update process and changing the ultimate evolution direction of the game. The adjustment method of the parameter is shown in Eq. (3.3).

$$W\_adjust = W \times (1 - full\_percent), \tag{3.3}$$

where

$$full\_percent = \frac{(Queue.length - PKT\_NUM)}{(Queue.L_{max} - PKT\_NUM)}, \tag{3.4}$$

and $Queue.length$ represents the current length of the node's buffer queue, $Queue.L_{max}$ represents the maximum acceptable length of the buffer queue.

Additionally, when the buffer queue length of a node is approaching the maximum acceptable length, it indicates that congestion is about to occur. At this time, in order to prolong the network survival time, cooperation between nodes needs to be further suppressed. Therefore, in such scenarios, it becomes necessary to manually modify and adjust the node's strategy before the data packet forwarding process (i.e., the begin of the game). Regardless of the strategy chosen by the node in that round of the game, we forcefully modify its strategy to defection, i.e., refusing to forward data packets from other nodes. This setting implies that when the buffer queue lengths of all nodes reach $(L_{max} - 5)$, none of the nodes will choose the cooperation strategy, resulting in a decrease in the number of successfully transmitted data packets. It is not difficult to discern that there exists a certain contradiction between the two optimization objectives we aim to achieve. We can only prioritize one optimization objective at different stages of the game and design routing methods accordingly.

3.1.2.4 Replicator Dynamics

The replicator dynamic equation [23] proposed by Taylor and Jonker is a model used to describe the evolution process of group strategies. According to Table 3.3, we can obtain the replicator dynamic equation in this scenario as follows:

$$F(x_i) = \frac{dx_i}{dt} = \dot{x_i} = x_i \left[\mathrm{U}(i, X) - \bar{\mathrm{U}}(X, X)\right], \tag{3.5}$$

where $\dot{x_i}$ represents the change in the proportion of drones choosing strategy $i$over time, reflecting the dynamic evolution of the strategy group over time and $i \in \{Cooperate, Defect\}$ is the strategy of UAV node X is the set of proportions of UAVs with different strategies selected. $\mathrm{U}(i, X)$ represents the expected utility of UAV nodes who select strategy i.By comparing individual returns and group average returns, the equation indicates that strategies with higher returns will gradually dominate, thereby affecting the overall evolution of the group. $\bar{\mathrm{U}}(X, X)$ indicates the average payoff of the whole UAV swarm:

$$\bar{\mathrm{U}}(X, X) = \sum_{i=1}^{k} x_i \mathrm{U}(i, X), \tag{3.6}$$

where k is the number of strategies. Equation (3.48) implies that if the utility of UAV node is higher than the average (i.e. $\mathrm{U}(i, X) > \bar{\mathrm{U}}(X, X)$), then the number of UAV nodes for selecting strategy i will increase (i.e. $\frac{dx_i}{dt} > 0$), and vice versa. In addition, if the utility of UAV node is equal to the average (i.e. $\mathrm{U}(i, X) = \bar{\mathrm{U}}(X, X)$), the number of UAV nodes for selecting strategy i will remain unchanged.

As we discussed before, there are two optional strategies for UAVs: *Cooperate* or *Defect*, which means $k = 2$ in this part. We assume that x is the proportion of UAVs that choose to cooperate, and $(1 - x)$ is the proportion of UAVs that choose to defect. Therefore, the expected utility of choosing the *Cooperate* strategy for UAVs is:

$$\begin{aligned} \mathrm{U}\left(Cooperate, X\right) &= \sum_{j=1}^{2} x_j \mathrm{u}_{1j} \\ &= x\left(R_T + R_C + W - C\right) \\ &+ \left(1 - x\right)\left(R_T + W - C - L\right), \end{aligned} \tag{3.7}$$

and the expected utility of choosing the *Defect* strategy for UAVs is:

$$\begin{aligned} \mathrm{U}\left(Defect, X\right) &= \sum_{j=1}^{2} x_j \mathrm{u}_{2j} \\ &= x\left(R_C + R_D\right) + \left(1 - x\right) R_D. \end{aligned} \tag{3.8}$$

Then, the average payoff of the whole UAV swarm can be calculated by:

$$\begin{aligned} \bar{\mathrm{U}}\left(X, X\right) &= \sum_{j=1}^{2} x_j \mathrm{U}\left(i, X\right) \\ &= x\mathrm{U}\left(Cooperate, X\right) \\ &+ \left(1 - x\right)\mathrm{U}\left(Defect, X\right). \end{aligned} \tag{3.9}$$

Therefore, the replicator dynamic equation of *Cooperate* strategy can be described as:

$$\begin{aligned} F(x) &= x[\mathrm{U}\left(Cooperate, X\right) - \bar{\mathrm{U}}\left(X, X\right)] \\ &= x[(1 - x)\mathrm{U}\left(Cooperate, X\right) - (1 - x)\mathrm{U}\left(Defect, X\right)] \\ &= x(1 - x)[(R_T + W - C - L - R_D) + Lx], \end{aligned} \tag{3.10}$$

similarly, the replicator dynamic equation of *Defect* strategy can be described as:

$$\begin{aligned} F(1 - x) &= (1 - x)[\mathrm{U}\left(Defect, X\right) - \bar{\mathrm{U}}\left(X, X\right)] \\ &= x(1 - x)[\mathrm{U}\left(Defect, X\right) - \mathrm{U}\left(Cooperate, X\right)] \\ &= x(1 - x)[(R_D - R_T - W + C + L) - Lx]. \end{aligned} \tag{3.11}$$

3.1.2.5 Evolutionary Stable Strategy

In this subsection, we discuss the ESS during the forwarding process. The ESS is a fixed point of the replicator dynamics, and reaching it indicates that the system has achieved a stable equilibrium state. At this point, the proportion of the population

Algorithm 4 UAV swarm routing optimization algorithm based on evolutionary game

```
for uav = k_1, k_2, ··· , k_N do
    Set the payoff matrix Payoff_Matrix and the buffer queue length of each node Queue.length
    if Queue.length ≥ (L_max − 5) then
        UAV node strategy uav.strategy = 'Defect'
    else
        full_percent = (Queue.length−PKT_NUM) / (Queue.L_max−PKT_NUM)
        W = W × (1 − full_percent)
    end if
    Dequeue the first packet packet
    Find the set of neighbor nodes neighbor_uav that are closer to the destination node of the packet
    if packet.dst_id in neighbor_uav then
        Transmit packet to the destination node dst_id
    else
        Remove the nodes that refuse to cooperate in neighbor_uav to get a new collection
        if neighbor_uav is empty then
            packet re-enters the buffer queue of uav
        else
            Randomly select a node another_uav in neighbor_uav
            packet enters the buffer queue tail of the another_uav
        end if
    end if
    Calculate the profit value of this game uav.profit according to the adjusted income matrix
end for
for uav = k_1, k_2, ··· , k_N do
    Find the neighbor node collection neighbor_uav_set of uav
    for neighbor_uav in neighbor_uav_set do
        if neighbor_uav.profit > uav.profit then
            uav.next_strategy = neighbor_uav.strategy
        end if
    end for
end for
for uav = k_1, k_2, ··· , k_N do
    Update node strategy:
    uav.strategy = uav.next_strategy
    Reset node payoff uav.profit = 0
end for
```

that chooses any strategy reaches stability and the payoff of each UAV node will no longer change (i.e., $\dot{x_i} = \frac{dx_i}{dt} = 0$).

As discussed above, we set Eq. (3.10) equal to 0 to derive the ESS point of the system:

$$
\begin{aligned}
F(x) = 0, \\
x(1-x)[(R_T + W - C - L - R_D) + Lx] = 0.
\end{aligned}
\tag{3.12}
$$

Therefore, we can obtain the three solutions of the replicator dynamic equation as follows:

$$\begin{cases} x_1 = 0, \\ x_2 = 1, \\ x_3 = (R_D + C + L - R_T - W)/L. \end{cases} \tag{3.13}$$

Based on the preceding discussion, we assign a large value to the parameter W in the initial state. Since all parameters in the payoff matrix are positive numbers, this setting also results in the third solution x_3 of the replicator dynamic equation being negative. However, due to the fact that the proportion of individuals choosing a certain strategy is $x \in [0, 1]$, x_3 is not a solution that satisfies the problem. For the other two solutions x_1 and x_2, we calculate the first derivative of Eq. (3.10) as follows:

$$\begin{aligned} F'(x) &= \frac{dF(x)}{dx} \\ &= (1-2x)(R_T + W - C - L - R_D + Lx) + (x - x^2)L, \end{aligned} \tag{3.14}$$

then, we substitute the two solutions x_1, x_2 to obtain $F'(x_1) = F'(0) > 0$ and $F'(x_2) = F'(1) < 0$. According to the judgment conditions of evolutionary stable strategy, it is easy to conclude that x_2 is a solution that satisfies the problem, i.e., in the initial state of the game between UAV nodes for data packet forwarding, the evolutionary stable strategy is that all UAV nodes choose the cooperation strategy, i.e., choose to forward data packets from other nodes.

In addition, to prevent the entire UAV swarm network from converging to a non-ESS Nash equilibrium, we have also introduced a mutation probability to avoid such a phenomenon. The introduction of mutation probability enables drone networks to have higher flexibility, adapt to uncertainties and changes in the environment, and thus enhance the dynamic adjustment capability of routing strategies. The relationship between ESS and the mutation strategy is expressed by

$$(1-\epsilon)U(\tilde{S}, \tilde{S}) + \epsilon U(\tilde{S}, S') > (1-\epsilon)U(S', \tilde{S}) + \epsilon U(S', S'), \tag{3.15}$$

where $\tilde{S}$ represents the evolutionary stable strategy, S$'$ represents a mutated strategy, ϵ represents the proportion of individuals in the population that select the mutated strategy, and $U(\tilde{S}, S')$ represents the expected payoff of the population when two behavioral agents select strategies $\tilde{S}$ and S' respectively. Assume ϵ approaches 0, then Eq. (3.15) can be converted into either

$$U(\tilde{S}, \tilde{S}) > U(S', \tilde{S}), \tag{3.16}$$

or

$$U(\tilde{S}, \tilde{S}) = U(S', \tilde{S}) \ and \ U(\tilde{S}, S') > U(S', S'), \tag{3.17}$$

which indicates that an ESS must be a Nash equilibrium [24]. Through (3.15), it can be seen that the stability of ESS depends to some extent on the proportion of mutated individuals *epsilon*, which contributes to the diversity of group strategies and prevents the system from falling into non-ESS optimal solutions. The process of a one-round game and data packet forwarding in the highly dynamic UAV swarm scenario is described in Algorithm 4.

3.1.3 Simulation Results and Discussions

In this section, we will design several simulation experiments for the UAV swarm routing optimization algorithm based on evolutionary game theory. The performance of the algorithm will be assessed through the results of these simulation experiments to validate its effectiveness.

3.1.3.1 Environment Settings

We use Python to conduct simulation experiments for the UAV swarm routing optimization algorithm based on evolutionary game theory. The hardware environment was an Intel(R) Core(TM) i7-8550U with 16GB RAM and an NVIDIA Geforce MX150. The software environment was PyCharm Community Edition 2021.1.3 x64 and Windows 11 Professional. The numerical values of the parameters in the system environment are presented in Table 3.4, and the parameter settings for the simulation experiments and evolutionary game method are shown in Table 3.5.

As shown in Table 3.5, we designed the experiment to give a higher payoff for choosing the cooperation strategy, and motivated cooperation between nodes by providing a higher initial payoff value. In this simulation experiment, we conducted a total of 200 episodes, each of which supports a maximum of 100 rounds of game. In each episode, if network congestion occurs, we immediately record the number

Table 3.4 System model parameter setting

Parameter	Value
Number of UAV nodes N	10, 15, 25
UAV node motion range boundary $P^{\alpha}_{max}, \alpha \in \{x, y, z\}$	10 m
Motion speed of UAV nodes v_{UAV}	[−1.0 m/s,1.0 m/s]
Velocity change probability of UAV nodes p	0.25
Initial number of packets per node PKT_NUM	10
Speed of packets generating per node PKT_SPD	1/s
Maximum length of a node buffer queue L_{max}	50

Table 3.5 Algorithm parameter setting

Parameter	Value
Maximum number of simulation episodes $Episodes$	200
Maximum number of games per phase $Periods$	100
Strategy mutation probability $Probability$	0.01
Payoff when a node chooses the cooperation strategy R_T	12.5
Payoff when a node chooses the defection strategy R_D	10
Additional payoff when the opponent chooses the cooperation strategy R_C	7
Loss when a node chooses the cooperation strategy but the opponent refuses to cooperate L	3
Additional payoff when a node chooses the cooperation strategy W	15
Cost paid by a node to forward a packet C	5

of completed games at the time of congestion, i.e., the network's survival time, and continue the current test until the maximum number of game round is reached before moving on to the next test.

3.1.3.2 Comparison Algorithms

In this part, we refer to the proposed algorithm as EGT for brevity. In order to verify the superiority of the EGT algorithm, we designed corresponding comparison experiments. In our scenario, we mainly designed two types of comparison experiments to comprehensively evaluate the performance of the algorithm:

(1) RANDOM: During the data packet forwarding process, each UAV node randomly selects a neighboring node within its communication range to transmit the first data packet in its own queue.
(2) RUS: At the end of each game process, when updating strategies, each UAV node randomly updates its strategy of the next round, i.e., it does not follow the principle of imitating the best performer in the original algorithm to update its strategy, and then continues to participate in the next round of games.

In the first comparison experiment designed, we weakened the concept of node strategy, emphasizing more on the randomness of data packet forwarding process. This is essentially a special case of our designed algorithm, where all UAV nodes choose the cooperation strategy in each round of the game. In the second comparison experiment designed, we weakened the concept of node payoff and did not adopt the principle of imitating the best performer based on node payoff in the original algorithm. Instead, we emphasized more on the randomness of node strategy updates.

In addition, we also set up two scenarios for the UAV nodes in the experiment: static and high dynamic. In the static network, all UAV nodes are in a hovering state, which can achieve the best algorithm performance and highlight the algorithm's efficiency. In the high dynamic network, the UAV nodes are in a moving state, which can more authentically simulate a high dynamic UAV swarm network. In the high dynamic scenario, we set some even-numbered UAVs, i.e., UAV k_i, where $i = 0, 2, 4, 6$ as the backbone nodes in a hovering state, while the remaining UAVs are all mobile.

3.1.3.3 Performance Analysis

We recorded the maximum number of games (i.e., network survival time) and the number of successfully transmitted data packets in each test episode, and plotted their trend lines as the number of test periods increases. In this scenario, we selected these two network indicators to analyze the performance of our algorithm.

(1) *Network survival time:* This network metric is characterized by the duration it takes for the UAV swarm network to transition from its initial state to a state of congestion. If our algorithm can extend the network's survival time, it means the algorithm can effectively mitigate and alleviate congestion.
(2) *Number of successfully transmitted packets:* This network metric is defined as the number of data packets that are successfully transmitted from the source node to the destination node within the network survival time. The designed algorithm is expected to not only extend the network's survival time but also optimize the efficiency of data packet transmission in the network.

As shown in Fig. 3.3, in the static scenario, compared to the RANDOM algorithm and the RUS algorithm, our designed algorithm has the best performance. Figure 3.3a demonstrates that our proposed algorithm EGT achieves the highest average network survival time of about 65, while the average network survival time of the RANDOM algorithm and the RUS algorithm is approximately between 40 and 55, with significant oscillation and fluctuations as the number of test episodes increases. This is because the RANDOM algorithm is a special case of our proposed algorithm, where all nodes choose the $Cooperate$ strategy. Although each node can frequently send packets from its own queue, it also frequently receives packets from other nodes, which can lead to a surge in the number of packets transmitted in the network and accelerate network congestion. While in the RUS algorithm, the selection of forwarding strategy is random, which means that the probability of nodes choosing to cooperate or not to cooperate is basically consistent. When there are many nodes in the network that choose the $Defect$ strategy, the transmission of data packets will be hindered, which can lead to congestion among nodes. Furthermore, the EGT algorithm exhibits the most stable performance among these three algorithms. This is because in the other two algorithms, there is a lack of learning and evolution between nodes, making it difficult to achieve stability.

In addition, as shown in Fig. 3.3b, our designed algorithm EGT has a transmission success rate of about 500 packets, which is much higher than the RANDOM algorithm and the RUS algorithm. This indicates that in RANDOM algorithm, although the transmission of data packets is not hindered due to the cooperation of nodes, a decrease in network lifetime will lead to a decrease in the number of

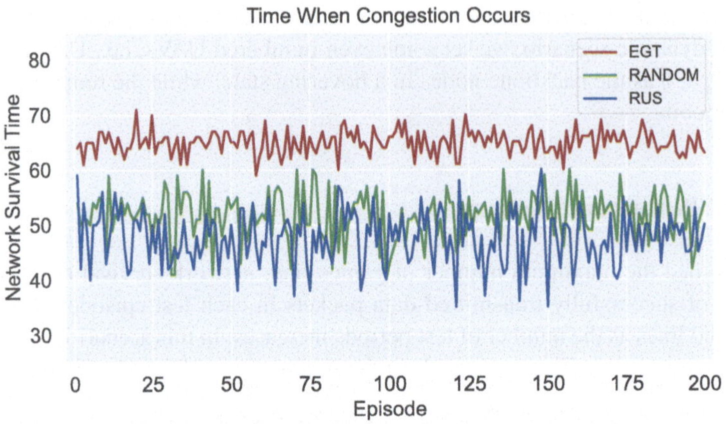

(a) network survival time

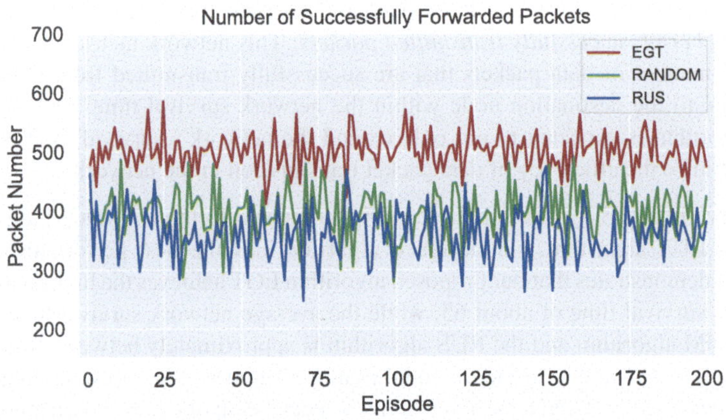

(b) number of successfully transmitted packets

Fig. 3.3 Network survival time and number of successfully transmitted packets in static scenario. (**a**) Network survival time. (**b**) Number of successfully transmitted packets

successfully transmitted data packets. Similarly, in the RUS algorithm, the early occurrence of network congestion also results in a decrease in the number of successfully transmitted packets. Therefore, in regard to both network survival time and the number of successfully transmitted data packets, our designed algorithm outperforms both the RANDOM algorithm and the RUS algorithm significantly, demonstrating higher stability.

Then, the performance of three algorithms in high dynamic environments can be found in Fig. 3.4. We can observe that in the high dynamic scenario, our proposed algorithm still outperforms the two algorithms in terms of network survival time

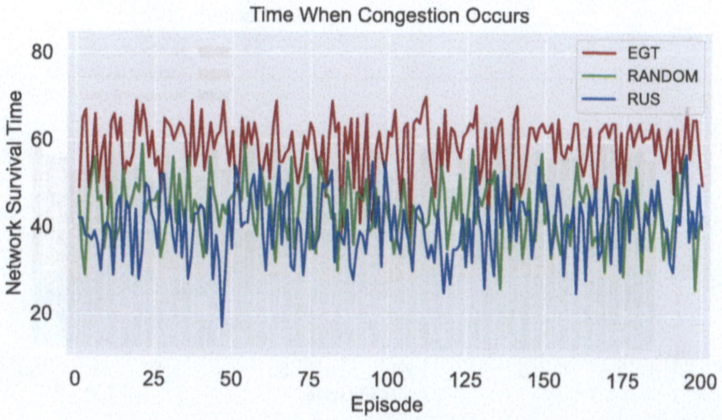

(a) network survival time

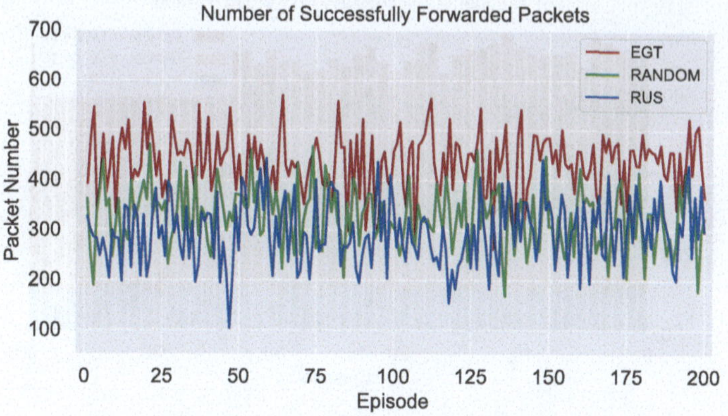

(b) number of successfully transmitted packets

Fig. 3.4 Network survival time and number of successfully transmitted packets in high dynamic scenario. (**a**) Network survival time. (**b**) Number of successfully transmitted packets

and number of successfully transmitted packets. Besides, in dynamic environments, the network performance curve fluctuates sharply due to the increase in network dynamics, and the network performance also decreases to a certain extent. This is because in dynamic environments, the neighbors of UAV nodes will change over time, and the strategies of nodes will also change frequently, which can cause instability and decrease in network performance. To sum up, our proposed evolutionary game-based UAV swarm routing method has a certain optimization effect in high dynamic scenarios. Figure 3.5 shows the performance of our proposed algorithm EGT in the static environment when the number of UAV nodes changes. In this situation, we calculate the average based on every set of 4 episodes out of

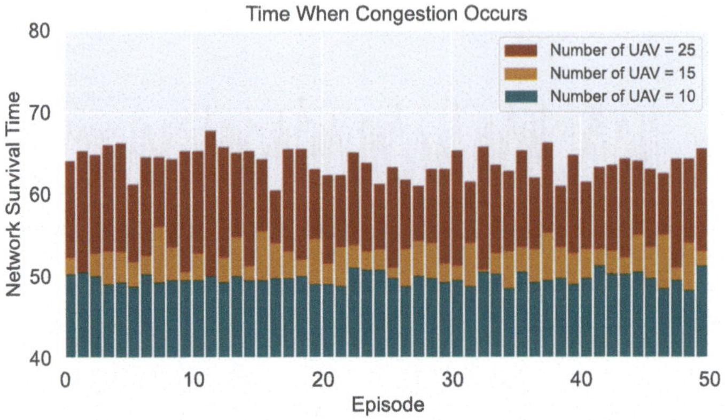

(a) Network survival time

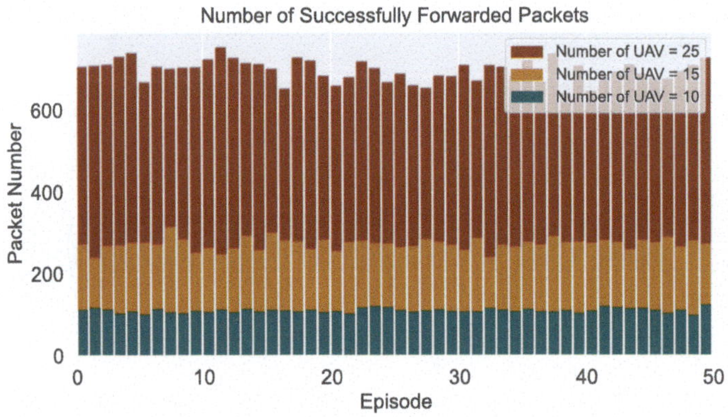

(b) Number of successfully transmitted packets

Fig. 3.5 Network survival time and number of successfully transmitted packets in static scenario v.s. the number of UAVs. (**a**) Network survival time. (**b**) Number of successfully transmitted packets

the total 200 to obtain the simulation results. As shown in Fig. 3.5a, the network survival time increases with the number of UAV nodes while the stability of the curve decreases. This is because the increase in the number of UAV nodes implies that, for a given UAV node, there are more neighboring nodes. Consequently, there are more UAVs available to serve as forwarding nodes, thereby extending the time before node congestion occurs. Therefore, the trend in the number of successfully transmitted packets, as depicted in Fig. 3.5b, aligns with the pattern observed in Fig. 3.5a. An increase in the number of UAV nodes with a longer network survival time leads to a higher count of successfully transmitted data packets.

3.2 Multi-Agent Reinforcement Learning-Based Routing Strategy for UAV Swarm Networks

The past few years have witnessed the compelling applications of the UAV swarm ranging from military to commercial domains. With the explosion of various relevant applications [25, 26], the expectations for reliable and efficient UAV swarm networks are more significant than ever [27]. However, compared to the fixed terrestrial networks, the high-speed mobility of UAV nodes has resulted in a new clan of networks known as FANETs. This new network presents many challenges in routing scheme design, such as dynamic topology, unstable link [28].

Recently, a rapidly growing number of studies have investigated designing the effective UAV swarm routing algorithm. In [29], Khaledi et al. proposed a distance-greedy routing algorithm, where each UAV node makes routing decisions based on the local observation. In [30], Wen et al. proposed a distributed routing method, in which the nodes transmit packets with the help of other cooperative nodes. These distributed schemes have severe non-convergence problems, especially when seeking a global optimum. In [31], Coelho et al. proposed a centralized routing solution that can select high-capacity paths among UAVs and avoid communications disruptions. In [32], Detti et al. proposed the Software-Defined Networking (SDN) based routing management method to increase the overall throughput of the network. However, the centralized solutions require collecting massive data from the controller, thus exhibiting delay to respond to network dynamics.

In this part, inspired by the recent success of multi-agent reinforcement learning, we propose a hybrid routing control algorithm. We adopt a centralized training and decentralized executing framework, where a centralized training platform is implemented to guide the policy updating of each UAV node. Note that compared to the centralized solution, the centralized platform in our approach only acts as a policy critic rather than the controller. Based on this, we design a multi-agent policy gradients-based UAV routing algorithm. By adopting the centralized training and distributed execution framework, the UAV nodes can learn the globally optimal policy cooperatively. Besides, we introduce a counterfactual baseline scheme to explicitly calculate each agent's contribution to the global reward to enhance the convergence speed. Extensive simulation results are provided to demonstrate the effectiveness and feasibility of our proposed algorithms.

Our main contributions are summarized as follows.

- We introduce a multi-agent reinforcement learning algorithm into the highly dynamic network environment. This method can form an efficient multi-hop routing mechanism by using the centralized training decentralized execution framework and a counterfactual baseline scheme.
- Our proposed dynamic routing method can jointly optimize network quality in terms of survival time, packet delivery rate, and throughput.
- We have conducted extensive experiments in different dynamic network environments to verify the effectiveness of our deployed methods.

3.2.1 System Model and Optimization Problem

In this section, we present our system model, routing model, and problem formulation.

3.2.1.1 System Model

As shown in Fig. 3.6, we consider a set of UAV nodes $K = [k_0, k_1, \ldots, k_n]$ distributed in a set of zones $Z = [zone_0, zone_1, \ldots, zone_m]$. We denote the system boundary as $(0, 0, 0)$, $(p^x_{max}, 0, 0)$, $(0, p^y_{max}, 0)$, $(0, 0, p^z_{max})$, where p^x_{max}, p^y_{max} and p^z_{max} are the maximum boundary values of the X, Y, and Z dimensions, respectively. At discrete-continuous time n, the position of node $k_i \in K$ can be described as: $\overrightarrow{P_{k_i}(n)} = (p^x_{k_i}(n), p^y_{k_i}(n), p^z_{k_i}(n))$, and its speed can be described as: $\overrightarrow{V_{k_i}(n)} = (v^x_{k_i}(n), v^y_{k_i}(n), v^z_{k_i}(n))$. The movement in each dimension can be formulated as follows:

$$p^{\varphi}_{k_i}(n+1) = \begin{cases} p^{\varphi}_{k_i}(n) + v^{\varphi}_{k_i}(n), \text{if } p^{\varphi}_{k_i}(n) + v^{\varphi}_{u_i}(n) \in [0, p^{\varphi}_{max}) \\ p^{\varphi}_{u_i}(n) - v^{\varphi}_{u_i}(n), \text{else (need speed change)} \end{cases} \tag{3.18}$$

where $\varphi \in (x, y, z)$ represents different dimensions in 3-D space. Moreover, we use the 3-D Mixed Mobility Model [33] to formulate the mobility of UAV nodes. This model is based on the traditional MANET (Mobile Ad Hoc Network) Mobility

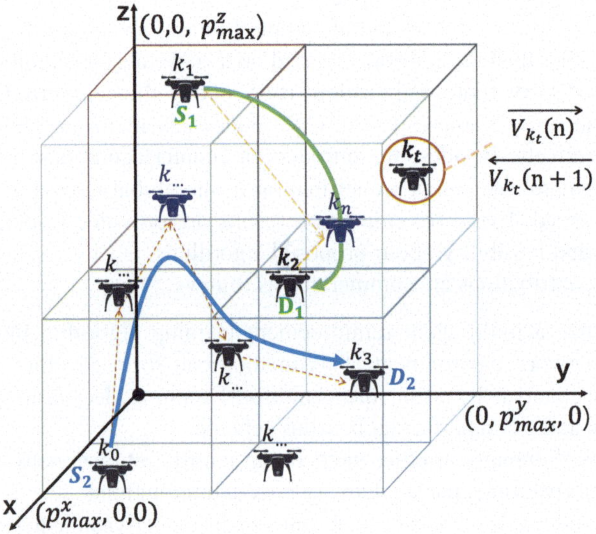

Fig. 3.6 System model

Models [34] and therefore can factually simulate the movement of UAV nodes. Firstly, the movement process for each UAV node on Z-axis is modeled as a Random Way Point Mobility Process, which includes the following steps:

(1) The UAV node randomly selects height h_t from $[p^l_{zone_i} \sim p^h_{zone_i}]$ as initial height, where $p^l_{zone_i}$ and $p^h_{zone_i}$ represent the lower bound and upper bound of $zone_i$.
(2) The UAV node at h_t randomly selects height h_{t+1} from $[p^l_{zone_i} \sim p^h_{zone_i}]$ and moves towards it with a speed choosing uniformly random from $[V_{min} \sim V_{max}]$.
(3) Once the node reaches h_{t+1}, it has a certain probability p_s of keep staying for a period of time T_s. The T_s is selected randomly from $[T_{min} \sim T_{max}]$, which corresponds to the minimum staying time and maximum staying time, respectively.
(4) The UAV nodes repeat the movement process from step (2) to step (3). In the above steps, the staying probability p_s can be calculated by the following equation:

$$p_s = \frac{\mathbb{E}[T_s]}{\mathbb{E}[T_s] + \mathbb{E}[T_m]}, \tag{3.19}$$

where

$$\mathbb{E}[T_m] = \frac{ln(V_{max}/V_{min})}{V_{max} - V_{min}} \times \frac{p^z_{max}}{3} \tag{3.20}$$

and $\mathbb{E}[T_s] = (T_{min} + T_{max})/2$ is the mean stay time.

Secondly, the movement process on the XY plane belongs to a Uniform Mobility Model. When the UAV node is hovering on the Z-axis in period T_s, it will choose the velocity component of the X-axis and Y-axis randomly from $[V_{min} \sim V_{max}]$, together forming a new velocity vector. The above process can be described as: $\overrightarrow{V_{k_i}(n+1)} = (v^x_{k_i}(n+1), v^y_{k_i}(n+1), v^z_{k_i}(n))$. Besides, in this part, we assume that the routing process is deployed in a limited 3-D space. Thus, the nodes will move in the opposite direction once they reach the region boundary.

3.2.1.2 Routing Model

In this part, we present the routing model in our system. We assume that the communication range of each UAV node is restricted to its own zone and neighbor zones. Besides, we assumed that the routing model is based on the hop by hop routing paradigm [19]. Two routing processes with a set of source-destination (SD) pairs are shown in Fig. 3.6: $\langle S_1, D_1\rangle$ from k_1 to k_2 and $\langle S_2, D_2\rangle$ from k_0 to k_3. The network environment has a set of packets $PKT(n)$, and the number of packets will increase as the new packet is injected into the network continuously.

Meanwhile, we introduce a packet dropout mechanism caused by transmission timeout. We designed the maximum transmission hop count as N_{hop}^{max}. For any data packet $pkt_i \in PKT(n)$, we consider the transmit process from $pkt_i^{S_k} \rightarrow pkt_i^{D_k}$: the packet will be discarded once $pkt_i^{N_{hop}(n)} \geq N_{hop}^{max}$, and set $pkt_i^{State} = -1$. Otherwise, if pkt_i is successfully transmitted, then set $pkt_i^{State} = 1$ and set $pkt_i^{N_{hop}(n)}$ to the constant value of the final hop count.

Moreover, we apply a queueing model for each node to smooth traffic fluctuation. For node $k_i \in K$, it has a buffer queue $Q_{k_i}(n) \in [0, Q_{k_i}^{max}]$, where $Q_{k_i}^{max}$ is the maximal capacity. We describe the input gate and the output gate as $g_{k_i}^{in}$ and $g_{k_i}^{out}$. Also, the increase of queue length can be formed as follows:

$$\Delta Q = v_{g_{k_i}^{in}}(n) - v_{g_{k_i}^{out}}(n), \tag{3.21}$$

$$\Delta Q_{k_i}(n+1) = \begin{cases} \Delta Q, & \text{if } 0 \leq Q_{k_i}(n) \leq Q_{k_i}^{max} \\ min(\Delta Q, 0), & \text{else} \end{cases}, \tag{3.22}$$

$$Q_{k_i}(n+1) = Q_{k_i}(n) + \Delta Q_{k_i}(n+1). \tag{3.23}$$

In the above equations, $v_{g_{k_i}^{in}}(n)$ and $v_{g_{k_i}^{out}}(n)$ represents the packet arrival rate from upstream and the packet service rate to downstream, respectively. We pre-cache ω packets in each UAV node $k_i \in K$ to maintain a high traffic load pressure in the network environment:

$$Q_{k_i}(0) = \omega. \tag{3.24}$$

Once the buffer queue is filled, the packet drop happens, and the network is considered as occurring the network congestion.

3.2.1.3 Problem Formulation

The dynamic routing process has two critical problems: How to increase the delivery ratio of the packet, and how to alleviate the traffic congestion. Therefore, the objective optimization function can be formulated as:

$$\begin{aligned} \lim_{n\to\infty} \mathbf{min}\{ & \sqrt{\frac{1}{|K|} * \sum_{k_i \in K} (Q_{k_i}(n) - \overline{\sum_{k_j \in K} Q_{k_j}(n)})^2} \\ & + \sum_{pkt_i \in PKT(n)} pkt_i^{N_{hop}^{(n)}} \}. \end{aligned} \tag{3.25}$$

Based on the above equation, the optimization objective includes two parts. The first part is to reduce the mean square error (MSE) of all buffer queue lengths, and hence it can relieve the traffic pressure among all the UAV nodes. The second part is to lower the mean hop count $pkt_i^{N_{hop}(n)}$ for all packets $pkt_i \in PKT(n)$ so that fewer packets will be dropped.

Moreover, the local observation space for each agent k_i at moment n can be described as follows:

$$[zone(D_k) \times 10 + pkt_j^{D_k}, zone(k_i) \times 10 + i, Q_{k_i}(n)], \tag{3.26}$$

which consists of three parts: the mixed id (zone id with node id) of packet pkt_j's destination node, the mixed id of node k_i, and the occupancy of its queue buffer. Also, the action space can be described as $[zone^{next}_{pkt_j}(n)]$, which denotes the next transmit zone of the current packet pkt_j in UAV node k_i. We designed the reward function as follow:

$$\begin{aligned} r_{k_i}(n) = \rho \times [Q_{k_i}(n) - &\overline{\sum_{k_t \in zone^{next}_{pkt_j}(n)} Q_{k_t}(n)]} \\ + \mu \times [&dis(zone^{next}_{pkt_j}(n), zone(pkt_j^{D_k})) \\ - &dis(zone(k_i), zone(pkt_j^{D_k}))], \end{aligned} \tag{3.27}$$

where $\overline{\sum_{k_t \in zone^{next}_{pkt_j}(n)} Q_{k_t}(n)}$ is the mean queue buffer length among the UAV nodes in pkt'_j next-hop zone area, function $dis(A, B)$ calculates the distance between A and B, and ρ, μ are the scale factors.

The above equation indicates that our routing algorithm has the following functions. The first part is to transmit packets among each UAV node to balance the occupancy of their queue buffer. The second part is to calculate whether the next-hop zone of pkt_j is closer to its destination zone, therefore decreasing the total hop count $pkt_j^{N_{hop}(n)}$ (i.e., shortest path).

3.2.2 *Adaptive Communication-Based UAV Swarm Routing Algorithm*

Inspired by the recent success of multi-agent reinforcement learning, we design a multi-agent reinforcement learning-based routing algorithm.

3.2.2.1 Decentralized Markov Decision Processes

Multiple UAV nodes form a decentralized Markov Decision Process (MDP) model. In the decentralized MDP, depending on the action they executed, each agent gained immediate reward. Meanwhile, the environment will change to a new state [35]. In the part, the current system state can be formulated as S. Each agent has their local observations, which can be represented by S: $[o_1, \ldots, o_n]$. Then, the agents choose action $[u_1, \ldots, u_n]$ by their local policies $[\pi_{\theta_1}, \ldots, \pi_{\theta_n}]$. In the system, the observations of all agents form a joint observation $\mathbf{O} = [o_1, \ldots, o_n]$, and the actions form a joint action $\mathbf{U} = [u_1, \ldots, u_n]$. The change of network state can be described as $S \times \mathbf{U} \to S'$. Each agent gets the same reward $r(S, \mathbf{U})$, also received a new local observation from the changed network state: $S' \to [o'_1, \ldots, o'_n]$. Besides, we denoted the τ_t^i as the action-observation history.

The goal of each agent is to maximize the cumulative discount reward:

$$R_t = \sum_{k=0}^{\infty} \gamma^k r^{t+k}, \tag{3.28}$$

where γ is the discount factor and $\gamma \in [0, 1]$. Besides, combining the joint policy $\pi = [\pi_{\theta_1}, \ldots, \pi_{\theta_n}]$, system state S_t, joint action $\mathbf{U}$ and the cumulative discount reward, a joint value function can be designed as: $Q^{\pi}(S_t, \mathbf{U}_t) = \mathbb{E}_{S_{t+1}:\infty; \mathbf{U}_{t+1}:\infty}[R_t | S_t, \mathbf{U}_t]$.

3.2.2.2 Multi-Agent Policy Gradients Based Routing Algorithm

The easiest way to solve our dynamic routing problem is to maintain multiple single-agent algorithms in parallel. In this section, we firstly propose a PG-based routing algorithm. PG is a traditional policy-gradient based single-agent reinforcement learning algorithm [36], the gradient can be described as:

$$\nabla_{\theta} J(\theta) = \mathbb{E}_{\pi_{\theta}}[\nabla_{\theta} log \pi_{\theta}(S, a) Q^{\pi}(S, a)], \tag{3.29}$$

where a is the action chosen by the single agent.

In order to enhance the cooperative learning ability of distributed UAV nodes, we further propose a multi-agent policy gradient-based routing algorithm. As shown in Fig. 3.7, it maintains a centralized training with a decentralized execution framework. In the training phase, we implement a training center to generate the joint value $Q(S, \mathbf{U})$ by inputting the environment state S and the joint action $\mathbf{U} = [u_1, \ldots, u_n]$ into the neural network, while the policy of each decentralized actor is trained by following a gradient depends on $Q(S, \mathbf{U})$. Then, in the execution

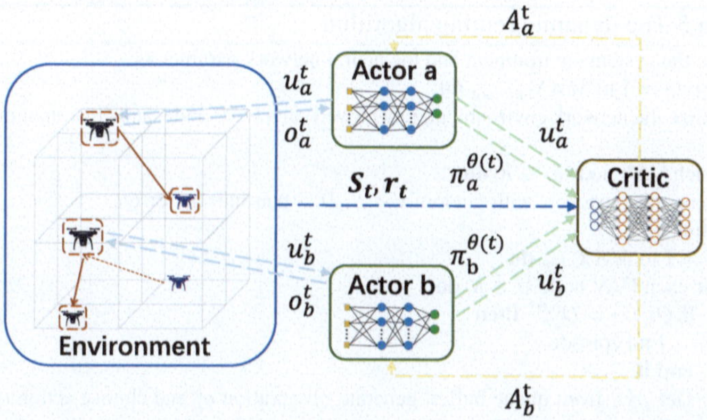

Fig. 3.7 The multi-agent actor critic reinforcement learning framework

phase, each agent chooses the action u_i based on their local observation o_i. Notice that we use the $\epsilon - greedy$ scheme during the action selection process to balance exploring and exploiting.

Moreover, we propose an advanced counterfactual baseline method to effectively evaluate each agent's contribution to the global reward, so as to enhance the convergence speed:

$$A^i(S, \mathbf{U}) = Q(S, \mathbf{U}) - \sum_{u'^i} \pi^i(u'^i|\tau^i)Q(S, (\mathbf{U}^{-i}, u'^i)). \tag{3.30}$$

The equation subtracted the average action Q-value of agent i from the global Q-value (calculated from the centralized training center). Therefore, each agent can obtain a precise, independent reward, and the algorithm can finally gain a better convergence ability. We use the following equation to calculate the gradient of agent i:

$$\nabla_{\theta_i} J(\theta_i) = \mathbb{E}_{\pi_{\theta_i}}[\nabla_{\theta_i} log\pi_{\theta_i}(u^i|\tau^i)A^i(S, \mathbf{U})]. \tag{3.31}$$

Besides, our routing algorithm trains the critic in an on-policy scheme, the loss function can be described as follows:

$$\mathcal{L}(\theta^c) = (\gamma^{(\lambda)} - f^c(\cdot_t, \theta^c)), \tag{3.32}$$

where $\gamma^{(\lambda)} = (1 - \lambda) \sum_{n=1}^{\infty} \lambda^{b-1} G_t(n)$. $G_t(n) = \sum_{l=1}^{\infty} \lambda^{l-1} r_{t+l} + \gamma^n f^c(\cdot_{t+n}, \theta^c)$, which can be calculated by the TD(λ) method using a mixture of n-step returns [37]. The dynamic routing algorithm is shown in Algorithm 5.

Algorithm 5 The dynamic routing algorithm

Initialize the system environment and the neural network parameters
for $episode = 1$ to $MAX_{episode}$ **do**
  Initialize the network environment, redeployed each UAV node, and clear their queue buffer
  for each UAV node $k_i \in K$ **do**
    Generate ω packets with randomly $< S, D >$ pair into buffer Q_{k_i}
  end for
  for $t = 1$ to MAX_{step} **do**
    for each UAV node $k_i \in K$ **do**
      if $Q_{k_i}(t) = Q_{k_i}^{max}$ **then**
        End episode
      end if
      Get pkt_i from queue buffer, generate observation o_i^t and choose action u_i^t based on policy $\pi_{\theta_i}^t$
      Transmit pkt_i to other UAV nodes according to u_i^t
    end for
    Form joint action $\mathbf{U}^t$, change the network environment $S_t \rightarrow S_{t+1}$, each agent get reward r_t and new observation o_i^{t+1}
    Generate new packets and handle the moving process for each UAV node
  end for
  Update critic network through Eq. (3.32)
  Update actor network for agents through Eq. (3.31)
end for

3.2.3 *Simulation Results and Discussions*

In this section, we firstly discuss network environment settings and the simulation settings. Secondly, we present the simulation results to evaluate the performance of our proposed algorithm. Our experiments simulate Ubuntu 16.04 with 32g RAM, Nvidia RTX 2060, and Intel i7-10875H.

3.2.3.1 Environment Settings

In the experiment, the boundary value of our experiment is: $p_{max}^x = 10m$, $p_{max}^y = 10m$, $p_{max}^z = 10m$. We set the total number of UAV nodes and space zones as $N = 10$ and $M = 8$, and the initial deployment process is described as follows: UAV nodes $[k_0, k_1, \ldots, k_7]$ are placed in zones $[zone_0, zone_1, \ldots, zone_7]$, respectively, and UAV nodes $[k_8, k_9]$ are randomly deployed in the whole zone area. This initialization process will be executed in each training episode. Besides, we set the speed of UAV nodes as: $V_{max} = 2.6$ m/s, $V_{min} = 1.2$ m/s. Furthermore, we set the minimum staying time T_{min} and maximum staying time T_{max} as 2 s and 6 s, respectively. The staying probability is $p_s \approx 0.5$, as calculated from Eq. (3.19). In each training step, our environment will generate a new packet in each UAV node, and the node can only transmit a single packet to the downstream in each training step (i.e., $v_{g_{k_i}^{out}}(n) = 1$). Other parameter settings include: $N_{hop}^{max} = 4$, $\omega = 30$, and $\forall k_i \in K, Q_{k_i}^{max} = 50$.

Finally, we configure the network environment at different dynamic levels to evaluate the performance of our proposed algorithm. For the UAV swarm, considering its high-speed mobility, it is necessary to deploy a backbone network architecture to ensure the basic communication quality [38]. In the highly-dynamic network scenario, UAV nodes $[k_0, k_2, k_4, k_6]$ are in the hover state to form the backbone nodes, while other nodes keep moving randomly. In a lowly-dynamic network scenario, all UAV nodes are in the hover state. However, the network still has an unstable topology structure between different training episodes due to the random deployment process of nodes k_8 and k_9. We also set three baseline algorithms to evaluate the relative performance of our proposed dynamic routing algorithm: the PG-based routing algorithm, the BackPressure routing algorithm (decides the routing policy based on congestion gradients) [39], and the Random routing algorithm.

3.2.3.2 Convergence Performance

In this section, we evaluate the convergence performance of our algorithm. Figure 3.8 shows the average reward in different dynamic network environments with different algorithms. Both the PG-based algorithm and our proposed algorithm can achieve convergence, while the former performs the worse. This is most likely caused by the purely distributed framework consisting of each single PG agent. This structure will cause each agent to focus only on maximizing their local reward instead of the global reward, forming a selfish routing policy. However, our proposed algorithm performs better because the centralized training distributed execution framework enhanced the cooperative routing ability among nodes.

Besides, the average reward of the PG-based routing algorithm and our proposed algorithm increase sharply before around 5000 episodes while tending to be flat after that. This is caused by the ϵ factor. Once ϵ reaches the maximum, the algorithm will become greedy and reduce the exploration of the environment. Also, we can find that the fluctuating range of those two algorithms' learning curves becomes higher as the dynamic level of our network increases. However, the BackPressure routing and Random routing method seem to be more stable, exposing their poor network state awareness ability. Nonetheless, the PG-based routing algorithm and our proposed intelligent routing algorithm can adjust their policy quickly once the network environment changes. Finally, our method performs the best among other algorithms, which means it can form a robust routing method in such a dynamic network environment.

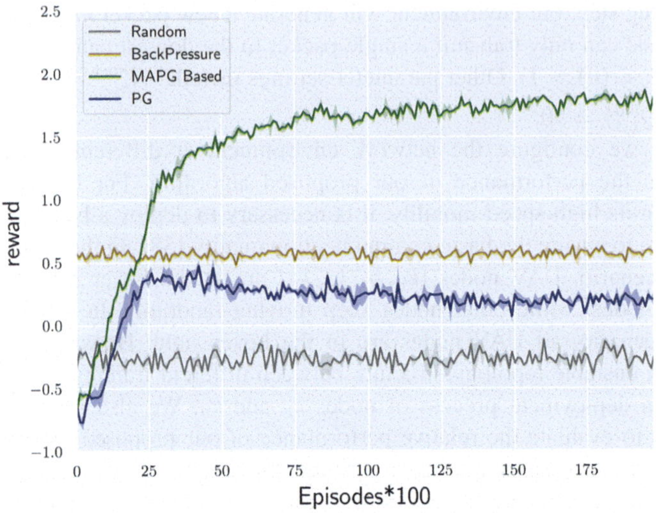

(a) lowly dynamic

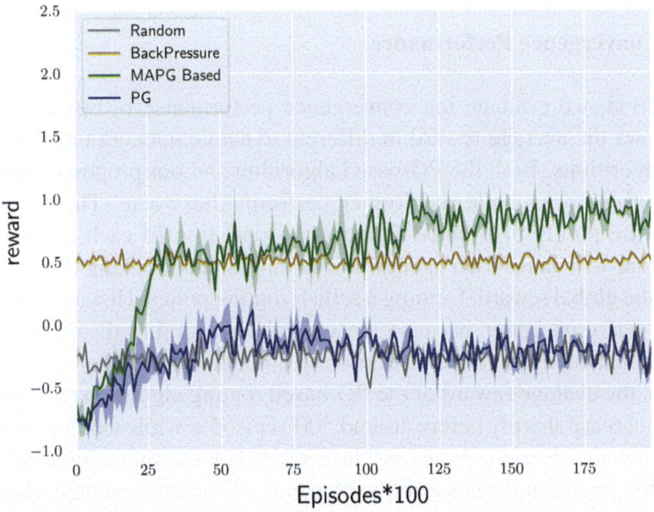

(b) highly dynamic

Fig. 3.8 Average reward. (**a**) Lowly dynamic. (**b**) Highly dynamic

3.2.3.3 Performance Analysis

To evaluate our algorithm performance, we will test the network quality from various indicators (e.g., survival time, delivery rate, and throughput) in this section.

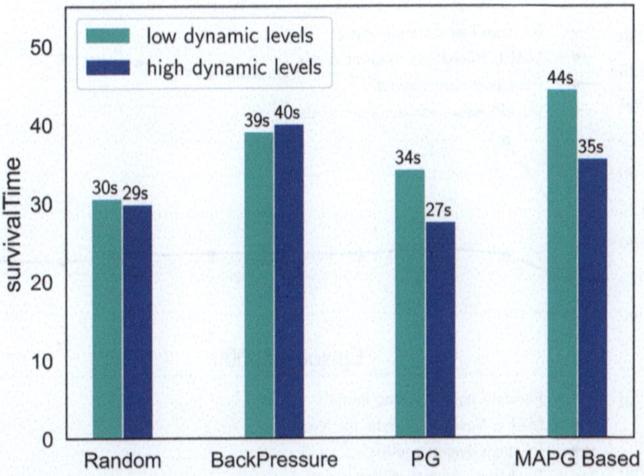

Fig. 3.9 Average survival time

(1) Average Survival Time Analysis: First of all, we analyze the average survival time with different networks dynamic. As shown in Fig. 3.9, the PG-based routing method cannot relieve the network congestion effectively. It even performs worse than the Random routing method in the highly-dynamic network environment. In contrast, our algorithm postpones the congestion occurrence time by 47% of Random algorithm, 13% of BackPressure algorithm, and 29% of PG algorithm. This demonstrates that our proposed algorithm can effectively prolong the network lifetime. However, we notice that the BackPressure method presents better performance than ours in the highly-dynamic network environment. This is probably because the BackPressure method only considers the congestion gradient as the routing decision basis, it built its survival time advantages at the expense of the transmission delay.
(2) Average Delivery Rate Analysis: Secondly, as shown in Fig. 3.10, we evaluate the average delivery rate. It shows that the average transmission success rate of our algorithm is greater than other baseline algorithms. The PG-based routing algorithm, on the other hand, has a similar delivery effect with the BackPressure method and Random method. Although the BackPressure algorithm has a better congestion control ability, it shows a lower packet delivery capability. It indicates that this method pays too much attention to the queue balancing problem among UAV nodes without considering how to transmit packets effectively. It is worth mentioning that the destination node cannot receive many packets because of a heavy network load environment. Hence, a packet delivery rate of about 20% maintained by our proposed algorithm is relatively acceptable under a highly dynamic network environment.
(3) Average Throughput Analysis: Finally, Fig. 3.11 demonstrates the throughput performance of our proposed algorithm. Generally, the throughput is defined as

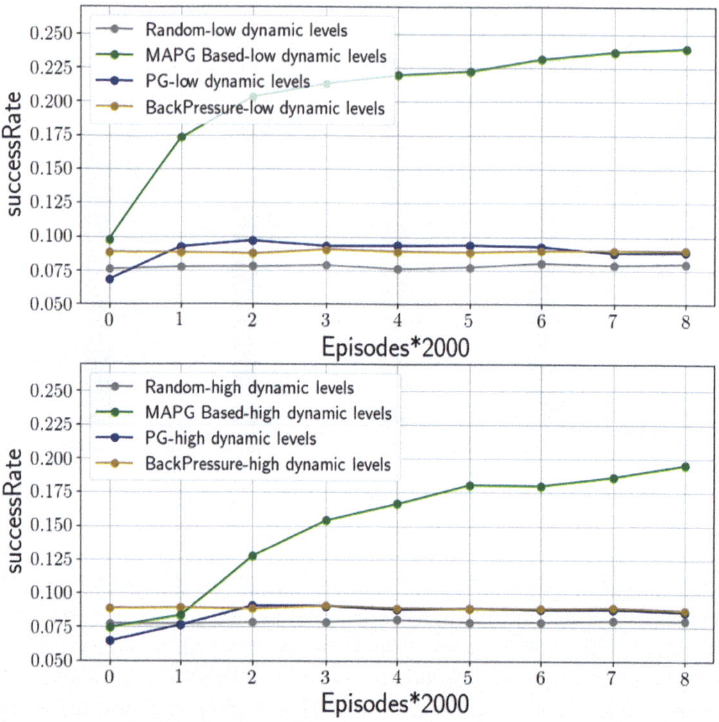

Fig. 3.10 Average delivery rate

the rate of successful message delivery over the whole network and a higher network throughput means a more robust packet processing ability. Similar to the packet delivery performance in Fig. 3.10, our proposed routing algorithm performs the best among all the other algorithms. However, the throughput capacity of the PG-based routing algorithm and BackPressure method only shows a weak advantage compared with the Random method.

From the above analysis results, we conclude the main advantage of our proposed algorithm as follows. By maintaining the centralized training and decentralized execution framework together with a counterfactual baseline scheme, our algorithm greatly enhances the cooperative among the UAV nodes. Overall, we find that it can effectively balance the network congestion and packet delivery while maintaining a high throughput ability. Finally, our proposed algorithm emerges with a swarm-intelligent capability and can converge into a robust routing policy within a dynamic network environment.

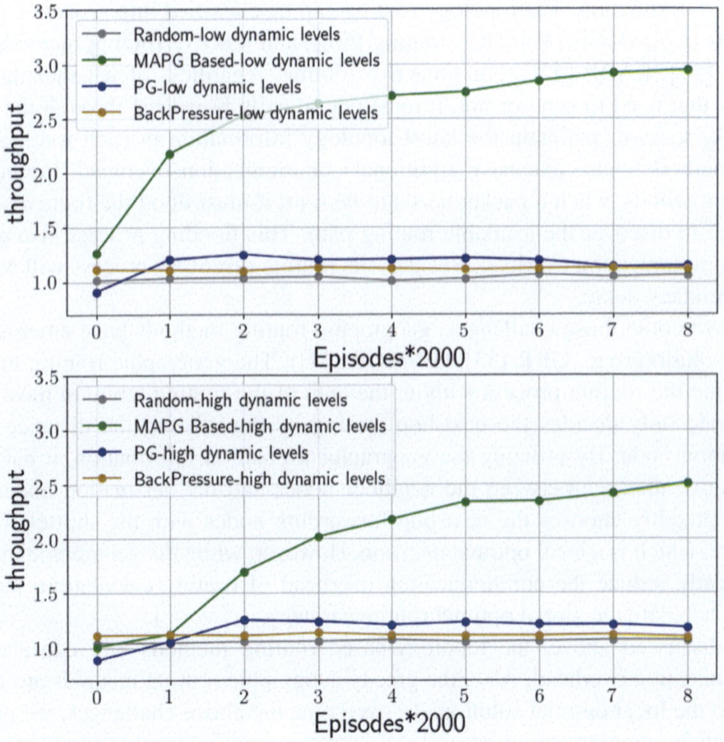

Fig. 3.11 Average throughput

3.3 Adaptive Communication-Based Routing Strategy for UAV Swarm Networks

Recently, the UAV swarm network has been widely used in various scenarios like agricultural production, intelligent transport, and disaster rescue [40–43]. The UAV swarm belongs to a new clan of networks named FANET [44]. In FANET, each UAV can assist the communication of other air stations [45]. Also, each UAV can directly communicate with other UAV nodes. In this part, we focus on the latter condition. The UAVs possess the characteristic of limited-energy nodes and bandwidth-constrained links [46]. Besides, the high mobility of UAVs also leads to time-varying network topology [47]. These features pose huge challenges for routing design in such a highly dynamic network.

In the past few decades, a rapidly growing number of studies have investigated effective FANET routing algorithms design [48]. The topology-based routing methods are the most widely used. It deploys a dynamically-updated routing table into each node, which records the single-hop forwarding policy. Based on the routing table, the topology-based routing methods decide the routing path from source to destination. The topology routing can be classified into proactive routing methods (e.g., OLSR [49], GCS-routing [50]), and reactive routing methods (e.g., AODV [51], RARP [52]). For proactive routing, regardless of whether there are packets that need to send or not, it must periodically broadcast through the whole network, so as to maintain the latest topology information in each routing table. Such methods cause extensive additional communication overhead. For reactive routing methods, when a packet needs to be sent, it must flood the route discovery message to discover the available routing path. This flooding process also poses a huge communication overload. Besides, the routing discovery process will bring an indeterminacy delay.

To overcome those challenges, geographic routing methods have emerged as a viable solution (e.g., GBR [53], GP-MOR [54]). The geographic routing methods determine the routing process without the help of the routing table. In most cases, each node only decides the next-hop routing by considering the distance to the destination node. By utilizing the geographical position information, it calculates the relative distance between the neighbor nodes and the destination node. After that, it greedily chooses the next-hop forwarding nodes with the shortest relative distance, which is a local optimal decision. However, while the geographic methods can greatly reduce the communication overhead of routing calculation, they can hardly calculate the global optimal routing strategy.

As discussed above, the topology-based routing methods will cause a huge communication overhead. Also, the greedy geographic routing methods are easy to fail into the local optimal solution. To overcome the above challenges, we propose an adaptive communication-based UAV swarm routing algorithm to enable UAVs routing efficiently under the partially observable environment. In our algorithm, we design the Multilayer Perceptron algorithm to learn when routing flooding is required among different UAVs (i.e., on-demand routing flooding). Each UAV determines whether it should participate in the flooding process by considering its local state, including the packet destination location, the network status of itself, and its neighbors. If needed, the selected UAVs will unite with the surrounding UAVs to share route information over the whole network. While other UAVs keep "silent" because their local observation information doesn't change obviously, which means they don't need to update their routing strategy at this time. Besides, we design a Gate Recurrent Unit to select the cooperative information and compress the volume of communication data. Moreover, we further design the multi-agent actor-critic algorithm to learn how to integrate shared information for cooperative routing decision-making. Each UAV can dynamically update its routing strategy according to the interaction with the environment and other UAVs. Simulation results demonstrated the effectiveness of our proposed algorithm.

The major contributions of the proposed work are summarized as follows:

(1) We design the Multilayer Perceptron algorithm to learn when routing flooding is required among different UAVs. It decides whether each UAV should flood routing information with others by considering its local network state. Multiple selected UAVs will share routing information with their surrounding agents, while other agents keep "silent" because their local network state hasn't

changed distinctly. This adaptive routing flooding mechanism greatly reduces communication overhead.

(2) We design the Gate Recurrent Unit algorithm to greatly compress the volume of communication data. The Gate Recurrent Unit constructs a communication channel to share routing information, which will further filter the information by selectively outputting the cooperative part.

(3) We design the multi-agent actor-critic algorithm to utilize compressed information for optimal routing decisions. For each agent, the compressed cooperative information promotes a comprehensive awareness of the network, which helps to adjust the training of its actor-critic network towards global optimal. Therefore, the actor-critic network can generate a robust routing strategy.

3.3.1 System Model and Problem Formulation

In this section, we first introduce the network system model. Secondly, we model the mobility and the energy of UAVs. Finally, we present the queue model and the optimization problem.

3.3.1.1 System Model

As shown in Fig. 3.12, we consider a UAV swarm includes a set of UAVs $\mathcal{K}$, which can be denoted as $\mathcal{K} = [1, \ldots, k]$. This UAV swarm is located in a bounded 3D space with the boundary points as $(x_{\max}, 0, 0)$, $(0, y_{\max}, 0)$, $(0, 0, z_{\max})$. Further,

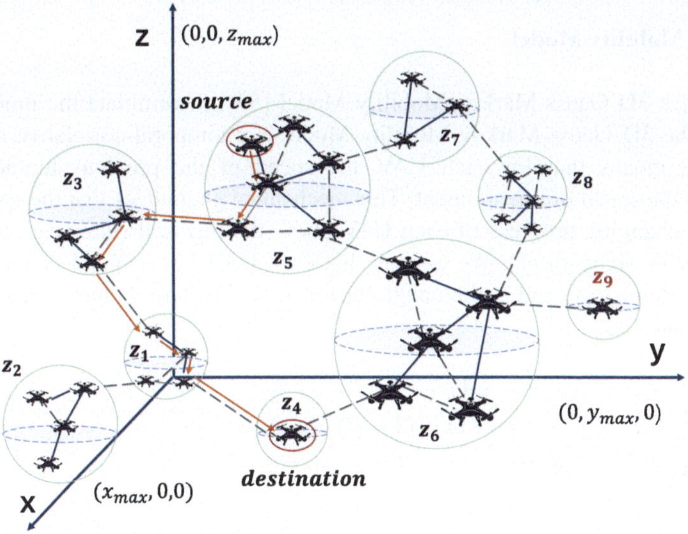

Fig. 3.12 UAV swarm network system model

we divide the UAV swarm 3D space into several zones $\mathcal{Z} = [z_1, \ldots, z_w]$. Each zone z_i has t gateway nodes and u common nodes. The gateway nodes are usually at the edge of their zones and can share information with other gateway nodes in different zones. Therefore, they take charge of inter-zone communication. Note that due to the highly dynamic network, the zones will be periodically re-divided according to network features, which will be introduced later.

Each UAV node i has a communication range r, and the position and speed at discrete-continuous time n can be formulated as:

$$\begin{cases} \overrightarrow{P_i(n)} = \left(\left(p_i^x(n), p_i^y(n), p_i^z(n)\right)\right. \\ \overrightarrow{S_i(n)} = \left(\left(s_i^x(n), s_i^y(n), s_i^z(n)\right)\right. \end{cases}. \tag{3.33}$$

Specifically, if the UAV node i reaches the boundary, it will move reversely to avoid out of bounds:

$$p_i^{\varphi}(n+1) = \begin{cases} p_i^{\varphi}(n) + s_i^{\varphi}(n), \text{ if } p_i^{\varphi}(n) + s_i^{\varphi}(n) \in [0, \varphi_{\max}) \\ p_i^{\varphi}(n) - s_i^{\varphi}(n), \text{ otherwise} \end{cases}, \tag{3.34}$$

where $\varphi \in (x, y, z)$, represents any dimension in the 3D space.

Besides, we assume that the high dynamic UAV swarm network contains a set of Source-Destination (SD) routing pairs, and the packet delivery process belongs to a multi-hop routing scheme. During the whole network system operation period from 0 to n, we assume that the system totally generates $C(n)$ packets, and $S(n)$ of them are successfully transmitted. Figure 3.12 presents a multi-hop routing process from the source node to the destination node, where the packet is transmitted across z_5, z_3, z_1, z_4.

3.3.1.2 Mobility Model

We adopt a 3D Gauss-Markov Mobility Model [55] to formulate the mobility of UAVs. The 3D Gauss-Markov Mobility Model is a temporal-correlative mobility model. It means that for each UAV, the speed of the previous moment will influence the speed of this moment. This mechanism guarantees that the speed and direction-changing progress of each UAV obeys the physical rules (e.g., avoiding unreasonable sharp turns). We assume that each UAV has a speed factor $v(n)$, a direction factor $d(n)$ and a pitch angle factor $a(n)$. The speed change process can be formulated as:

$$\begin{cases} v(n) = \beta \times v(n-1) + (1-\beta) \times \bar{v} + \sqrt{1-\beta^2} \times v_g \\ d(n) = \beta \times d(n-1) + (1-\beta) \times \bar{d} + \sqrt{1-\beta^2} \times d_g \\ a(n) = \beta \times a(n-1) + (1-\beta) \times \bar{a} + \sqrt{1-\beta^2} \times a_g \end{cases}, \tag{3.35}$$

where $\beta \in [0, 1]$ is the tuning parameter. It determines whether the node mobility is more random or more time-dependent. v_g, d_g, a_g are random variables obeying a Gaussian distribution. Based on Eq. (3.35), we further calculate the speed component of node i at time n:

$$\begin{cases} s_i^x(n) = v(n) \times \cos(d(n)) \times \cos(\mathrm{a}(n)) \\ s_i^y(n) = v(n) \times \sin(d(n)) \times \cos(\mathrm{a}(n)) \\ s_i^z(n) = v(n) \times \sin(a(n)) \end{cases}. \tag{3.36}$$

3.3.1.3 Energy Consumption Model

In this part, we consider the UAV's energy consumption caused by packet transmission. For each packet, its hop-by-hop forwarding energy consumption includes transmission energy consumption E_T and receiving energy consumption E_R [56]. The transmission energy consumption is generated by the radio electronics and the power amplifier of the sender node, while the receiving energy consumption is only caused by the radio electronics of the receiver node. We also consider the wireless channel fading problem and introduce the free space model and the multipath fading model. The above two fading models will switch according to the distance change between the sender and the receiver. We assume a packet pkt_u $(u \in C(n))$ with the sender UAV node i and the receiver UAV node j. The communication energy consumption in each hop can be formulated as:

$$E_T = \begin{cases} l \times E_{\text{elec}} + l \times \eta_{fs} \times \psi(i,j)^2, & \psi(i,j) < \tau \\ l \times E_{\text{elec}} + l \times \eta_{mp} \times \psi(i,j)^4 & \psi(i,j) \geq \tau \end{cases}, \tag{3.37}$$

$$E_R = l \times E_{\text{elec}}, \tag{3.38}$$

$$E_{\text{tot}} = E_T + E_R, \tag{3.39}$$

where E_T presents the energy consumption of the sender UAV node i, and E_R presents the energy consumption of the receiver UAV node j. l is the bits number of packet pkt_u, and E_{elec} is the energy consumption of radio electronics per bit. η_{fs} and η_{mp} are the energy consumption factors of the power amplifier, corresponding to the free space model and the multipath fading model, respectively. $\psi(i, j)$ is the 3D Euclidean distance between node i and j, and τ is the distance switching threshold of the above two models. $\psi(i, j)$ can be formulated as:

$$\psi(i,j) = \sqrt{\left(p_i^x(n) - p_j^x(n)\right)^2 + \left(p_i^y(n) - p_j^y(n)\right)^2 + \left(p_i^z(n) - p_j^z(n)\right)^2}. \tag{3.40}$$

We assume the total hop count of packet pkt_u is H_u. Then, the total communication energy consumption of the complete multi-hop transmission process can be calculated as:

$$E_{pkt_u} = \sum_{k=1}^{H_u} E_{tot}. \tag{3.41}$$

3.3.1.4 Queue Model

The traffic fluctuation of the high-speed UAV swarm network is extremely unstable [57]. Therefore, we apply a queue for each UAV to smooth the traffic fluctuation. The queue can cache the packets during the peak flow and release them at other times, thereby alleviating the network congestion. We assume that at time n, the queue capacity of node i is $q_i(n)$, and the maxsize is q_i^{max}. Beside, for UAV node i, the number of packets flowing into it and forwarding from it is $\lambda_i^{in}(n)$ and $\lambda_i^{out}(n)$, respectively. The queue length changing process can be formulated as:

$$\Delta q_i(n+1) = \begin{cases} \lambda_i^{in}(n) - \lambda_i^{out}(n), \text{if } 0 \le q_i(n) \le q_i^{\max} \\ \min\left(\lambda_i^{in}(n) - \lambda_i^{out}(n), 0\right), \text{otherwise} \end{cases}, \tag{3.42}$$

$$q_i(n+1) = q_i(n) + \Delta q_i(n+1). \tag{3.43}$$

Equation (3.42) indicates the change of queue length for UAV node i from time n to $n+1$. If the queue $q_i(n)$ is not full, then the change of queue length equals the difference between $\lambda_i^{in}(n)$ and $\lambda_i^{out}(n)$ (i.e., the difference between incoming and outgoing packets). Otherwise, the queue $q_i(n)$ is already full (i.e., local congestion), and the length can only be reduced if the outgoing packets are more than the incoming packets. Equation (3.43) indicates the queue length change result at time $n+1$, which equals the basic length value at time n plus the change length value.

Moreover, the queue of each UAV node is essentially a typical $M/M/1/N/+\infty$ queuing system [58]. Therefore, the average waiting time W_s of each packet in the queue can be formulated as:

$$\begin{aligned} W_s &= \frac{1}{a-d} - \frac{L(\eta)^{L+1}}{a\left(1-(\eta)^L\right)} \\ &= \frac{1 - \frac{2a^L}{d^L} + \frac{La^{(L+1)}}{d^{(L+1)}}}{d - \frac{a^L}{d^{(L-1)}} - a + \frac{a^{(L+1)}}{d^L}} \\ &\approx \frac{1}{d-a}, \end{aligned} \tag{3.44}$$

where $a = \mathbb{E}(\lambda_i^{in}(n))$, $d = \mathbb{E}(\lambda_i^{out}(n))$ obeys the Poisson distribution, $\eta = a/d$ and $L = q_i^{max}$.

To evaluate the congestion control ability of our proposed routing algorithm, we maintain a high congestion pressure level in the UAV swarm network by pre-cached ω packets in each UAV node. During the UAV swarm network system operation, once the buffer queue of a UAV node is filled, the local network congestion occurred.

3.3.1.5 Problem Formulation

At this stage, we can formulate the multiobjective optimization problem of UAV swarm routing. We wish our routing method has the congestion control ability and the shortest path delivery ability. The former aims to alleviate traffic fluctuation so that the routing method can prolong the network lifetime under an unstable FANET environment. And the latter corresponds to improving the communication quality. The above optimization problem can be formulated as:

$$\lim_{n\to\infty} \mathbf{min}\{ \sqrt{\frac{1}{K} \times \sum_{i=1}^{k} \left(q_i(n) - \frac{1}{k} \times \sum_{j=1}^{k} q_j(n) \right)^2}, \\ + \frac{1}{S(n)} \times \sum_{u=1}^{S(n)} dis(pkt_u)\}. \tag{3.45}$$

where dis (pkt_u) presents the total transmission distance of packet pkt_u. The above equation is subject to the following constraints:

$$0 \leq q_i(n) \leq q_i^{\max}, \forall i \in [1, \ldots, k] \tag{3.46}$$

$$0 \leq S(n) \leq C(n). \tag{3.47}$$

Equation (3.45) states that as the UAV swarm network system is stabilized, we hope our proposed routing method can minimize the mean square error (MSE) of the queue length among UAVs. It indicates that our proposed routing algorithm can prolong network congestion effectively. We also hope to minimize the average transmission distance of successful delivery packets. Therefore, the proposed routing algorithm can deliver packets efficiently. Equation (3.46) ensures that each UAV node can operate normally without congestion. And Eq. (3.47) guarantees that the total number of the successfully transmitted packet is less than the total number of packets generated in the network.

3.3.2 Multi-Agent Policy Gradients-Based Routing Algorithm

In this part, we propose an adaptive communication-based UAV swarm routing algorithm. We first model the routing process among multiple UAV nodes as a decentralized partially observable Markov decision process, then we propose a UAV swarm zone division method. Based on the above, we lastly introduce the proposed routing algorithm.

3.3.2.1 Decentralized POMDP

We model the UAV swarm network routing process as a multi-agent decentralized partially observable Markov decision process (Dec-POMDP). Without loss of generality, the Dec-POMDP can be formalized as a 5-tuple $< N, S, A_i, R, O_i >$ [59]. N is the set of agents, S is the set of global states, A_i is the action space of agent i, R is the global reward, and O_i is the local observation space of agent i. As a multi-agent reinforcement learning system, we define the system environment state as the combination of all agent's local observations:

$$s(n) = [o_1(n), \ldots, o_k(n)] . \tag{3.48}$$

For each UAV node i at each time n, we assume that $\lambda_i^{out}(n) = 1$. It means that at each moment, each UAV can get a packet pkt_u from their queue and forward it to one neighbor. Therefore, the local observation of agent i can be formulated as:

$$o_i(n) = [geo(D\,(pkt_u)), info(i), info\,(N_r(i))], \tag{3.49}$$

where $N_r(i)$ is the set of the neighbor nodes of node i, which can be formulated as:

$$N_r(i) = \{j \mid \psi(i, j) \leq r, i \neq j, \forall j \in [1, \ldots, k]\}, \tag{3.50}$$

and $\psi(i, j)$ is the distance between node i and node j, which can be calculated by Eq. (3.40).

Equation (3.49) includes three-part. The first is the geographic information of the packet pkt_u's destination node, the second is the information of node i, and the last is the information of node i's neighbor nodes. Detailed information includes location, ID number, and queue length.

We define the action of each agent as:

$$a_i(n) = j, j \in N_r(i), \tag{3.51}$$

which means that each node can only forward the packet to one of its neighbor nodes within its communication range.

Corresponding to the optimization problem in Sect. 3.3.1.5, the reward function should simultaneously consider congestion control and shortest path delivery. Therefore, the reward function can be described as:

$$\begin{aligned} r_i(n) =& \rho \times \left[q_i(n) - q_{a_i(n)}(n)\right] \\ &+ (1-\rho) \times \left[\psi(a_i(n), D(pkt_u)) - \psi(i, D(pkt_u))\right] \end{aligned}, \tag{3.52}$$

where $D(pkt_u)$ is the destination node of pkt_u and $\psi(x, y)$ is the distance between node x and node y, which can be calculated by Eq. (3.40). ρ is the balance factor between congestion control ability and shortest path delivery ability. By adjusting ρ, we can jointly optimize the multi-objective problems.

3.3.2.2 Zone Division

We divide the UAV swarm 3D space into several zones to reduce the large-scale routing flooding. Our proposed zone division method for the 3D space UAV swarm network extends a traditional approach in [60]. It abstracts the network topology as multiple trees, while each tree corresponds to a non-overlapping zone with an independent ID. The zone-based trees have different sizes, and each zone can communicate with the adjacent zones by gateway nodes. Also, the proposed method periodically re-divides the zones to balance the routing accuracy and computing overhead. The zone division method is composed of three sub-processes.

(1) Optimal Neighbor Selection: We assume that the degree of UAV node i is the number of neighbors within the communication range r, which can be formulated as:

$$deg(i) = |N_r(i)|. \tag{3.53}$$

Moreover, node i constructs the optimal neighbor set by seeking the neighbor nodes with the maximum degree:

$$BN(i) = \{j \mid deg(j) = \max(deg(t)), t \in N_r(i)\}. \tag{3.54}$$

(2) Tree Structure Construction: For UAV node i, once it selects the neighbor node j as the optimal neighbor, it will broadcast this result to other neighbors. After that, the proposed method will construct logical edges between each node and its optimal neighbor. Based on this logical connection relationship, the proposed method further builds multiple trees. Note that some isolated nodes may not have neighbors, but can still form a single-point tree. In each tree, nodes with lower degrees are located in the root or leaf, which usually forward the packets across different zones as a gateway node.

(3) Zone Division: We assume that each tree corresponds to a zone and number the zones based on their spatial location distribution characteristics. Specifically,

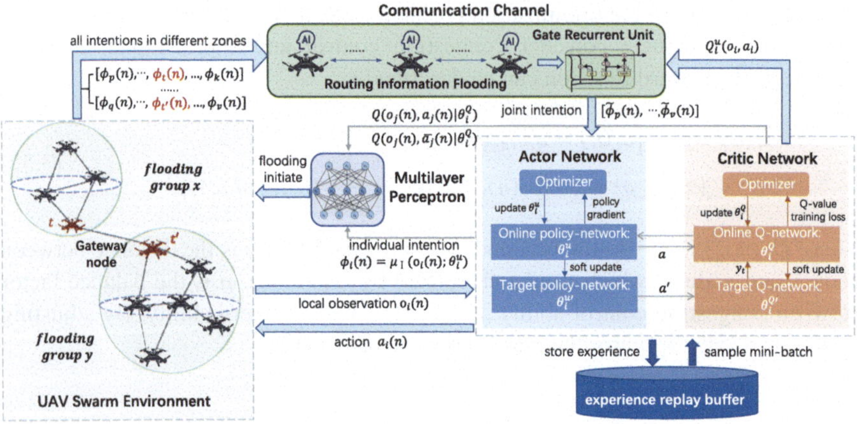

Fig. 3.13 Adaptive communication-based UAV swarm routing algorithm

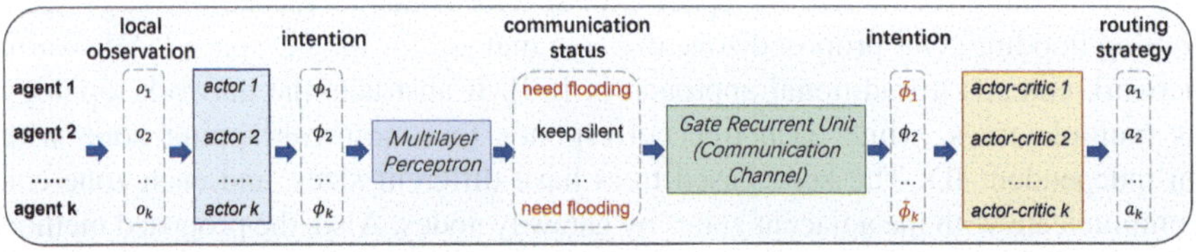

Fig. 3.14 The logic flow of our proposed algorithm

we sort the distance of each zone from the center point to its origin, and number each zone in ascending order of distance. For example, as shown in Fig. 3.12, we divide nine zones from z_1 to z_9 in the UAV swarm network environment (the blue line in each zone represents the logical connection in the corresponding tree). Also, we may re-divide the zones periodically according to some network features such as node density/mobility, rate of network connection/disconnection, and transmission power.

3.3.2.3 Adaptive Communication-based Routing Algorithm

As shown in Fig. 3.13, we propose the adaptive communication-based routing algorithm. It includes k actor-critic network, a Multilayer Perceptron, and a Gate Recurrent Unit (i.e., Communication Channel), where k is the number of the agent. Figure 3.14 presents the logic flow of our proposed algorithm. The previous part of the actor network first converts the local observation of each UAV into the independent intention. The intentions are further inputted to the Multilayer Perceptron to learn whether the route information is required to flood among different UAVs. After that, the Gate Recurrent Unit compresses these data and forms the joint intention (i.e., shared cooperative information). Then, the algorithm combines the latter part of the actor network with the critic network to learn the cooperative routing strategy. Note that our proposed routing algorithm only considers UAVs that operate normally, thus the destroyed nodes will exit the UAV swarm network, and the algorithm will leave out them. We will detail the algorithm design as follows.

As discussed above, our proposed algorithm includes four parts: the actor network of each agent, the critic network of each agent, the Multilayer Perceptron, and the Gate Recurrent Unit (i.e., communication channel). These parts can be parameterized as $\theta_i^{\mu}, \theta_i^{Q}, \theta^{d}, \theta^{g}$, respectively. Firstly, for agent i at time n, the actor network inputs the local observation $o_i(n)$ and outputs a hidden layer variable as the intention. The intention essentially compresses the coding of the observation $o_i(n)$. It indicates how much the agent wants to communicate, which can be described as:

$$\phi_i(n) = \mu_I\left(o_i(n); \theta_i^{\mu}\right), \tag{3.55}$$

where μ_I represents the anterior layers of the actor network. After that, the Multilayer Perceptron inputs the intentions of each agent to determine whether they need to route flooding.

(1) Multilayer Perceptron: We design the Multilayer Perceptron algorithm to learn when routing flooding is required among different UAVs. The Multilayer Perceptron is essentially a binary classifier. At each step, the algorithm inputs the intention of each agent $\phi_i(n)$ into the Multilayer Perceptron to decide whether this agent should flood routing information among nodes within the same zone:

$$\chi_i(n) = g\left(\phi_i(n); \theta^{d}\right). \tag{3.56}$$

If $\chi_i(n) > \chi_0$, the agent needs to flood routing information with the intra-zone nodes. Note that in each zone, if there exists more than one agent that satisfied the threshold judgment condition, the Multilayer Perceptron only selects the agent with the highest intention value to initiate communication. Therefore, at each step n, the network builds multiple zone-based flooding groups $[C_1, \ldots, C_m]$. Each group can flood routing information not only in their own zone but also with other neighbor groups. For some zones where none of the agents in it meet the condition, the zone will keep "silent" and not participate in the route information flooding at this step. This adaptive routing flooding mechanism greatly reduces communication overhead. Also, with the operation of the network system, the flooding groups dynamically change, thus the network status information can be gradually diffused.

(2) Gate Recurrent Unit: We adopt the Gate Recurrent Unit algorithm to greatly compress the volume of communication data [61]. It integrates the independent intentions of agents and guides the agents to learn cooperative routing strategies. The Gate Recurrent Unit constructs a communication channel, which firstly combines the intentions of agents in flooding group C_j:

Algorithm 6 Adaptive communication-based UAV swarm routing algorithm

Initialize the UAV swarm network environment, the hyper-parameter and the experience replay buffer R
for $episode = 1$ to MAX_{eps} **do**
  Reset the UAV swarm network environment
  for UAV node $i \in [1, \ldots, k]$ **do**
    Reset the position and clear the queue
    Generate ω packets into the queue with the source node i and a random destination node
  end for
  Re-divide the zones according to Sect. 3.3.2.2, and number the results as $\mathcal{Z} = [z_1, \ldots, z_w]$
  for $t = 1$ to MAX_{step} **do**
    Forward packets by Algorithm 7
    Change the environment state $S_t \rightarrow S_{t+1}$, get reward $r_i(t)$ and next observation $o_i(t+1)$ for each UAV node $i \in [1, \ldots, k]$
    Store the observation, action, reward, and next observation of each agent into the experience replay buffer R
    Update the hyper-parameter by Algorithm 8
  end for
end for

$$\Phi_{C_j}(n) = \left[\phi_p(n), \ldots, \phi_k(n)\right], p, k \in C_j. \tag{3.57}$$

Secondly, the Gate Recurrent Unit selectively filters the intentions that promote cooperation, and compresses the joint intention to share within C_j:

$$g(\Phi_{C_j}(n); \theta^g) = \left[\tilde{\phi}_p(n), \ldots, \tilde{\phi}_k(n)\right]. \tag{3.58}$$

Moreover, in a large-scale UAV swarm network, multiple adjacent flooding groups will further share routing information by gateway nodes. For example, as shown in Fig. 3.13, agent t and t' are the gateway nodes of flooding groups x and y, respectively. Because node t and t' are within each other's communication range, the Gate Recurrent Unit further filters and compresses the joint intentions of the flooding groups:

$$\begin{aligned} g[\Phi_{C_x}(n), \Phi_{C_y}(n); \theta^g] = [&\tilde{\phi}'_p(n), \ldots, \tilde{\phi}'_t(n), \ldots, \tilde{\phi}'_k(n), \\ &\tilde{\phi}'_q(n), \ldots, \tilde{\phi}'_{t'}(n), \ldots, \tilde{\phi}'_v(n)] \end{aligned}, \tag{3.59}$$

where node $p, t, k \in C_x$ and $q, t', v \in C_y$.

Note that in the real UAV swarm scene, the routing information-sharing process is much more complex than in the above example, which involves multiple adjacent zones. Also, the zone division algorithm will periodically re-divide the zone. Therefore, the Gate Recurrent Unit will repeatedly compress the routing intentions data and only selects the cooperative information for routing flooding.

Algorithm 7 The single-hop routing process

for UAV node $i \in [1, \ldots, k]$ **do**
  if $q_i(t) = q_i^{\max}$ **then**
    End episode
  end if
  if $q_i(t) = 0$ **then**
    Skip this UAV node
  end if
  Get packet pkt_u from the queue, and get observation $o_i(t)$ by Eq. (3.49)
  Get independent intension $\phi_i(t)$ from Eq. (3.55)
  Decide whether to flood the routing information by Eq. (3.56)
  if Node i needs to flood **then**
    if The zone of node i still keeps "silent" **then**
      Build the zone-based flooding group as C_i
      Node i becomes the leader node of C_i
    else
      Compare the intention value between node i and the origin leader node of this zone-based flooding group
      Update the node with the higher value as the leader node
    end if
  end if
end for
for Flooding group $C_j \in [C_1, \ldots, C_m]$ **do**
  Generate the joint intention by Eq. (3.57) and Eq. (3.58)
  if Existing adjacent flooding groups **then**
    Further flooding routing information between adjacent zones, and output the joint intention by Eq. (3.58)
  end if
end for
for UAV agent $i \in [1, \ldots, k]$ **do**
  if Node i is in any flooding groups **then**
    Generate action $a_i(t)$ based on Eq. (3.60)
  else
    Generate action $\bar{a}_i(t)$ based on Eq. (3.61)
  end if
end for

(3) Multi-agent Actor-Critic: We design the multi-agent actor-critic algorithm to utilize the shared intention information to learn a cooperative routing strategy [62]. Note that the agents with an empty queue don't participate in the packet forwarding process at this moment. Therefore, for any other agent i that belongs to a flooding group, the proposed algorithm inputs its joint intention into the actor network to decide the next-hop forwarding strategy:

$$a_i(n) = \mu_{II}\left(\tilde{\phi}_i(n); \theta_i^{\mu}\right), \tag{3.60}$$

where μ_{II} represents the posterior layers of the actor network.

While for those "silent" agents that don't belong to any flooding group and whose queue is not empty, they output the actions based on their independent intentions:

$$\bar{a}_i(n) = \mu_{II}\left(\phi_i(n); \theta_i^{\mu}\right). \tag{3.61}$$

Algorithm 8 The hyper-parameter updating process

1: **for** $C_j \in [C_1, \ldots, C_m]$ **do**
2: Calculate independent action $\bar{a}_z(t)$ for each agent $z \in C_j$ by Eq. (3.61)
3: Calculate ΔQ_j in C_j by Eq. (3.68)
4: **end for**
5: Store $D(n)$ into experience replay buffer R
6: Randomly select b transitions from buffer R
7: **for** UAV agent $i \in [1, \ldots, k]$ **do**
8: Update the critic network by minimizing $\mathcal{L}\left(\theta_i^{Q}\right)$ calculated from Eq. (3.64)
9: Update the actor network using policy gradients $\nabla_{\theta_i^{\mu}} J\left(\theta_i^{\mu}\right)$ calculated from Eq. (3.65)
10: Soft update the target networks by Eq. (3.66)
11: **end for**
12: Update the Gate Recurrent Unit by Eq. (3.67)
13: Get $\Delta\hat{Q}_j$ by normalize each ΔQ_j in sampled $D(n)$
14: Update the Multilayer Perceptron by minimizing the loss calculated from Eq. (3.69)

Note that the communication agents can also calculate the independent action based on Eq. (3.61), which is not a real action but is only used for algorithm training. Besides, the hyperparameter ϵ is used in the action selection process, so as to fully explore the action space and avoid falling into the local optimal solution [63].

Lastly, the critic network is essentially a RL neural network based on the Q-value function [64]. It inputs the local observation $o_i(n)$ and action $a_i(n)$ of agent i, then outputs $Q_i^{\mu}\left(o_i, a_i; \theta_i^{Q}\right)$ to guide the update of the actor network. Note that the output Q-value will also be used to update the Gate Recurrent Unit and the Multilayer Perceptron, which will be described later.

As for the algorithm training, an experience replay buffer R is applied to store the historical experience of all agents [65]. At each step n, the buffer stores the following data:

$$(O(n), A(n), R(n), O(n+1), D(n)), \tag{3.62}$$

where

$$\begin{cases} O(n) = (o_1(n), \ldots, o_k(n)) \\ A(n) = (a_1(n), \ldots, a_k(n)) \\ R(n) = (r_1(n), \ldots, r_k(n)) \\ O(n+1) = (o_1(n+1), \ldots, o_k(n+1)) \end{cases}, \tag{3.63}$$

and $D(n)$ records the relevant components at time n for training the Multilayer Perceptron, which will be introduced later. Note that our proposed algorithm belongs to a step-by-step update scheme [66]. Therefore, at each step, it extracts b pieces of experiences from the experience replay buffer to update the neural network. Also, each single-step training forms a new piece of experience, which will be stored in the experience replay buffer [67].

According to the above settings, the critic network can be updated by the training loss [68]:

$$\mathcal{L}\left(\theta_i^Q\right) = \mathbb{E}_{o_i,a_i,r_i,o_i' \sim R}\left[\left(Q_i^{\mu}\left(o_i, a_i\right) - y_i\right)^2\right], \tag{3.64}$$

where $y_i = r_i + \gamma \times Q_i^{\mu'}\left(o_i', a_i'\right)\Big|_{a_i'=\mu'(o_i')}$.

The policy gradient of the actor network can be described as:

$$\begin{aligned}\nabla_{\theta_i^{\mu}} J(\theta_i^{\mu}) =& \mathbb{E}_{o_i,a_i \sim R}[\nabla_{\theta_i^{\mu}} \mu(a_i \mid o_i) \times \\ & \nabla_{a_i} Q_i^{\mu}(o_i, a_i)|_{a_i=\mu(o_i)}]\end{aligned}. \tag{3.65}$$

Moreover, the target network of the actor-critic structure adopts a soft-update mechanism with a factor σ:

$$\begin{cases} \theta_i^{Q'} = \sigma \times \theta_i^Q + (1-\sigma) \times \theta_i^{Q'} \\ \theta_i^{\mu'} = \sigma \times \theta_i^{\mu} + (1-\sigma) \times \theta_i^{\mu'} \end{cases}. \tag{3.66}$$

The Gate Recurrent Unit (i.e., communication channel) can be updated by the gradient of the joint intentions, which can be calculated by the chain rule:

$$\begin{aligned}\nabla_{\theta^g} J(\theta^g) =& \frac{1}{k} \times \sum_{i \in k} \mathbb{E}_{o_i,a_i \sim R}[\nabla_{\theta^g} g(\widetilde{\phi}_i \mid \Phi) \times \\ & \nabla_{\widetilde{\phi}_i} \mu(a_i, \widetilde{\phi}_i) \nabla_{a_i} Q_i^{\mu}(o_i, a_i)|_{a_i=\mu(o_i)}]\end{aligned}. \tag{3.67}$$

Lastly, the update process of the Multilayer Perceptron is as follows. For each leader node j (i.e., the node with the highest intention value at this step) and the corresponding flooding group C_j, the proposed algorithm calculates the average difference between the cooperative Q-value and independent Q-value:

$$\begin{aligned}\Delta Q_j = & \frac{1}{|C_j|} \times \Big(\sum_{z \in C_j} Q(o_z, a_z \mid \theta_z^Q) - \\ & \sum_{z \in C_j} Q(o_z, \bar{a}_z \mid \theta_z^Q)\Big)\end{aligned}, \tag{3.68}$$

where $\bar{a_z}$ means the independent action of agent z, which can be calculated by Eq. (3.61). ΔQ_j indicates the influence of the routing information flooding. $\Delta Q_j > 0$ means the agent can form a better routing decision benefit from routing information sharing. Otherwise the opposite. Note that due to the unstable training feature, ΔQ_j may jitter during the early training process. But it will gradually increase as the algorithm tends to be stable. At step n, supposed that there exist m flooding groups, and the algorithm will store $D(n) = (\Delta Q_1, \ldots, \Delta Q_m)$ into the experience replay buffer according to Eq. (3.62). In the training phase, the algorithm normalizes $\Delta Q_1, \ldots, \Delta Q_m$ from b pieces of experiences to get $\Delta \hat{Q}_1, \ldots, \Delta \hat{Q}_m$ with a normalized-value between [0,1]. The normalized values are used as the classifier label to train the Multilayer Perceptron:

$$\begin{aligned}\mathcal{L}(\theta^d) = & \frac{1}{m} \times \sum_{j \in m} [\Delta \hat{Q}_j \times \log(g(\phi_j \mid \theta^d)) \\ & -(1 - \Delta \hat{Q}_j) \times \log(1 - g(\phi_j \mid \theta^d))]\end{aligned} \tag{3.69}$$

The detail of our proposed routing algorithm can be found in Algorithms 6, 7, and 8.

3.3.3 Simulation Results and Discussions

In this section, we construct several experiments to evaluate the performance of our proposed routing method by comparing it with other baseline algorithms. We use OMNET++ to build the network simulation environment. Our experiment environment is Ubuntu 20.04 with 64g RAM, Nvidia RTX 3060, and Intel i7-10875H. The environment setting can be found in Table 3.6, and the hyper-parameter setting of the proposed algorithm can be found in Table 3.7.

3.3.3.1 Baseline Algorithm

To evaluate our algorithm, we adopt some other routing algorithms as the baseline algorithms. The first one is the backpressure routing algorithm [69]. The backpressure routing algorithm forwards the packet based on the queue congestion difference between neighbor nodes. Therefore, it can effectively prolong the network survival time and postpone the local congestion in a high-load pressure network. The second one is the multi-agent reinforcement learning routing algorithm without routing information flooding. This method is a simplified version of our algorithm, which only retains the multi-agent actor-critic algorithm part. Therefore, no flooding groups are built to share routing information, and each agent decides the routing strategy only based on their local observation. The last method is the random routing method. Moreover, to further consider the influence caused by the network

Table 3.6 Environment parameter setting

Parameter	Value
Number of UAV nodes k	20
System boundary $x_{max}/y_{max}/z_{max}$	100 m
Initial speed factor $v(0)$ and initial average speed factor $\bar{v}$	1 m/step $\sim$ 5m/step
Initial direction factor $d(0)$ and initial average direction factor $\bar{d}$	random unit vector
Initial pitch angle factor $a(0)$ and initial average pitch angle factor $\bar{a}$	$0° \sim 90°$
Tuning parameter (β)	0.5
Bits of each packet l	500 bits
Energy consumption of radio electronics per bit E_{elec}	$1.5 * 10^{-4}$ J/bit
Energy consumption factor of the power amplifier in the free space model η_{fs}	$25 * 10^{-9}$ J/(bit $*$ m^2)
Energy consumption factor of the power amplifier in the multipath fading model η_{mp}	$4 * 10^{-12}$ J/(bit $*$ m^4)
Distance switch threshold between the free space model and the multipath fading model τ	100 m
Max queue capacity q_i^{max} of each UAV node i	50
Number of packets forwarded by each UAV node at each step $\lambda_i^{out}(n)$	1
Number of pre-cached packets ω	10
Communication range r of each UAV node i	20 m

Table 3.7 Algorithm hyper-parameter setting

Parameter	Value
Discount factor γ of the critic network	0.95
Soft update factor σ of the target network	0.02
Max training episodes MAX_{eps}	20,000
Max training step MAX_{step} in each episode	100
Max capacity of the experience replay buffer R	10,000
Batch size b	256
Initial ϵ factor	0.9
The Minimum value of ϵ	0.05
Learning rate of $\theta_i^{\mu}, \theta_i^{Q}, \theta^{d}, \theta^{g}$	0.001
Balance factor of the optimization problem ρ	0.5

dynamics, we set three network environments with different degrees of mobility: the low dynamic network environment, the middle dynamic network environment, and the high dynamic network environment. In the low-dynamic network environment, all the UAVs stay hover. In the middle dynamic network environment, we randomly select ten UAVs to move while the others stay hover. As for the highly dynamic network environment, all UAVs move continuously, so the network environment is the most unstable.

3.3.3.2 Convergence Analysis

Figure 3.15 presents the reward convergence curve in the UAV swarm networks with different dynamic levels. We can find that the convergence curves of the random and backpressure methods have no obvious fluctuation. The reward value of these two methods keeps at 0.9 and 1.2 no matter how the network dynamic changes. This is because they are unintelligent and cannot flexibly adjust their routing strategy by interacting with the UAV swarm environment. As for the proposed routing method, the convergence reward value under different network scenarios is 5.0, 4.7, and 4.1, respectively. And the convergence reward value of the non-communication MARL routing method is 4.5, 4.2, and 3.8, respectively. We can find that although the two intelligent algorithms suffer from a dynamic network environment, they perform better than the random routing algorithm and the backpressure routing algorithm. Moreover, our proposed adaptive communication-based UAV swarm routing algorithm has the highest reward in all situations. It shows that the dynamic flooding group can enhance routing information sharing among multiple UAVs, therefore improving convergence performance. Besides, although the convergence curve of the two intelligent algorithms rises rapidly at about 3000 steps, the non-communication MARL routing method spends more time to achieve stable. In contrast, our proposed routing method reaches the optimal convergence state quickly. It indicates that the dynamic flooding among multiple zones promotes a robust routing strategy.

3.3.3.3 Complexity Analysis

In this section, we analyze the time complexity of the proposed routing algorithm. During the training process, the proposed algorithm updated k actor networks according to Eq. (3.65), k critic networks according to Eq. (3.64), one Gate Recurrent Unit according to Eq. (3.67), and one Multilayer Perceptron according to Eqs. (3.68) and (3.69). Therefore, the time complexity of training can be described as $O(k * (O_i)^2 * A_i)$, where k represents the agent number, O_i represents the local observation space of agent i, and A_i represents the action space of agent i.

3.3.3.4 Performance Analysis

In this section, we provide a detailed performance analysis to further evaluate the proposed algorithm.

(1) Queue Utilization Evaluation: Figure 3.16 shows the change process of max queue occupancy under different UAV swarm network environments. The curve gradient reflects the congestion control ability of the routing method. The smaller the gradient, the stronger ability. Besides, we count the maximum queue length among all the UAVs at each step. Therefore, the curve may not grow

Fig. 3.15 Average reward. (**a**) Lowly dynamic. (**b**) Middle dynamic. (**c**) Highly dynamic

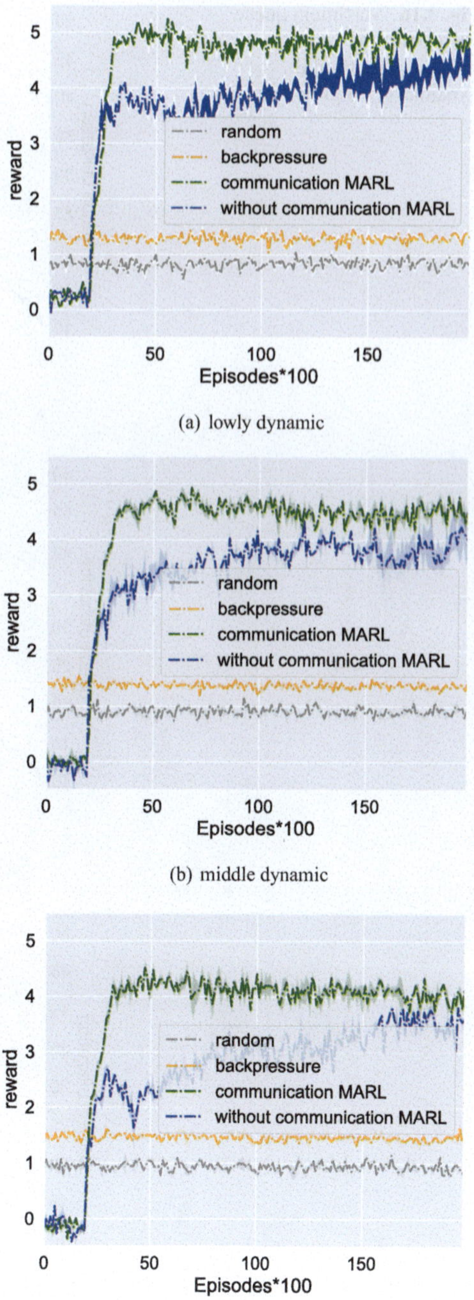

(a) lowly dynamic

(b) middle dynamic

(c) highly dynamic

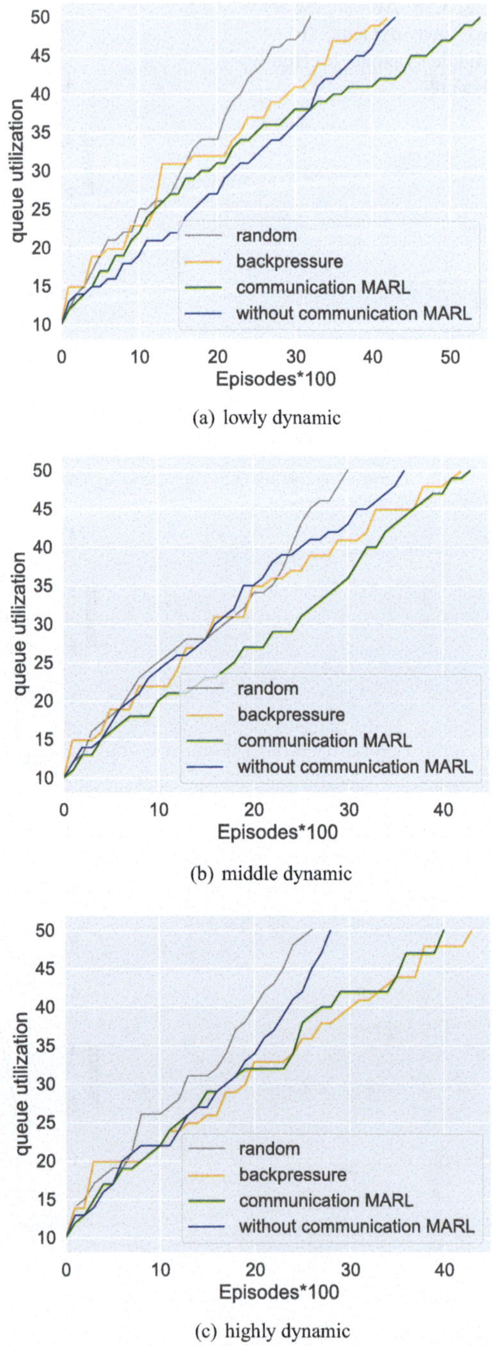

Fig. 3.16 Maximum queue occupancy. (**a**) Lowly dynamic. (**b**) Middle dynamic. (**c**) Highly dynamic

in a certain period, which means multiple UAVs may reach the same queue utilization successively. For the lowly dynamic network environment and the middle dynamic network environment, the queue occupancy growth of our proposed method is the slowest. In a highly dynamic environment, our proposed method's performance is slightly worse than the backpressure method. This is because the backpressure routing method only decides the routing based on the queue length differences of neighbor UAVs. Therefore is not affected by the network dynamics. Besides, the non-communication MARL routing method performs slightly better than the backpressure routing algorithm in the lowly dynamic network environment, but only better than the random routing method as the environment dynamic increases. It indicates that a UAV swarm network without necessary routing information sharing is difficult to form a cooperative routing strategy, therefore unable to postpone the occurrence of local congestion.

(2) Residual Energy Evaluation: Based on the energy consumption model in Sect. 3.3.1.3, we analyze the residual energy of the UAV swarm network in three different dynamic environments. As shown in Fig. 3.17, the horizontal axis is the residual energy of the UAV swarm network, and the vertical axis is the frequency of the residual energy. If the residual energy of each UAV is relative concentration, then the frequency histogram will have a higher peak with a narrower distribution, otherwise the opposite. Although we assume that each UAV can only forward one packet per step, the number of received packets from upstream is uncertain, thus will cause a different receiving energy consumption. Similar to the queue utilization evaluation, a robust routing method can not only balance the queue load but also balance the residual energy. This ability will be reflected in the concentration of the frequency histogram curve. Besides, a good routing method can relieve local congestion and prolong the network system operation time, therefore should have a lower average energy. In conclusion, a robust routing algorithm will present a concentrated frequency histogram curve with a low mean residual energy value. As shown in Fig. 3.17, both our proposed routing algorithm and the backpressure routing algorithm have the above characteristics in the lowly dynamic network environment. However, our proposed routing algorithm has lower average energy, which means a longer network system operation time. Although the non-communication MARL routing method has lower average energy than the random routing method, both of them have a scattered frequency distribution. As for the middle dynamic network environment, the backpressure routing method presents the most concentrated energy distribution, while our proposed routing method still has the lowest average energy. It indicates that the backpressure routing algorithm averages the residual energy among all the UAVs with an excellent load balancing capability. Yet, it cannot effectively prolong the network survival time. Besides, the without-communication MARL method only performs better than the random routing method in this environment. Finally, as for the highly dynamic network environment, the difference between algorithms is reduced,

Fig. 3.17 Residual energy distribution. (**a**) Lowly dynamic. (**b**) Middle dynamic. (**c**) Highly dynamic

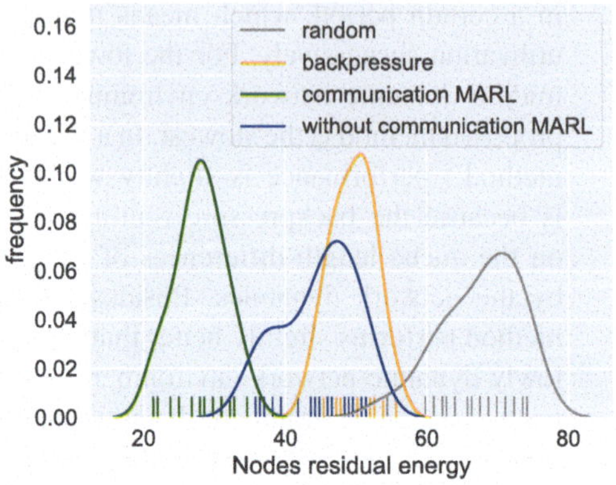

(a) lowly dynamic

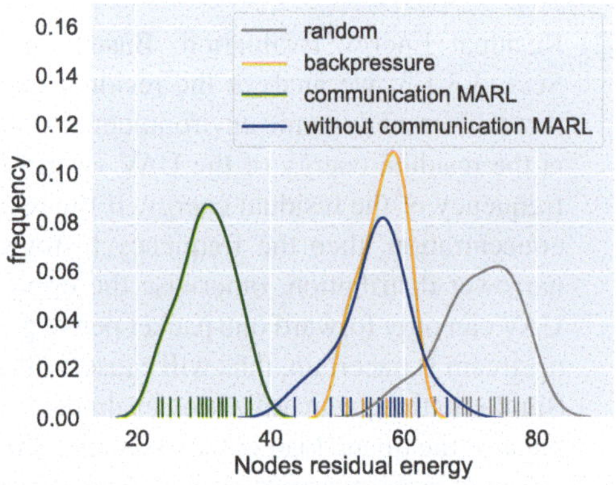

(b) middle dynamic

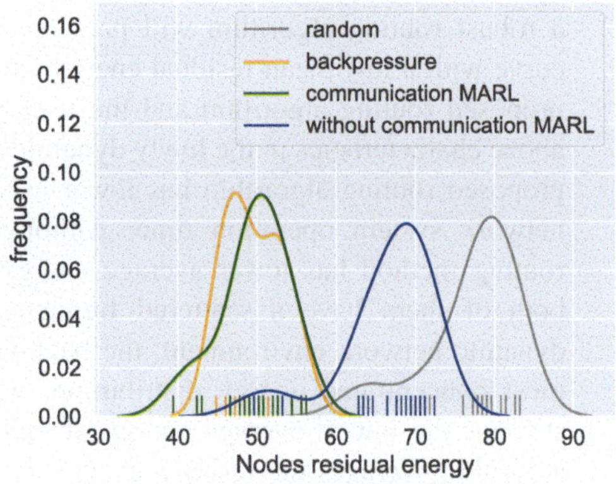

(c) highly dynamic

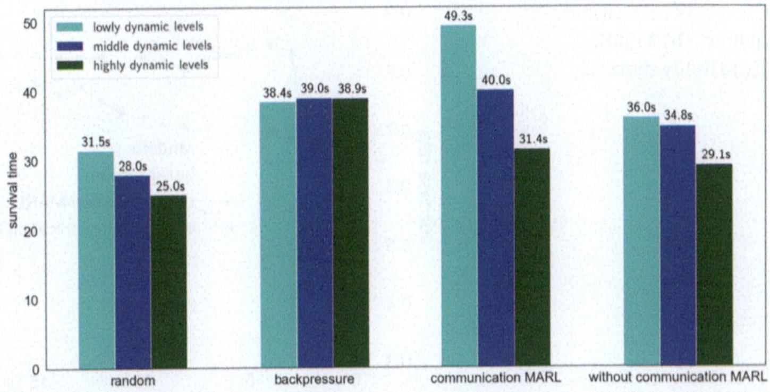

Fig. 3.18 Network survival time

but the backpressure algorithm and our proposed algorithm maintain a relative advantage. The above analysis shows that by on-demand routing flooding between multiple zones, our proposed routing method can balance the energy of each UAV in the dynamic network environment.

(3) Survival Time Evaluation: The above analysis reflects the network survival time problem through queue utilization and residual energy evaluation. In this part, we directly analyze it through Fig. 3.18. Our proposed routing method presents the longest network survival time in low and middle-dynamic network environments. In the low dynamic network environment, our method improves the performance by 56% of the random routing method, by 28% of the backpressure routing method, and by 37% of the without-communication MARL routing method. In the middle dynamic network environment, the improvement percentage changed to 42%, 3%, and 15%, respectively. As for the high dynamic network environment, our proposed routing algorithm is still better than the non-communication MARL and random routing methods, but the backpressure routing method performs best. We can find that the backpressure routing method is least affected by the network dynamics. The reason for this advantage will be further discussed.

(4) Success Rate Evaluation: Figure 3.19 shows the routing success rate under different UAV swarm network environments. Unlike the above performance, the backpressure routing method cannot deliver packets effectively and even performs similarly to the random routing method. Considering the characteristics of the backpressure routing algorithm, we find that it built its survival time advantages at the expense of packet delivery. Therefore, the backpressure routing algorithm can only perform better in some relevant performance indicators of the network survival time. By comparison, our proposed algorithm performs the best, although the packet transmission success rate gradually decreases as the network dynamic increases. Also, the non-communication

Fig. 3.19 Success rate. (**a**) Lowly dynamic. (**b**) Middle dynamic. (**c**) Highly dynamic

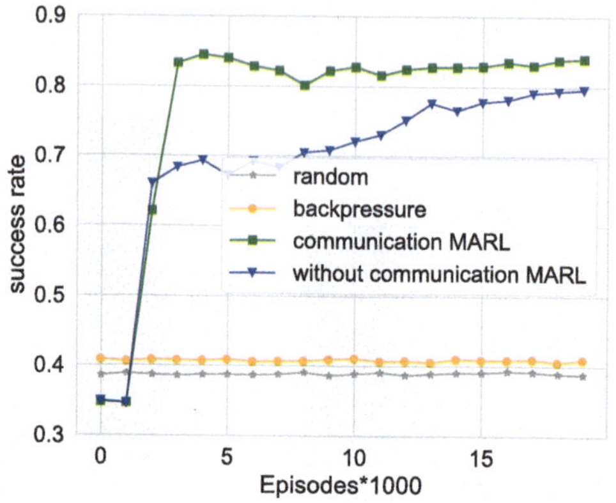

(a) lowly dynamic

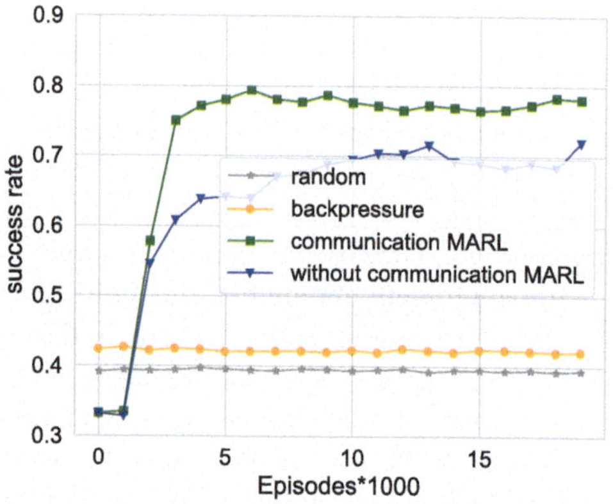

(b) middle dynamic

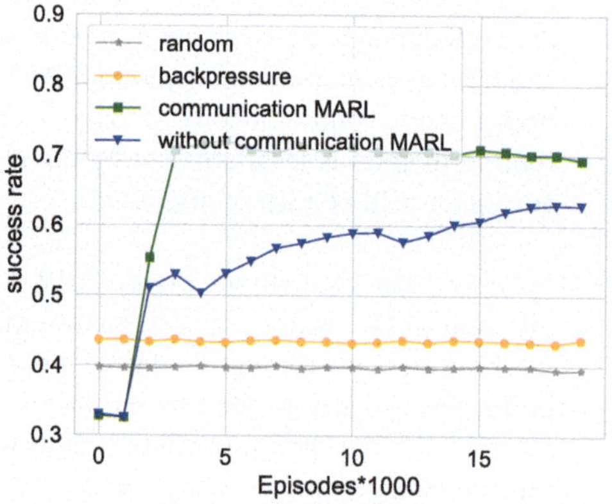

(c) highly dynamic

MARL routing method transmits packets more efficiently than the other two methods. However, compared with our method, it has worse convergence and greater fluctuation, thus indicating the necessity of adaptive routing flooding.

(5) Throughput Evaluation: The throughput performance of different routing methods in multiple network environments can be found in Fig. 3.20. We can find that the throughput performance is similar to the success rate. It is because the throughput reflects the number of successfully transmitted packets per step. Compared with the random routing method, the backpressure routing method only shows a slight advantage, further proving our previous analysis. In the low dynamic network environment, the convergence value of the without-communication routing method and our proposed routing method are similar, while the former converges more slowly. Also, the throughput curve of the without-communication routing method fluctuates sharply as the network dynamic increases. Conversely, our proposed algorithm achieves the best throughput performance in different environments. It shows that by forming flooding groups and sharing routing information within multiple zones, our proposed routing algorithm can transmit packets more efficiently, and construct a stable routing strategy.

3.4 Summary

This chapter primarily addresses the routing problem in dynamic UAV swarm networks, with the goal of enhancing routing efficiency, network resilience, and network lifetime. This chapter first proposes an evolutionary game-based UAV swarm routing optimization method to reduce computational complexity and extend the network's operational lifetime. This method enables the evolution of routing strategies in highly dynamic environments, ensuring the stability and resilience of routing decisions. Additionally, we employ a MARL framework to optimize routing decisions. A MLP algorithm is designed to determine when routing flooding is necessary, and a GRU algorithm is used to compress communication data, thus reducing communication overhead. Finally, we introduce a Multi-Agent Actor-Critic algorithm that facilitates cooperative routing decision-making through shared routing information. This approach enables efficient routing in partially observable distributed environments and optimizes network quality in terms of network lifetime, packet delivery rate, and throughput.

Fig. 3.20 Throughput. (**a**) Lowly dynamic. (**b**) Middle dynamic. (**c**) Highly dynamic

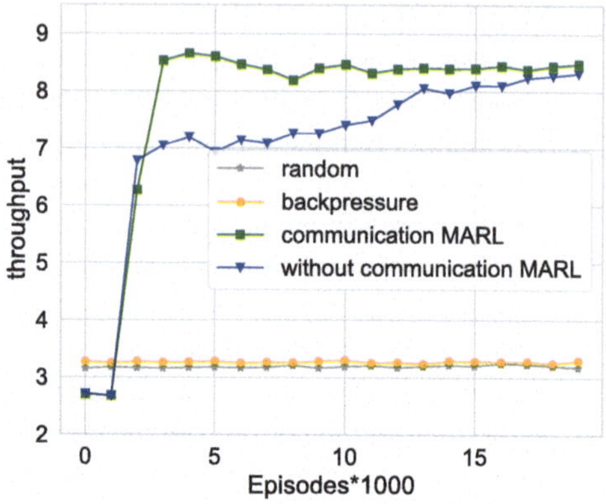

(a) lowly dynamic

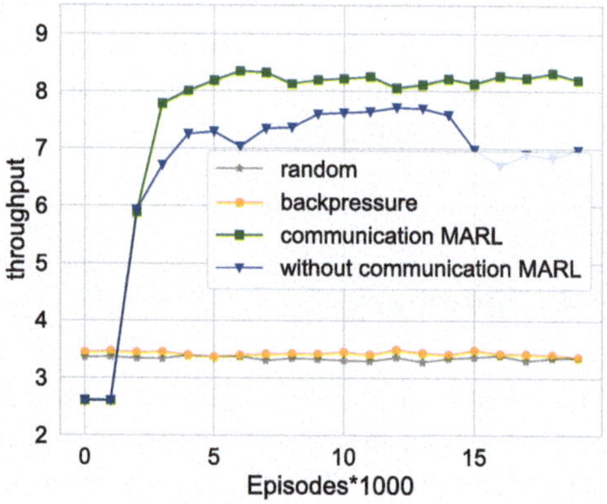

(b) middle dynamic

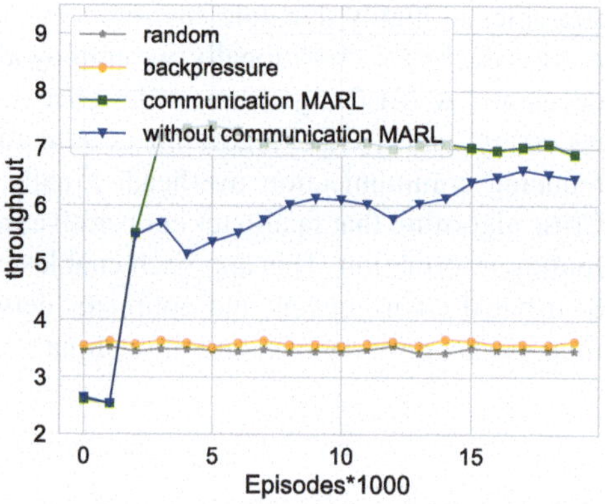

(c) highly dynamic

References

1. D. Wu, C. Jin, H. Yao, T. Mai, X. Xin, Dynamic routing optimization method for UAV swarm networks: An evolutionary game approach. IEEE Trans. Sustain. Comput. **10**(4), 616–627 (2025)
2. Z. Wang, H. Yao, T. Mai, Z. Xiong, X. Wu, D. Wu, S. Guo, Learning to routing in UAV swarm network: A multi-agent reinforcement learning approach. IEEE Trans. Veh. Technol. **72**(5), 6611–6624 (2022)
3. Z. Wang, H. Yao, T. Mai, Z. Xiong, F.R. Yu, Cooperative reinforcement learning aided dynamic routing in UAV swarm networks, in *ICC 2022-IEEE International Conference on Communications* (IEEE, Piscataway, 2022), pp. 1–6
4. H. Shakhatreh, A.H. Sawalmeh, A. Al-Fuqaha, Z. Dou, E. Almaita, I. Khalil, N.S. Othman, A. Khreishah, M. Guizani, Unmanned aerial vehicles (UAVs): A survey on civil applications and key research challenges. IEEE Access **7**, 48572–48634 (2019)
5. L. Gupta, R. Jain, G. Vaszkun, Survey of important issues in UAV communication networks. IEEE Commun. Surv. Tutorials **18**(2), 1123–1152 (2016)
6. M.A. Khan, A. Safi, I.M. Qureshi, I.U. Khan, Flying ad-hoc networks (FANETs): A review of communication architectures, and routing protocols, in *2017 First International Conference on Latest trends in Electrical Engineering and Computing Technologies (INTELLECT)* (2017), pp. 1–9
7. M.A. Khan, I.U. Khan, A. Safi, I.M. Quershi, Dynamic routing in flying ad-hoc networks using topology-based routing protocols. Drones **2**(3), 27 (2018)
8. Y.-B. Wang, T.-Y. Wu, W.-T. Lee, C.-H. Ke, A novel geographic routing strategy over VANET, in *2010 IEEE 24th International Conference on advanced information networking and Applications Workshops* (IEEE, Piscataway, 2010), pp. 873–879
9. A. Rovira-Sugranes, A. Razi, F. Afghah, J. Chakareski, A review of AI-enabled routing protocols for UAV networks: Trends, challenges, and future outlook. Ad Hoc Netw. **130**, 102790 (2022)
10. M.E. Mkiramweni, C. Yang, J. Li, Z. Han, Game-theoretic approaches for wireless communications with unmanned aerial vehicles. IEEE Wirel. Commun. **25**(6), 104–112 (2018)
11. B. Benmammar, F. Krief, Game theory applications in wireless networks: A survey, in *Applications of Information Systems in Engineering and Bioscience* (2014), pp. 208–215
12. M.E. Mkiramweni, C. Yang, J. Li, W. Zhang, A survey of game theory in unmanned aerial vehicles communications. IEEE Commun. Surv. Tutorials **21**(4), 3386–3416 (2019)
13. M. Zhou, Y. Guan, M. Hayajneh, K. Niu, C. Abdallah, Game theory and machine learning in UAVs-assisted wireless communication networks: A survey. Preprint. arXiv:2108.03495 (2021)
14. J. Tanimoto, *Fundamentals of Evolutionary Game Theory and Its Applications* (Springer, Tokyo, 2015)
15. D. Greiner, J. Periaux, J.M. Emperador, B. Galván, G. Winter, Game theory based evolutionary algorithms: a review with Nash applications in structural engineering optimization problems. Arch. Comput. Methods Eng. **24**, 703–750 (2017)
16. P. Du, M. Gerla, An evolutionary multi-player game model for two-hop routing in delay tolerant networks, in *2017 IEEE 14th International Conference on Mobile Ad Hoc and Sensor Systems (MASS)* (IEEE, Piscataway, 2017), pp. 108–116
17. Z. Wang, H. Yao, T. Mai, Z. Xiong, F.R. Yu, Cooperative reinforcement learning aided dynamic routing in UAV swarm networks, in *ICC 2022-IEEE International Conference on Communications* (IEEE, Piscataway, 2022), pp. 1–6
18. Z. Wang, H. Yao, T. Mai, Z. Xiong, F.R. Yu, Cooperative reinforcement learning aided dynamic routing in UAV swarm networks, in *ICC 2022-IEEE International Conference on Communications* (IEEE, Piscataway, 2022), pp. 1–6

19. M. Li, F.R. Yu, P. Si, R. Yang, Z. Wang, Y. Zhang, UAV-assisted data transmission in blockchain-enabled M2M communications with mobile edge computing. IEEE Netw. **34**(6), 242–249 (2020)
20. M.A. Nowak, R.M. May, Evolutionary games and spatial chaos. Nature **359**(6398), 826–829 (1992)
21. C.A. Kamhoua, N. Pissinou, J. Miller, S.K. Makki, Mitigating routing misbehavior in multi-hop networks using evolutionary game theory, in *2010 IEEE Globecom Workshops* (IEEE, Piscataway, 2010), pp. 1957–1962
22. Y. Li, X. Wu, Packet forwarding strategies in multiagent systems: An evolutionary game approach. Wirel. Commun. Mob. Comput. **2018**, 4965343 (2018)
23. P.D. Taylor, L.B. Jonker, Evolutionary stable strategies and game dynamics. Math. Biosci. **40**(1–2), 145–156 (1978)
24. D. Fudenberg, D.K. Levine, *The Theory of Learning in Games*, vol. 2 (MIT Press, 1998)
25. R. Ch, G. Srivastava, T.R. Gadekallu, P.K.R. Maddikunta, S. Bhattacharya, Security and privacy of UAV data using blockchain technology. J. Inf. Secur. Appl. **55**, 102670 (2020)
26. J. Wang, C. Jin, Q. Tang, N. Xiong, G. Srivastava, Intelligent ubiquitous network accessibility for wireless-powered MEC in UAV-assisted B5G. IEEE Trans. Netw. Sci. Eng. **8**, 2801 (2020)
27. X. Pang, M. Liu, N. Zhao, Y. Chen, Y. Li, F.R. Yu, Secrecy analysis of UAV-based mmWave relaying networks. IEEE Trans. Wirel. Commun. **20**, 4990 (2021)
28. D. Zhai, H. Li, X. Tang, R. Zhang, Z. Ding, F.R. Yu, Height optimization and resource allocation for NOMA enhanced UAV-aided relay networks. IEEE Trans. Commun. **69**, 962 (2020)
29. M. Khaledi, A. Rovira-Sugranes, F. Afghah, A. Razi, On greedy routing in dynamic UAV networks, in *2018 IEEE International Conference on Sensing, Communication and Networking (SECON Workshops)* (IEEE, Piscataway, 2018), pp. 1–5
30. S. Wen, C. Huang, X. Chen, J. Ma, N. Xiong, Z. Li, Energy-efficient and delay-aware distributed routing with cooperative transmission for internet of things. J. Parallel Distrib. Comput. **118**, 46–56 (2018)
31. A. Coelho, E.N. Almeida, P. Silva, J. Ruela, R. Campos, M. Ricardo, Redefine: Centralized routing for high-capacity multi-hop flying networks, in *2018 14th International Conference on Wireless and Mobile Computing, Networking and Communications (WiMob)* (IEEE, Piscataway, 2018), pp. 75–82
32. A. Detti, C. Pisa, S. Salsano, N. Blefari-Melazzi, Wireless mesh software defined networks (wmSDN), in *2013 IEEE 9th International Conference on Wireless and Mobile Computing, Networking and Communications (WiMob)* (IEEE, Piscataway, 2013), pp. 89–95
33. P.K. Sharma, D.I. Kim, Coverage probability of 3-D mobile UAV networks. IEEE Wirel. Commun. Lett. **8**(1), 97–100 (2019)
34. R.R. Roy, *Handbook of Mobile Ad Hoc Networks for Mobility Models* (Springer, New York, 2011)
35. M.T. Spaan, Partially observable Markov decision processes, in *Reinforcement Learning* (Springer, Berlin, 2012), pp. 387–414
36. R.S. Sutton, D.A. McAllester, S.P. Singh, Y. Mansour et al., Policy gradient methods for reinforcement learning with function approximation, in *NIPs*, vol. 99 (Citeseer, 1999), pp. 1057–1063
37. H. Yao, T. Mai, C. Jiang, L. Kuang, S. Guo, AI routers & network mind: A hybrid machine learning paradigm for packet routing. IEEE Comput. Intell. Mag. **14**(4), 21–30 (2019)
38. M. Gerla, K. Xu, Minuteman: Forward projection of unmanned agents using the airborne internet, in *Aerospace Conference Proceedings (2002) IEEE* (IEEE, Piscataway, 2002)
39. M.J. Neely, Stochastic network optimization with application to communication and queueing systems. Synthesis Lect. Commun. Netw. **3**(1), 1–211 (2010)
40. G.A. Kakamoukas, P.G. Sarigiannidis, A.A. Economides, FANETs in agriculture - a routing protocol survey. Internet Things **18**, 100183 (2022)

41. H. Fatemidokht, M.K. Rafsanjani, B.B. Gupta, C.-H. Hsu, Efficient and secure routing protocol based on artificial intelligence algorithms with UAV-assisted for vehicular ad hoc networks in intelligent transportation systems. IEEE Trans. Intell. Transp. Syst. **22**(7), 4757–4769 (2021)
42. M.Y. Arafat, S. Moh, Location-aided delay tolerant routing protocol in UAV networks for post-disaster operation. IEEE Access **6**, 59891–59906 (2018)
43. H. Zhang, W. Quan, Networking automation and intelligence: A new era of network innovation. Engineering **17**(10), 13–16 (2022)
44. S. Bhandari, X. Wang, R. Lee, Mobility and location-aware stable clustering scheme for UAV networks. IEEE Access **8**, 106364–106372 (2020)
45. W.Y.B. Lim, S. Garg, Z. Xiong, Y. Zhang, D. Niyato, C. Leung, C. Miao, UAV-assisted communication efficient federated learning in the era of the artificial intelligence of things. IEEE Netw. **35**(5), 188–195 (2021)
46. M.J. Sousa, A. Moutinho, M. Almeida, Decentralized distribution of UAV fleets based on fuzzy clustering for demand-driven aerial services, in *2020 IEEE International Conference on Fuzzy Systems (FUZZ-IEEE)* (IEEE, Piscataway, 2020), pp. 1–8
47. M.Y. Arafat, S. Moh, Localization and clustering based on swarm intelligence in UAV networks for emergency communications. IEEE Internet Things J. **6**(5), 8958–8976 (2019)
48. D.S. Lakew, U. Sa'ad, N.-N. Dao, W. Na, S. Cho, Routing in flying ad hoc networks: A comprehensive survey. IEEE Commun. Surv. Tutorials **22**(2), 1071–1120 (2020)
49. T. Clausen, P. Jacquet, RFC3626: Optimized link state routing protocol, Internet Engineering Task Force (IETF) Request For Comments, http://ietf.org/rfc/rfc3626.txt, (2003).
50. J. Lee, K. Kim, S. Yoo, A.Y. Chung, J.Y. Lee, S.J. Park, H. Kim, Constructing a reliable and fast recoverable network for drones, in *2016 IEEE International Conference on Communications (ICC)* (IEEE, Piscataway, 2016), pp. 1–6
51. C. Perkins, E. Belding-Royer, S. Das, RFC3561: Ad hoc on-demand distance vector (AODV) routing. Internet Engineering Task Force (IETF) Request For Comments, http://ietf.org/rfc/rfc3626.txt, (2003).
52. G. Gankhuyag, A.P. Shrestha, S.-J. Yoo, Robust and reliable predictive routing strategy for flying ad-hoc networks. IEEE Access **5**, 643–654 (2017)
53. S.-C. Choi, H.R. Hussen, J.-H. Park, J. Kim, Geolocation-based routing protocol for flying ad hoc networks (FANETs), in *2018 Tenth International Conference on Ubiquitous and Future Networks (ICUFN)* (IEEE, Piscataway, 2018), pp. 50–52
54. L. Lin, Q. Sun, J. Li, F. Yang, A novel geographic position mobility oriented routing strategy for UAVs. J. Comput. Inf. Syst. **8**(2), 709–716 (2012)
55. D. Broyles, A. Jabbar, J.P. Sterbenz et al., Design and analysis of a 3-D gauss-Markov mobility model for highly-dynamic airborne networks, in *Proceedings of the International Telemetering Conference (ITC), San Diego, CA* (2010)
56. W.B. Heinzelman, A.P. Chandrakasan, H. Balakrishnan, An application-specific protocol architecture for wireless microsensor networks. IEEE Trans. Wirel. Commun. **1**(4), 660–670 (2002)
57. F. Ashtiani, J.A. Salehi, M.R. Aref, Mobility modeling and analytical solution for spatial traffic distribution in wireless multimedia networks. IEEE J. Sel. Areas Commun. **21**(10), 1699–1709 (2003)
58. D. Gross, J.F. Shortle, J.M. Thompson, C.M. Harris, Fundamentals of queuing theory. J. R. Stat. Soc. **35**(5), 76–77 (1986)
59. T. Mai, H. Yao, N. Zhang, L. Xu, M. Guizani, S. Guo, Cloud mining pool aided blockchain-enabled Internet of Things: An evolutionary game approach. IEEE Trans. Cloud Comput. **11**, 692 (2021)
60. N. Nikaein, H. Labiod, C. Bonnet, DDR-distributed dynamic routing algorithm for mobile ad hoc networks, in *2000 First Annual Workshop on Mobile and Ad Hoc Networking and Computing. MobiHOC (Cat. No. 00EX444)* (IEEE, Piscataway, 2000), pp. 19–27
61. K. Cho, B. Van Merriënboer, C. Gulcehre, D. Bahdanau, F. Bougares, H. Schwenk, Y. Bengio, Learning phrase representations using RNN encoder-decoder for statistical machine translation. Preprint. arXiv:1406.1078 (2014)

62. Z. Qin, H. Yao, T. Mai, D. Wu, N. Zhang, S. Guo, Multi-agent reinforcement learning aided computation offloading in aerial computing for the Internet-of-Things. IEEE Trans. Serv. Comput. **16**, 1976 (2022)
63. M. Tokic, G. Palm, Value-difference based exploration: adaptive control between epsilon-greedy and softmax, in *Annual Conference on Artificial Intelligence* (Springer, Berlin, 2011), pp. 335–346
64. X. Ren, C. Qiu, X. Wang, Z. Han, K. Xu, H. Yao, S. Guo, AI-bazaar: A cloud-edge computing power trading framework for ubiquitous AI services. IEEE Trans. Cloud Comput. **11**, 2337 (2022)
65. Y. Gong, H. Yao, J. Wang, M. Li, S. Guo, Edge intelligence-driven joint offloading and resource allocation for future 6G industrial Internet of Things. IEEE Trans. Netw. Sci. Eng. **11**(6), 5644–5655 (2024)
66. T. Mai, H. Yao, N. Zhang, W. He, D. Guo, M. Guizani, Transfer reinforcement learning aided distributed network slicing resource optimization in industrial IoT. IEEE Trans. Ind. Inf. **18**, 4308 (2021)
67. W. Quan, M. Liu, N. Cheng, X. Zhang, D. Gao, H. Zhang, Cybertwin-driven DRL-based adaptive transmission scheduling for software defined vehicular networks. IEEE Trans. Veh. Technol. **71**(5), 4607–4619 (2022)
68. C. Yu, W. Quan, D. Gao, Y. Zhang, K. Liu, W. Wu, H. Zhang, X. Shen, Reliable cybertwin-driven concurrent multipath transfer with deep reinforcement learning. IEEE Internet Things J. **8**(22), 16207–16218 (2021)
69. M.J. Neely, R. Urgaonkar, Optimal backpressure routing for wireless networks with multi-receiver diversity. Ad Hoc Netw. **7**(5), 862–881 (2009)

Chapter 4
Intelligent UAV Swarm Network Resource Scheduling

UAV swarm networks face inherent constraints due to limited resources. Effectively managing and coordinating these resources to meet diverse service demands presents a significant research challenge. This chapter explores intelligent methods for UAV Network Resource Scheduling. Firstly, we employ virtualization techniques to abstract physical resources into virtual resources, enabling more flexible allocation of network resources to meet the varying demands of UAV nodes. Traditional heuristic methods for mapping virtual network requests to physical networks typically yield suboptimal solutions, as they often rely on manually designed features and constraints, which lack both flexibility and efficiency. To address these limitations, we leverage Graph Transformer technology to assist in the embedding of resource virtualization. Specifically, we propose a self-supervised graph neural network that effectively captures both local and global structural information of the graph, and rapidly adapts to dynamic scenarios, thus addressing the inherent dynamism of UAV networks [1]. Additionally, to mitigate the challenges posed by insufficient energy resources in UAV Named Data Networking (UNDN), we propose a resource-efficient content-sharing mechanism. This mechanism incentivizes content producers to share content and disclose private information. Building on this framework, we introduce a contract-based approach and apply the Gale-Shapley (GS) algorithm to achieve stable matching between content consumers and producers, thereby enhancing overall system efficiency and resource utilization [2].

4.1 Graph Transformer-Based Resource Virtualization Embedding for UAV Swarm Networks

Over the past decade, UAV swarm network has been extensively utilized in the field of military, agriculture, and commercial applications. In the scene of 5G and

H. Yao, T. Mai, *UAV Swarm Cooperation*, Wireless Networks,
https://doi.org/10.1007/978-3-031-96444-2_4

emerging 6G, diverse services require networks to have high bandwidth, low latency and large-scale connectivity. However, the limited capabilities of UAV networks (e.g. limited computational capacity) restrict their potential to be applied in 5G and 6G network [3–5]. Therefore, solving the problems of resource management and coordination to integrate diverse types of services into a unified UAV network, effectively meeting complicated services requirements simultaneously [6], represents a significant challenge. In the face of the diverse requirements from emerging 5G vertical domains and applications, technologies such as network function virtualization and network slicing have been widely embraced as key enabling technologies for future upgrades. By adopting virtualization techniques, physical resources are abstracted into virtual resources, enabled more flexible allocation of network resources to fulfill users' requests. To accommodate as many virtual network services as possible on the underlying physical infrastructure, developing efficient algorithms for virtual holds significant importance. Among these research areas, virtual network embedding (VNE) has received extensive attention as a core direction, and it is known to be NP-hard [7].

To reduce computational complexity and runtime, the classical VNE problem has employed heuristic methods [8, 9] to obtain suboptimal solutions. However, these methods typically rely on manually designed features and constraints, which are often sparse and intricate. Consequently, they exhibit low efficiency and lack flexibility in problem-solving, while also neglecting time constraints. In contrast, offline training and online inference methods based on deep reinforcement learning (DRL) ensure policy performance and alleviate the online computational burden of VNE mapping [10, 11]. However, these methods face the challenge of sparse rewards, resulting in training times that can extend to several tens of hours. Moreover, the dynamism of UAVs and the scarcity of node energy make it more difficult for traditional methods to solve VNE problems deployed on UAV networks.

In this part, we investigate a new VNE method based on graph transformer technique to counter existing difficulties. Specifically, we design an self-supervised graph neural network that partitions the physical UAV network based on service capabilities such as resource capacity and network status to reduce the search space. Then, during the embedding process, the corresponding service partition is embedded based on the different capability requirements of the virtual network requests. Our contributions can be summarized as follows:

- We design an self-supervised graph transformer model to encode both local and global structural information, enabling a more effective capture of the graph's structural and dynamic characteristics, thereby achieving better service partitioning performance through multi-scale feature fusion and providing good generalization.
- We define a end-to-end optimization function that enables our model to quickly aggregate physical node features and information about the interconnected network. After simple pre-training, the model can perform rapid service partition adjustment in the dynamic scenario of UAV nodes.

- Finally, we conduct extensive and thorough simulations to evaluate the performance of our proposed algorithm.

4.1.1 System Model and Problem Formulation

In this section, we present our system mobility model, VNE problem model, and evaluation metrics.

4.1.1.1 System Model

Based on the problem we investigated, we have designed a service-oriented UAV network system model, as shown in Fig. 4.1, We consider a group of UAVs $K = [k_0, k_1, \ldots, k_n]$ distributed in a set of zones $Z = [z_0, z_1, \ldots, z_n]$. We denote the system boundary as $(0, 0, 0)$, $(x_{k_i}^{max}, 0, 0)$, $(0, y_{k_i}^{max}, 0)$, $(0, 0, z_{k_i}^{max})$ where $x_{k_i}^{max}$, $y_{k_i}^{max}$ and $z_{k_i}^{max}$ are the maximum boundary values of the X, Y, and Z dimensions, X_{buffer}, Y_{buffer},Z_{buffer} are the buffer zone of the boundary,respectively. The trajectory of each UAV node can be approximated as a sequence of waypoints at discrete time, in discrete continuous time n, the position of node $k_i, \in, K$ can be described as: $\overrightarrow{P_{k_i}(n)} = (x_{k_i}(n), y_{k_i}(n), z_{k_i}(n))$, and its speed can be described as: $\overrightarrow{V_{k_i}(n)} = (v_{k_i}^x(n), v_{k_i}^y(n), v_{k_i}^z(n))$. The movement of node k_i can be represented as follows:

$$p_{k_i}^{\partial}(n) = \begin{cases} p_{k_i}^{\partial}(n-1) + v_{k_i}^{\partial}(n-1), \\ \qquad \text{if } p_{k_i}^{\partial}(n-1) + v_{k_i}^{\partial}(n-1) \in \left[0, p_{k_i}^{\max} - \partial_{u_i}\right) \\ p_{k_i}^{\partial}(n-1) - \theta v_{k_i}^{\partial}(n-1), \text{ else (need speed change)} \end{cases} \tag{4.1}$$

where $\partial \in (x, y, z)$ represents different dimensions in 3-D space.

Within the specified boundaries, we utilize the particle swarm mobility model to generate the mobile scenarios for UAV networks [12]. The movement of UAV nodes adheres to both boundaries and dynamic constraints, rather than pure random motion, ensuring smooth trajectories for the nodes. Moreover, the model considers the collective movement from the perspective of particle swarm [13], ensuring a certain level of connectivity among nodes during the flight process. Therefore the speed $V_{k_i}^{\partial}(n)$ at time n is determined by the previous speed $V_{k_i}^{\partial}(n-1)$ and the guidance of center location $V_{k_i}^{\partial} \cdot gCenter(n-1)$:

$$\begin{aligned} v_{k_i}^{\partial}(n) &= (1-\lambda) v_{k_i}^{\partial}(n-1) \\ &\quad + \lambda\varphi\left(\text{gCenter}(n-1) - p_{k_i}^{\partial}(n-1)\right), \end{aligned} \tag{4.2}$$

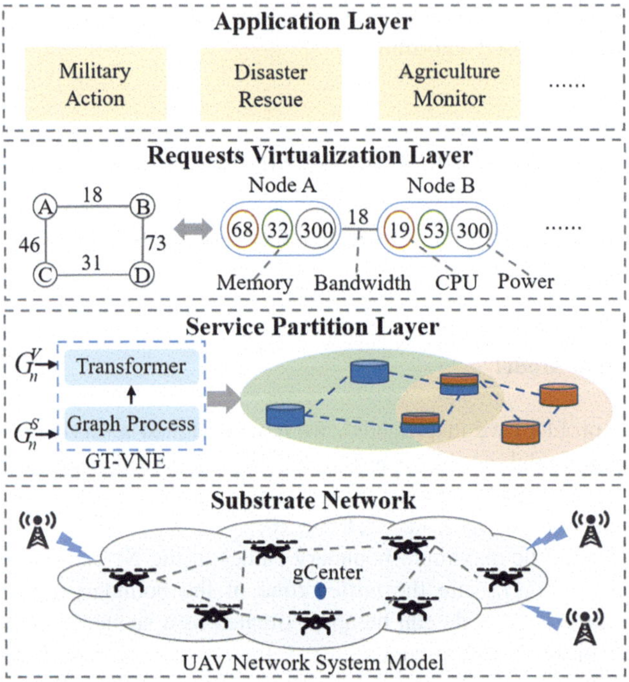

Fig. 4.1 System model

$\lambda \in (0, 1)$ is a random variable that follows a Gaussian distribution, it reflects the correlation between the current state and the previous state of the UAV node, $gCenter(n-1)$ represents the centroid position of the entire swarm at time $n-1$.Vector $gCenter(n-1) - p_{k_i}^{\partial}(n-1)$ is used to ensure that the movement speed of UAV nodes is guided by the central position. On the other hand, $\partial \in (0, 1)$ is a random variable that follows a Gaussian distribution and represents the randomness of speed.The following equation provides the swarm's center position at time n, where $|K|$ represents the number of nodes in the swarm:

$$\mathrm{gCenter}(n) = \frac{\sum_{i=1}^{N} p_{k_i}^{\partial}(n)}{|K|}. \tag{4.3}$$

Based on the UAV mobility model mentioned above, we model the swarm snapshot at time n as a weighted undirected dynamic graph G_n^S. The existence of links between nodes is determined by a certain distance from each UAV.

4.1.1.2 Problem Formulation

In this part, we present the problem formulation. At time n, we assume the network has K_n^S nodes and L_n^S links. Therefore the substrate network can be formulated as $G_n^S = (K_n^S, L_n^S, A_K^S, A_L^S)$, A_K^S and A_L^S represent the resource capacity of nodes and links at time n. Similarly, we model the virtual network requests (VNR) as $G^V = (K^V, L^V, R_K^V, R_L^V)$. R_K^V and R_L^V are the resource requirement of VNRs. Therefore, the network resource scheduling problem can be modeled as a embedding problem: $G^V = (K^V, L^V) \rightarrow G_n^{S'} = (K_n^{S'}, L_n^{S'})$, $K_n^{S'} \subset K_n^S$, $L_n^{S'} \subset L_n^S$. The network embedding process can be described in two stages: (1) Node embedding phase: mapping virtual nodes to substrate nodes with enough resource, represented by $X = \{x_{ij} \mid k_i \in K^V, k_j \in K_n^S\}$. If

$$\sum_{j=1}^{|K_n^S|} x_{ij} = 1, \tag{4.4}$$

node k_i is successfully embedded into node k_j. (2) Link embedding phase: allocating virtual links to a loop-free path consisting of one or multiple physical links with enough bandwidth. Similarly, the allocation can be represented by $Y = \{y_{ij} \mid l_i \in L^V, l_j \in L_n^S\}$. If

$$\sum_{j=1}^{|L_n^S|} y_{ij} \geq 1, \tag{4.5}$$

it indicates that l_i is embedded successfully.

It is worth noting that virtual nodes from different virtual network requests can share the same substrate node. A virtual link can not only share a substrate link with other virtual links but can also traverse multiple substrate links, forming a substrate path between the source and destination nodes.

4.1.1.3 Evaluation Metrics

The ultimate goal of VNE is to maximize the number of serviced VNRs and achieve greater VNE benefits. In order to assess whether a VNE algorithm can achieve such excellent performance, it is common to define and establish evaluation metrics for the algorithm. In our research, we define and formulate evaluation metrics for VNE in UAV swarm network, where the service resource requirement is associated with node computing resource $CPU(k_V)$, node storage resource $MEM(k_V)$, node energy consumption $POW(k_V)$, and link bandwidth requirement $BW(l_V)$. The revenue of successfully deploying VNR services can be expressed as follows:

$$\begin{aligned}&\text{Revenue}\left(G^V, n_d\right) = \\ &n_d\left[\alpha_R \sum_{k^V \in K^V} CPU\left(k^V\right) + \beta_R \sum_{k^V \in K^V} MEM(k^V)\right. \\ &\left. +\chi_R \sum_{k^V \in K^V} POW(k^V) + \gamma_R \sum_{l^V \in L^V} BW\left(l^V\right)\right],\end{aligned} \tag{4.6}$$

n_d represents the duration of the service request, while α_R, β_R, χ_R and γ_R are weighting factors aimed at balancing different virtual resources. Similarly, the resource consumption of G^V can be represented by:

$$\begin{aligned}&\text{Cost}\left(G^V, n_d\right) = \\ &n_d\left[\alpha_C \sum_{k^V \in K^V} CPU\left(k^V\right) + \beta_C \sum_{k^V \in K^V} MEM(k^V)\right. \\ &\left. +\chi_C \sum_{k^V \in K^V} POW(k^V) + \gamma_C \sum_{l^V \in L} \sum_{l^S \in L_n^S} BW\left(f_{l^S}^{l^V}\right)\right],\end{aligned} \tag{4.7}$$

where $BW(f_{l^S}^{l^V})$ represents the bandwidth consumed by l^V in the substrate link. It is important to note that completed tasks release computing, storage, and bandwidth resources, but energy consumption is a permanent loss. Based on this, we have designed the following three evaluation metrics. The first one is the utilization rate of node resources, including $\frac{\sum_{VNR} CPU(k^V)}{CPUk^S}$, $\frac{\sum_{VNR} MEM(k^V)}{MEMk^S}$ and $\frac{\sum_{VNR} POW(k^V)}{POWk^S}$. Then, we designed the revenue-cost ratio (Rev2Cos) metric, which is calculated by:

$$Rev2Cos = \frac{\sum_{n=0}^{N} Revenue(G^V, n_d)}{\sum_{n=0}^{N} Cost(G^V, n_d)}. \tag{4.8}$$

Meanwhile, the embedded ratio of VNRs is another significant indicator which can be formulated as:

$$Acp = \frac{\sum_{n=0}^{N} Acp(G^V, n_d)}{\sum_{n=0}^{N} All(G^V, n_d)}. \tag{4.9}$$

4.1.2 Graph Transformer-Based Resource Embedding Algorithm

In this section, we propose a design for a scalable method of a Transformer-based self-supervised graph neural network to address the problem of VNE. Our algorithm effectively integrates local and global structural information into vector representations, enabling long-distance interaction between nodes within a single layer. This approach solves the issues of low computational efficiency and oversmoothing, and captures the dynamic nature of the graph through self-attention mechanisms.

This method computes representations for each node, incorporating the following information: (1) the resource capacity of UAV nodes, (2) the resource capacity of neighboring nodes within the maximum reachable distance, and (3) the UAV network topology and link bandwidth at the current time. Subsequently, we partition these physical nodes based on the obtained node representations. Nodes within the same service partition exhibit similar capabilities. Based on the characteristics of virtual network service requests, we select the node with the lowest cost within the corresponding partition as the starting point for virtual network service mapping. Additionally, we define freshness criteria, updating the partitioning of service areas only when topology changes due to node movements or significant changes in node capabilities caused by service acceptance occur.

4.1.2.1 Input Representation

At time n, giving a graph G with node set K and edge set l, for each target node k_i, we sample associated nodes within k-hop or structurally similar nodes as k_j. The neighbours set of k_i is defined as $E_i = E_i^{k-hop} \cup E_i^{sim}$, including neighbours within k-hop distance and the set of structurally similar nodes. And $I_{k-hop}(k_i)$ represents an ordered degree sequence of k_i in 1 to k-hop. By comparing the ordered degree sequences between k_i and k_j, we can apply a score to measure the structural distance defined as:

$$\begin{aligned} S = exp(-\sqrt{DTW[I_{k-hop}(k_i), I_{k-hop}(k_j)]}) \\ + exp(-\frac{1}{\sqrt{d_{k_i} + d_{k_i}}}), \end{aligned} \tag{4.10}$$

where d_{k_i} represents the degree of k_i. And the node pair with a higher S are closer in latent space. Accordingly, the feature h_i of k_i can be mapped to a d-dimensional hidden feature representation through a linear transformation:

$$\hat{x}_i^0 = W_0 x_i + b_0, \tag{4.11}$$

where W_0 and b_0 are parameters of the linear transformation. Since we aim to enable the Graph Transformer can learn the global positional context of each node in the entire graph, we incorporate a linear transformed positional embedding λ_i of dim n into the node features as:

$$\hat{\lambda}_i = W_1 \lambda_i + b_1, \tag{4.12}$$

$$h_i^0 = \hat{x}_i^0 + \hat{\lambda}_i, \tag{4.13}$$

where W_1 and b_1 are parameters of the positional embeddings, and we use eigenvectors of the graph Laplacian matrix.

4.1.2.2 Graph Transformer Layer

We propose a self-attention bias to encode both local and global structural information. To capture the contextual nodes within a k-hop, self-attention needs to understand the k-hop distance and structural distance between node pairs. Formally, we define the self-attention between k_i and k_j for the head μ at layer l as:

$$a_{ij}^{\mu,l} = \frac{\left(Q^{\mu,l} y_i^l\right) \cdot \left(K^{\mu,l} y_j^l\right)}{\sqrt{d_{\text{hid}}}} + D_{ij}^{\mu,l} + P_{ij}^{\mu,l}, \tag{4.14}$$

K and V represent the query vectors and key vectors of multi head attention mechanism [14], respectively. d_{hid} represents the dimension of the hidden layer, y_i and y_j are the features of node k_i and k_j, D_{ij} and P_{ij} are sequences of linear transformation for the structural distance dis_{ij} and transition probability p_{ij} between k_i and k_j. We define the structural identities of each target node based on hierarchical information, including the maximum, minimum, mean, and standard deviation of the contextual degree of each node k_i within a k-hop distance. This can be formulated as:

$$J_i = \{d_{k_i}, T_1, T_2, \ldots, T_{k-hop}\}, \tag{4.15}$$

$$dis_{ij} = Ln\left(\frac{1}{\left\| J_i^{(n)} - J_j^{(n)} \right\|}\right), \tag{4.16}$$

where $T_{k-hop} = \{max_{k-hop}, min_{k-hop}, avg_{k-hop}, \delta_{k-hop}\}$, $J_i^{(n)}$ represents the $n-th$ element of J_i, Ln donates the linear transformation. Besides, to capture the paths between two structurally similar nodes and perceive the connectivity between them from a probabilistic perspective. In this way, GT layer can capture both local and global structural information, so the probability distance between nodes k_i and k_j can be defined as:

$$p_{ij} = Ln\left(A_{ij}^1 A_{ij}^2, \cdots, A_{ij}^{\tau}\right), \tag{4.17}$$

where A_{ij}^{τ} represents the transition probability from k_i to k_j at step τ. Then, we concatenate the outputs of self-attention into a vector representation and perform a linear transformation. The feature of node k_i can be updated in the $l-th$ layer as follows:

$$\hat{y}_i^{l+1} = \|_{\mu=1}^{M} \left(\sum_{k_j} softmax(a_{ij}^{\mu,l}) V^{\mu,l} y_j^l \right), \tag{4.18}$$

V represents the value vectors in the multi-head attention mechanism. Afterwards, the output is passed along with layer normalization to the feed-forward network:

$$y_i^{l+1} = \mathrm{LN}\left[\mathrm{LN}\left(y_i^l + \hat{y}_i^{l+1}\right) + \omega_1^l \mathrm{SeLU}\left(\omega_0^l \hat{y}_i^{l+1}\right)\right], \tag{4.19}$$

where ω_0^l and ω_1^l refer to learnable parameters, while LN represents layer normalization.

4.1.2.3 GT-VNE Algorithm

We adopt an end-to-end differentiable approach to optimize the service partitioning process, integrating the entire process from extracting the raw features to obtaining the final results into model training, in order to enhance empirical performance. Our method consists of the encoding structure mentioned in the previous section and an objective function for optimizing the allocation, Which can be described as:

$$L_{GT} = \alpha \sum_{\tau} \left\| A'^{(\tau)} - Z^{(\tau)} \right\|_F^2 + \frac{\beta}{|K|} \sum_{k_i \in K} \|x_i - FFN(z_i)\|_2, \tag{4.20}$$

where $Z^{(\tau)}$ is a score matrix constructed by calculating the cosine similarity between vector embeddings, and $x_i = \{cpu_i, mem_i, pow_i\}$ represents the original features of node k_i. $A'^{(\tau)}$ refers to the logarithmically scaled probability matrix for the τ-step which can be calculated by:

$$A'^{(\tau)} = \log \left(\frac{A_{i,j}^{(\tau)}}{\sum_t A_{t,j}^{(\tau)}} \right) - \log \left(\frac{|N_s|}{|V|} \right). \tag{4.21}$$

When the VNR G_n^V arrives at time n in the substrate network G_n^S, the model representation Z and the slicing result C are passed to the GT-VNE algorithm together.

As shown in Algorithm 9, GT-VNE selects the best node based on cost and benefit utility within different service partitions to handle service requests, according to the requirements of the service request. This selected nodes serve as the starting points for the embedding process, the Floyd [15] algorithm is used to find the closest path among them. The algorithm also limits the number of traversed nodes and the maximum search depth. Once the optimal mapping scheme is found, the service is initiated, and the changes in available resources are checked. When the changes in available resources or topology exceed a certain threshold due to the movement of unmanned cluster nodes, the model representation matrix Z and the partitioning result C are updated to adapt to the new network state.

Algorithm 9 GT-VNE: graph transformer aided embedding

Initialize the graph transformer network parameters
Train GT-VNE model **GT-VNE**(G_n^V, G_n^S, Z, C)
for $request \in requestSet$ **do**
    for $k_i^V \in request$ **do**
        for $k_i^S \in slice\ C$ **do**
            $queue \leftarrow start\_node(C)$
        end for
    end for
    if $Resource\_Free[queue] < Request$ **then**
        BREAK
    end if
    Node Embedded Successfully;
    for $l_i^V \in request$ **do**
        Floyd(link)
    end for
    if All links have enough resource **then**
        Store G_n^V, calculate Revenue and Cost
    else
        Delete G_n^V
    end if
    $num++$
end for
if Resource changed more than θ_1 or Topology changed more than θ_2 **then**
    $Z, C =$re GT-VNE(G_n^S, Z)
end if

4.1.3 Simulation Results and Discussions

In this section, we give the experiment results of our GT-VNE algorithm. We conducted extensive simulations to demonstrate that our proposed method outperforms other algorithms in terms of service requests acceptance rate, revenue-to-cost ratio, and resource utilizations.

4.1.3.1 Environment Setup

In the experiment, we simulate the movement of UAV nodes using the mobility model proposed above, and represent the FANET network topology at time n as a graph G_n^S using the Networkx tool [16]. The initial experiment consists of 50 UAV nodes. Based on line-of-sight communication, the simulation ensures that the links between nodes remain within a range of (550, 650) during their movement. The computation, storage resources, and bandwidth capacity of each node are randomly distributed within range (100, 200). The initial energy capacity of each node is set to 2000. Additionally, we generated a set of VNRs using a random method [17], which includes 1000 service requests. Specifically, the number of nodes in each VNR is uniformly distributed in (3, 10). The required computation, storage resources, and bandwidth capacity are between 10 and 80. The link connection probability for the links is set to 0.7. The arrival time of virtual requests follows a Poisson distribution, with an average of 20 service requests arriving within every 100 time units. The duration of requests follows an exponential distribution, with an average duration of 800 time units. Each node consumes 1 unit of energy per time unit when providing services.

All experiments are implemented in Python 3.9.13. Using a computer equipped with an AMD Ryzen 9 5950X 16-Core Processor, 128GB of memory, and Nvidia GeForce RTX 3070Ti GPU.

4.1.3.2 Convergence Analysis

In this part, we evaluate the convergence of the algorithm proposed. As shown in Fig. 4.2, the loss gradually decreases as the training time progresses. At the beginning of training, we notice that the loss rapidly decreases. This is because the randomization of parameters provides enough optimization space. As the training continues, the rate of loss reduction becomes slower. Around 90–100 epochs, the loss becomes almost constant. The simulation results show that our GT-VNE algorithm is able to effectively learn during the training process and converge to the optimal point at a relatively fast rate.

4.1.3.3 Performance Analysis

In the experiment, we compared the performance of our algorithm with three baseline algorithms using the same set of virtual network requests. In addition to GT-VNE, we implemented the following three baseline algorithms:

The first is BestFit, which selects the physical node with the largest resource capacity and fills it with the virtual node demand. The Second baseline is NodeRank presented in [8], which considers the ranking of nodes. And the last one is RLVNE [10], using a policy network-based CNN to optimize embedding decisions. As shown in Figs. 4.3, 4.4, 4.5, and 4.6, we observe that the trends of the four

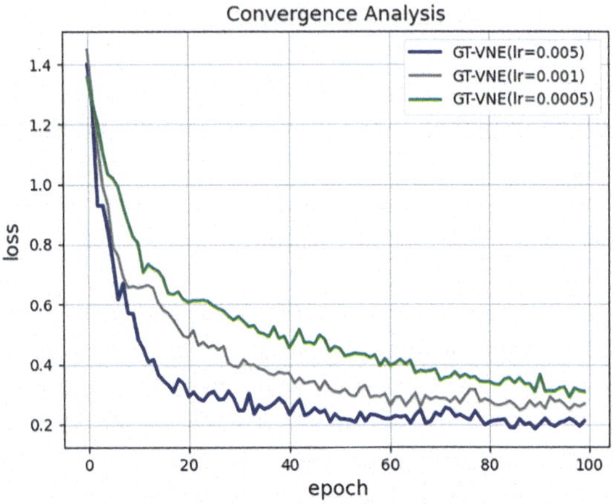

Fig. 4.2 Convergence analysis

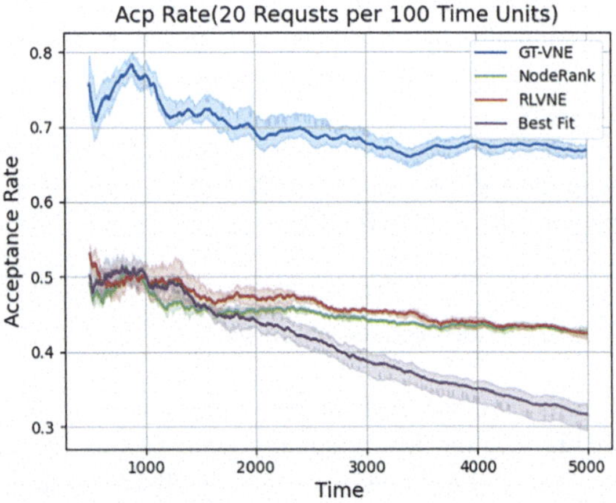

Fig. 4.3 Acceptance rate

algorithms are generally consistent. Specifically, at the beginning of the test, the acceptance rate is high due to the sufficient resources of the substrate network. As the embedding continues, service requests are correspondingly rejected. In contrast, the revenue-cost metric doesn't fluctuate during the test [18]. So we present the metric under various arrival rates of requests. In the test, we can see that our algorithm consistently outperforms the other algorithms.

4.2 Resource-Efficient Content Sharing Mechanism for UAV Named Data Networking

Recently, the widespread application of UAV swarms has brought tremendous benefits and changes to our lives, industry, and military affairs [19]. The members of the UAV swarm utilize a new architecture for collaboration to complete complex

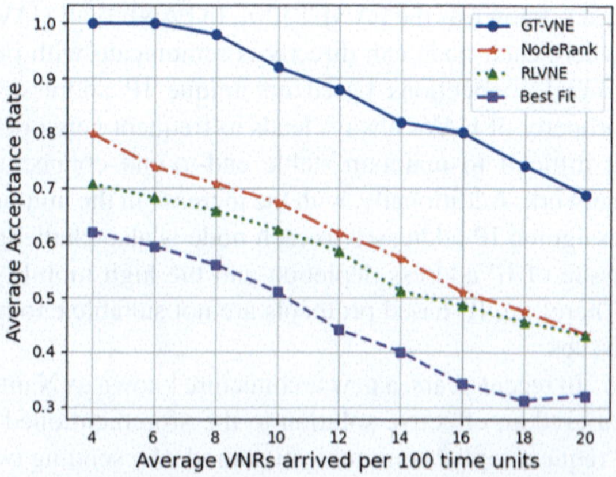

Fig. 4.4 Acceptance rate under different VNRs arrival speed

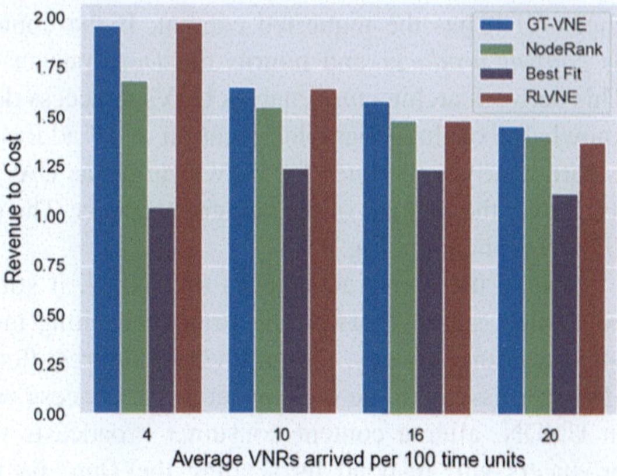

Fig. 4.5 Revenue to cost

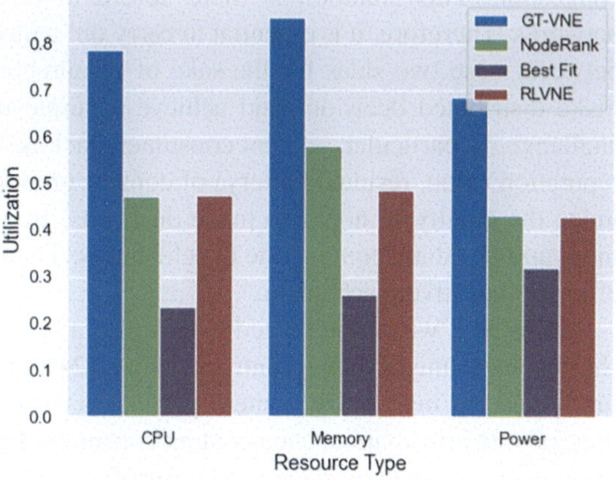

Fig. 4.6 Resource utilization

tasks, known as the FANET [20]. In FANET, all UAV nodes form an ad hoc network, where each node can directly communicate with other nodes by establishing end-to-end connections based on unique IP addresses. However, the high mobility property of UAVs always leads to frequent network topology changes [21], making it difficult to maintain stable end-to-end connectivity in such a highly dynamic network. Additionally, with the increase in the number of UAV nodes in the network, assigning IP addresses to each node is also challenging, especially considering the issue of IP address depletion and the high mobility characteristics of UAVs [22]. Therefore, IP-based protocols are not suitable enough for FANET due to the above issues.

In recent years, a new architecture known as Named Data Networking (NDN) has offered an effective solution to the aforementioned problems [23]. NDN utilizes a "request-reply" communication model by sending two types of packets. Specifically, when a UAV node intends to request data, it broadcasts an *Interest* packet with content name as *content consumer*. Upon receiving the *Interest* packet, a UAV node checks if it has the requested content. If the content is present, the node serves as *content producer* and returns the *Data* packet back to the content consumer. This network architecture enables UAVs to access desired content without requiring knowledge of the geographic location or IP address of the content producer. This is particularly well-suited for UAV swarms, as UAVs prioritize the required content itself over the identity of the content producer. The communications architecture of UNDN is shown in Fig. 4.7.

Despite the above advantages of UNDN, it still has some issues that need to be studied, such as forwarding strategy, caching mechanism, naming scheme, etc. In forwarding strategy, the problem of *Interest* flooding has been widely studied. However, issues in the content returning process are rarely involved. Specifically, in UNDN, after a content consumer broadcasts the *Interest* packet, all content producers will respond by sending the *Data* packet back to content consumer, which can cause serious resource waste, data redundancy, and even network congestion. This situation is more severe in resource-constrained UAV swarm networks. Therefore, it is essential to carry out a more precise matching mechanism between these two sides for the sake of resource efficiency. Nevertheless, how to make distributed decisions and achieve accurate matching in UNDN is the core challenge. In particular, content consumers lack access to private information (e.g., reputation value, residual battery) of content producers, which makes it difficult to judge the quality of nodes to make decisions. In addition, content producers lack motivation to share content due to selfishness. Therefore, it is essential to design an efficient incentive mechanism.

In this part, we are motivated to establish a resource-efficient two-stage content-sharing mechanism using contract theory [24] and matching theory [25]. In the first phase, to incentivize content producers to send content while also exposing their private information, each content consumer designs a set of contracts acting as principal for content producers, who function as agents. Each contract item contains bundled content amounts and corresponding rewards. Then, content consumers broadcast the contract items with *Interest* packets and each content producer selects

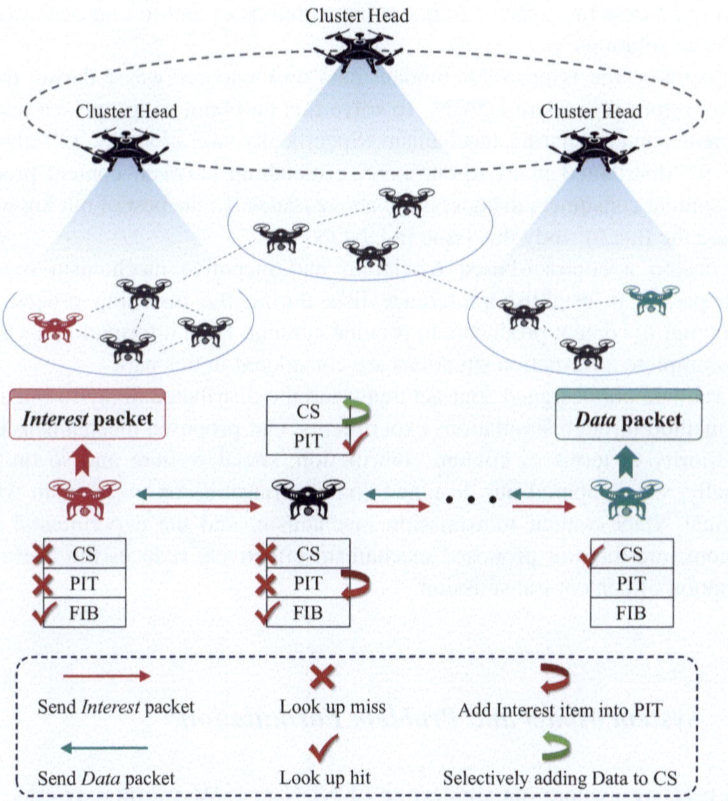

Fig. 4.7 Communication paradigm in UAV named data networking UNDN

the item that can maximize its utility. The application of contract theory offers two distinct advantages in such scenarios. Firstly, it can motivate content producers in situations of asymmetric information. Secondly, it can reveal the types of content producers (i.e., private information), ensuring that content producers must be honest and cannot misrepresent their types. In the second stage, a precise distributed many-to-one matching game is conducted between two sides, which implies that one content consumer can only match with one content producer, while one content producer can provide content to multiple content consumers. During the matching process, both content producers and content consumers first recalculate their utilities based on the results of the selected contracts from the first stage to formulate preference lists. Subsequently, the matching process commences in a distributed manner.

The major contributions of the proposed work are summarized as follows:

- We introduce the NDN network architecture into UAV swarm networks, utilizing a "request-reply" communication model and a content-centric routing mechanism to address the issue of frequent interruptions of end-to-end connections in IP-based solutions.

- We focus on the issues of redundant data and resource waste during the data packet return process in UNDN. To solve this problem, we propose a resource-efficient content sharing mechanism. Specifically, we adopt the GS algorithm [26] for distributed many-to-one precise matching between content producers and content consumers to address the above issues. To the best of our knowledge, we are the first to study this issue in UNDN.
- We design a contract-based revelation and incentive mechanism to enable both parties to establish preference lists during the matching process while encouraging content producers to provide content. Both information asymmetry and complete information situations are considered in this part.
- We validate our designed contract items and the distributed many-to-one matching method through simulations experiments. Our proposed mechanisms exhibit superiority in terms of content contribution, social welfare and so on. Additionally, we compared our designed content transmission mechanism with the original NDN content transmission mechanism, and the experimental results demonstrate that our proposed mechanism effectively reduces the energy consumption of content transmission.

4.2.1 System Model and Problem Formulation

In this part, we explore the application of NDN in UAV swarm networks, where UAVs are organized with a cluster structure for effective management. We consider dividing continuous time into multiple discrete time slots and regard the entire system as quasi-static during each time slot. And in each time slot, we assume the presence of M content consumers and N content producers for a specific type of content. The set of content consumers and content producers can be denoted as $\mathcal{M} = \{1, \ldots, j, \ldots, M\}$ and $\mathcal{N} = \{1, \ldots, i, \ldots, N\}$, respectively. As shown in Fig. 4.8, two stages are contained in our system model. In the first stage, each content consumer (principal) designs multiple contract items, which are disseminated along with *Interest* packets. Upon receiving the *Interest* packet, each UAV node first checks if it possesses the corresponding content. A successful content name match signifies that the node contains the relevant content and can serve as a content producer. Each potential content producer (agent) then preselects an item from the contract sets and sends their preselection results back to the cluster head. The cluster head, in turn, transmits this information to neighboring cluster heads, ultimately conveying it back to the content consumer. After the contract preselection phase is completed, the matching between content consumers and content producers begins. In this phase, each content consumer requires only one content producer to provide the content, while each content producer can serve multiple content consumers. Therefore, this phase can be modeled as a many-to-one matching game.

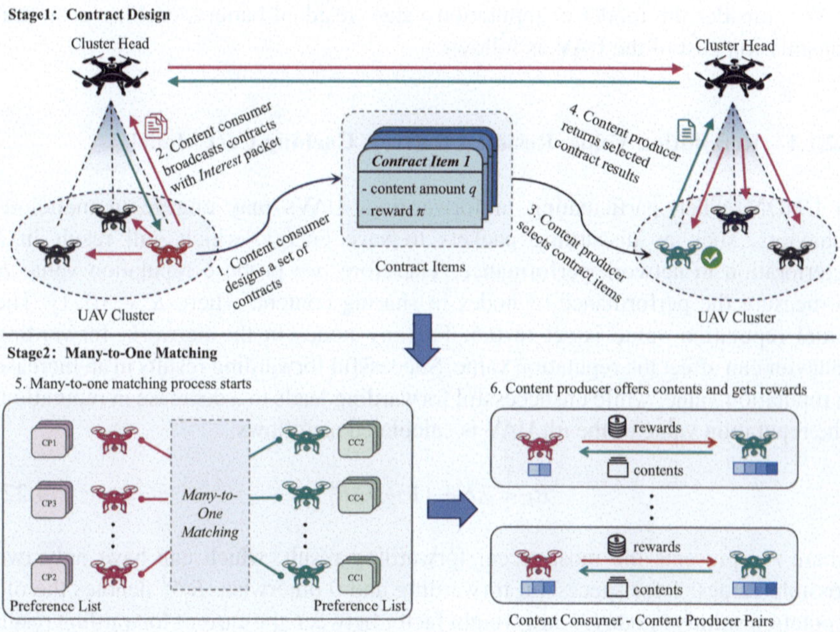

Fig. 4.8 System model

The two phases in our system model are stated below:

(1) Content Consumers Design Contract
In UNDN, content producers lack motivation to share content with other UAVs without additional incentive mechanisms due to their selfishness. The types of content producers, such as reputation value and remaining electricity, constitute private information that content consumers are unaware of. Therefore, each content consumer designs a set of contract items to stimulate content producers while revealing their corresponding types, thus facilitating the matching process in the second stage.

(2) Many-to-One Matching Between Content Consumers and Content Producers
For a content consumer, multiple content producers may have accepted the contracts. If all these nodes send data packets to the content consumer, it can result in content redundancy and lead to a waste of transmission resources. Besides, each content consumer tends to choose a high-quality producer. Therefore, after completing the first phase of contract pre-selection, both sides establish their preference lists based on their utilities. Then, the GS algorithm is employed to achieve many-to-one stable matching between content consumers and content producers.

We consider the model of reputation value, residual battery, caching cost, and transmission cost of the UAV as follows.

4.2.1.1 Reputation Value, Residual Battery, Caching Cost Model

In UNDN, when participating in forwarding, UAVs may engage in malicious behaviors, such as discarding packets to save energy, which will result in a deterioration in network performance. Therefore, we propose reputation value R to measure the performance of nodes in sharing content, where $R \in (0, 1]$. The initial reputation value is set to 0.5. For any nodes in the network, forwarding behavior can affect the reputation value. Successful forwarding results in an increase in reputation value, while unsuccessful forwarding leads to a decrease in reputation. The reputation value of the ith UAV is calculated as follows:

$$R_i = \lambda r + (1 - \lambda) R_i^{old}, \tag{4.22}$$

where r represents the most recent forwarding result, which can have only two possible values: 1 for successful forwarding and 0 otherwise. R_i^{old} denotes the old reputation value of node i. λ is a weight factor between the current forwarding result and the old reputation value.

Residual battery B is another parameter to evaluate the quality of UAVs. A UAV with high residual power is more inclined to forward and share content with other nodes, whereas a UAV with low residual power tends to conserve energy for its own survival.In this part, the residual battery B of ith UAV is calculated by the ratio of the current battery capacity to the maximum battery capacity. Thus, $B \in (0, 1]$. The definition of B is given as follows:

$$B_i = \frac{B_i^{res}}{B_{\max}}. \tag{4.23}$$

The caching cost of UAV is incurred due to cache contents, which depends on the size of cached contents. The caching cost of content producer i when caching q_{ij} contents for content consumer j is defined as follows:

$$E_{ij}^{c} = c \times q_{ij}, \tag{4.24}$$

where c is the unit caching cost.

4.2.1.2 Energy Consumption Model

This part considers the energy consumption of UAVs due to content transmission. The transmission energy consumption E^T stems from the transmitter's circuitry and power amplifier. Regarding the issue of wireless channel fading, we are taking into account both the free space model and the multi-path fading model simultaneously. The energy consumed by content producer i when transmitting q_{ij} content to content consumer j can be formulated as [27]:

$$E_{ij}^{T} = \begin{cases} q_{ij} \times [E_{elec} + \eta_{fs} \times d(i,r)^2], & d(i,r) < \tau \\ q_{ij} \times [E_{elec} + \eta_{mp} \times d(i,r)^4], & d(i,r) \geq \tau \end{cases}, \tag{4.25}$$

where r is the next-hop node of content producer i. Due to the high mobility nature of UAVs, transmitting data packets to content consumers along the *Interest* packet's reverse paths is no longer applicable in UNDN. Consequently, in this part, the content producer initially transmits the *Data* packet to the cluster head, which then sequentially relays it to the adjacent cluster heads, ultimately delivering it to the content consumer. E_{elec} represents the unit energy consumption. η_{fs} and η_{mp} represent the energy consumption factors for power amplifiers in the free space model and multi-path fading model, respectively. $d(i,r)$ is the distance between content producer i and the next-hop node r, and τ denotes the distance switching threshold of the two aforementioned models:

$$\tau = \sqrt{\frac{\eta_{fs}}{\eta_{mp}}}. \tag{4.26}$$

4.2.2 *Contract-Based Incentive Mechanism*

In this part, a contract-based disclosure and incentive mechanism is designed to expose private information of content producers to content consumers while motivating content producers to share content. We first present the type model of the content producer. Then, we model the utility functions of both content consumers and content producers, and formulate the optimization problem. Lastly, we derive the optimal contracts both with and without information asymmetry.

4.2.2.1 Content Producer Type Modeling

During the content sharing process in UNDN, sending content will cause power consumption, and the trustworthiness of the node also requires attention. As a result, type θ is introduced to denote the willingness and ability of nodes to share content as content producers. The two parameters mentioned above, reputation value R and residual battery B, are the confidential information of the node, which is only known to itself. Inspired by [28], the type of content producer is determined by both reputation value R and residual battery B:

$$\theta_i = \alpha_1 R_i + \alpha_2 B_i, \tag{4.27}$$

where $\alpha_1, \alpha_2 \in (0, 1)$ are weight factors that regulate the importance of R and B, and $\alpha_1 + \alpha_2 = 1$. A higher θ indicates that the corresponding content producer has a greater reputation value and remaining battery capacity, thus is more willing to share content and also deserves more rewards. We arrange the types of content producers in ascending order and assume their types are discretely finite, which is represented by

$$\theta_1 < \ldots < \theta_i < \ldots < \theta_N. \tag{4.28}$$

In this part, the expression "content producer i" is equivalent to "type θ_i content producer". Moreover, the probability of a content producer belonging to type θ_i is expressed as p_i, and $\sum_{i=1}^{N} p_i = 1$.

4.2.2.2 Contract Formulation

In the contract-based model, for a specific content, each content consumer designs a set of contribution-reward bundle contracts $(\boldsymbol{q}, \boldsymbol{\pi})$ for different types of content producers. The contract collection designed by content consumer j is $(\boldsymbol{q}_j, \boldsymbol{\pi}_j)$. Each contract item is (q_{ij}, π_{ij}), where q_{ij} represents the amount of content provided by content producer i to content consumer j, π_{ij} represents the corresponding reward. These contracts are spread with *Interest* packets. For content producers, they can choose to accept or refuse the contracts.

Next, we define the utility functions of content producers and content consumers, each comprising two parts: revenue and cost.

Definition 4.1 (Utility of Content Producer) The utility of θ_i-type content producer when provides q_{ij} content to content consumer j can be expressed as reward minus caching cost:

$$U^P(\theta_i, q_{ij}) = \theta_i W(\pi_{ij}) - E_{ij}^c. \tag{4.29}$$

In (4.29), $W(\pi_{ij})$ is a reward valuation function for the content producer, and it must follow: $W(0) = 0$, $\frac{dW}{d\pi} > 0$, $\frac{d^2W}{d\pi^2} < 0$, which means the reward valuation function is strictly increasing and concave. Without loss of generality and inspired by Miglani and Kumar [22], Su et al. [29], Wang et al. [30], we set $W(\pi) = ln(1 + \pi)$ for the characteristic of decreasing marginal utility.

Definition 4.2 (Utility of Content Consumer) The utility of content consumer j when receives content from content producer i can be expressed as the value of content minus the reward:

$$U^C(\theta_i, q_{ij}) = V(q_{ij}) - b\pi_{ij}, \tag{4.30}$$

In (4.30), $V(q_{ij})$ denotes the gain from content q_{ij}, and there exits $V(0) = 0$, $\frac{dV}{dq} > 0$, $\frac{d^2V}{dq^2} < 0$, which means content gain function is strictly increasing and concave. Furthermore, inspired by [31], $V(q)$ is defined as $V(q) = \sqrt{aq}$ for the characteristic of decreasing marginal utility, in which the positive factor a measures the value of the content to content consumers. b represents the unit cost of content consumers on offering rewards. For simplicity's sake, we assume $b = 1$.

4.2.2.3 Contract Conditions Analysis and Problem Formulation

In this part, the type of content producer θ is private information, and some nodes may misreport their types to obtain higher rewards. To truly reveal the hidden information (types) of selfish content producers, the feasible contract needs to meet two constraints: Individual Rationality (IR) and Incentive Compatibility (IC).

Definition 4.3 (IR Constraints) The IR constraints indicate that the contract chosen by a content producer must ensure its utility is non-negative, i.e.,

$$U^P(\theta_i, q_{ij}) = \theta_i ln(1 + \pi_{ij}) - cq_{ij} \geq 0. \tag{4.31}$$

Definition 4.4 (IC Constraints) The objective of each content producer is to maximize its utility. The IC constraints indicate that for a specific θ_i type content producer, its utility can be maximized only when it selects the contract corresponding to its type, i.e.,

$$U^P(\theta_i, q_{ij}) \geq U^P(\theta_i, q_{i'j}), \forall j, \ \forall i' \neq i, \tag{4.32}$$

$$\theta_i ln(1 + \pi_{ij}) - cq_{ij} \geq \theta_i ln(1 + \pi_{i'j}) - cq_{i'j}, \ \forall j, \ \forall i' \neq i. \tag{4.33}$$

Therefore, the contract-based incentive mechanism inherently reveals the true information of the participants, where content consumers can infer the type of content producer by observing the contract it chooses to accept.

In contract theory, the goal is to maximize the contract principal's expected utility while meeting the constraints of IR and IC. Thus, the problem is formulated as follows:

$$\mathbf{P1}: \quad \max_{q_{ij}, \pi_{ij}} \sum_{i=1}^{N} p_i(\sqrt{aq_{ij}} - \pi_{ij}), \tag{4.34}$$

$$\text{s.t.} \quad q_j^{\min} \leq q_{ij} \leq q^{\max}, \tag{4.34a}$$

$$IR: \theta_i ln(1 + \pi_{ij}) - cq_{ij} \geq 0, \tag{4.34b}$$

$$IC: \theta_i ln(1 + \pi_{ij}) - cq_{ij} \geq \theta_i ln(1 + \pi_{i'j}) - cq_{i'j}. \tag{4.34c}$$

Equation (4.34a) indicates that for a specific type of content, content consumer j can accept a minimum amount of content q_j^{min}, and the total data volume of that content is $q^{\max}$. Equation (4.34b) is IR constraints, which ensures that each type of content producer can benefit from data sharing. Equation (4.34c) is IC constraints, which means that each content producer can only maximize its utility by selecting the contract item that corresponds to its type. Equations (4.34b) and (4.34c) jointly guarantee the feasibility of contracts.

4.2.2.4 Properties of Feasible Contract

The original contract-based optimization problem **P1** presented in Sect. 4.2.2.3 contains N IR constraints and $N(N-1)$ IC constraints. Because of the complexity of the constraints, the problem becomes too complicated to solve directly. Hence, we employ feasibility conditions to simplify the constraint conditions.

Lemma 4.1 $\forall j \in \mathcal{M}, \forall i, i' \in \mathcal{N}$ *and* $i \neq i'$, *any feasible contract items satisfy the following conditions [32]:*

1. *If* $\theta_i > \theta_{i'}$, *then* $\pi_{ij} \geq \pi_{i'j}$;
2. *If* $\pi_{ij} > \pi_{i'j}$, *then* $\theta_i \geq \theta_{i'}$.

Proof

1. $\theta_i > \theta_{i'} \rightarrow \pi_{ij} \geq \pi_{i'j}$.

   According to the IC constraints, we have:

$$\theta_i ln(1+\pi_{ij}) - cq_{ij} \geq \theta_i ln(1+\pi_{i'j}) - cq_{i'j},\ \forall i' \neq i, \tag{4.35}$$

$$\theta_{i'} ln(1+\pi_{i'j}) - cq_{i'j} \geq \theta_{i'} ln(1+\pi_{ij}) - cq_{ij},\ \forall i' \neq i. \tag{4.36}$$

Then, add (4.35) and (4.36), we have:

$$\begin{aligned} &\theta_i ln(1+\pi_{ij}) + \theta_{i'} ln(1+\pi_{i'j}) \\ &\geq \theta_i ln(1+\pi_{i'j}) + \theta_{i'} ln(1+\pi_{ij}), \forall i' \neq i. \end{aligned} \tag{4.37}$$

After transforming, we have:

$$ln(1+\pi_{ij})(\theta_i - \theta_{i'}) \geq ln(1+\pi_{i'j})(\theta_i - \theta_{i'}),\ \forall i' \neq i. \tag{4.38}$$

As $\theta_i > \theta_{i'}$, we can easily derive:

$$\theta_i - \theta_{i'} > 0. \tag{4.39}$$

By simultaneously eliminating $(\theta_i - \theta_{i'})$ on both sides, we can obtain $ln(1+\pi_{ij}) \geq ln(1+\pi_{i'j})$. Since $ln(\cdot)$ is a strictly increasing function, $\pi_{ij} \geq \pi_{i'j}$ holds.

2. $\pi_{ij} > \pi_{i'j} \rightarrow \theta_i \geq \theta_{i'}$.
   By representing (4.37) in another form, we can obtain:

$$\begin{aligned} &\theta_i[ln(1+\pi_{ij}) - ln(1+\pi_{i'j})] \\ &\geq \theta_{i'}[ln(1+\pi_{ij}) - ln(1+\pi_{i'j})],\ \forall i' \neq i. \end{aligned} \tag{4.40}$$

Due to $\pi_{ij} > \pi_{i'j}$ and $ln(\cdot)$ is strictly increasing, it is easy to obtain

$$ln(1+\pi_{ij}) - ln(1+\pi_{i'j}) > 0. \tag{4.41}$$

Therefore, we can get $\theta_i \geq \theta_{i'}$.

The proof is finished. □

Lemma 4.1 indicates that π_{ij} increases with θ_i, which means that a higher type content producer can receive no less rewards than a lower type content producer. We summarize this property to Proposition 4.1.

Proposition 4.1 (Monotonicity of Rewards) *For any feasible contract, the reward satisfies:*

$$0 \leq \pi_{1j} \leq \ldots < \pi_{ij} \leq \ldots \leq \pi_{Nj}. \tag{4.42}$$

Similar to Lemma 4.1 and Proposition 4.1, we can obtain Lemma 4.2 and Proposition 4.2 as follows.

Lemma 4.2 $\forall j \in \mathcal{M}, \forall i, i' \in \mathcal{N}$ *and* $i \neq i'$, *any feasible contract items satisfy* $\pi_{ij} > \pi_{i'j}$ *if and only if* $q_{ij} > q_{i'j}$, *and* $\pi_{ij} = \pi_{i'j}$ *if and only if* $q_{ij} = q_{i'j}$.

Proof

1. *Sufficiency:* $q_{ij} > q_{i'j} \rightarrow \pi_{ij} > \pi_{i'j}$.
   Utilizing the inequality in (4.35), we can obtain:

$$\theta_i[ln(1+\pi_{ij}) - ln(1+\pi_{i'j})] \geq c(q_{ij} - q_{i'j}) > 0,\ \forall i' \neq i. \tag{4.43}$$

   Since $c > 0$, $q_{ij} > q_{i'j}$ and $ln(\cdot)$ is strictly increasing, then $\pi_{ij} > \pi_{i'j}$.
2. *Necessity:* $\pi_{ij} > \pi_{i'j} \rightarrow q_{ij} > q_{i'j}$.
   Then, we can transform the inequality (4.36) into the following form:

$$\theta_{i'}[ln(1+\pi_{i'j}) - ln(1+\pi_{ij})] \geq c(q_{i'j} - q_{ij}),\ \forall i' \neq i. \tag{4.44}$$

   Since $\pi_{i'j} < \pi_{ij}$ and $ln(\cdot)$ is strictly increasing, we can acquire:

$$0 > \theta_{i'}[ln(1+\pi_{i'j}) - ln(1+\pi_{ij})] \geq c(q_{i'j} - q_{ij}),\ \forall i' \neq i. \tag{4.45}$$

And there exits $c > 0$, therefore we have:

$$q_{i'j} - q_{ij} < 0, \ \forall i' \neq i. \tag{4.46}$$

Thus when $\pi_{ij} > \pi_{i'j}, q_{ij} > q_{i'j}$ holds.

From the above proof, it can be seen that content producers who contribute more will receive greater rewards. Therefore, content producers who receive the same rewards must have equivalent contributions, and vice versa. In other words, $\pi_{ij} = \pi_{i'j}$ if and only if $q_{ij} = q_{i'j}$. Therefore, the proof is finished. □

We can summarize this property to Proposition 4.2.

Proposition 4.2 (Monotonicity of the Contributions) *As a strictly increasing function of π, the contribution amount of content q satisfies:*

$$0 \leq q_{1j} \leq \ldots \leq q_{ij} \leq \ldots \leq q_{Nj}. \tag{4.47}$$

Likewise, the third property of feasible contracts can be summarized in Lemma 4.3.

Lemma 4.3 $\forall j \in \mathcal{M}, \forall i, i' \in \mathcal{N}$ *and* $i \neq i'$, *any feasible contract items satisfy* $U^P(\theta_i, q_{ij}) > U^P(\theta_{i'}, q_{i'j})$ *if and only if* $\theta_i > \theta_{i'}$.

Proof

1. *Sufficiency:* $\theta_i > \theta_{i'} \rightarrow U^P(\theta_i, q_{ij}) > U^P(\theta_{i'}, q_{i'j})$.
   According to IC constraints, we have:

$$\begin{aligned} U^P(\theta_i, q_{ij}) &= \theta_i ln(1 + \pi_{ij}) - cq_{ij} \\ &\geq \theta_i ln(1 + \pi_{i'j}) - cq_{i'j} \\ &> \theta_{i'} ln(1 + \pi_{i'j}) - cq_{i'j} = U^P(\theta_{i'}, q_{i'j}). \end{aligned} \tag{4.48}$$

   Thus, $U^P(\theta_i, q_{ij}) > U^P(\theta_{i'}, q_{i'j})$ holds.
2. *Necessity:* $U^P(\theta_i, q_{ij}) > U^P(\theta_{i'}, q_{i'j}) \rightarrow \theta_i > \theta_{i'}$.
   Due to $U^P(\theta_i, q_{ij}) > U^P(\theta_{i'}, q_{i'j})$ and IC constraints, we can easily obtain:

$$\begin{aligned} \theta_i ln(1 + \pi_{ij}) - cq_{ij} &> \theta_{i'} ln(1 + \pi_{i'j}) - cq_{i'j} \\ &\geq \theta_{i'} ln(1 + \pi_{ij}) - cq_{ij}. \end{aligned} \tag{4.49}$$

   Therefore, $\theta_i > \theta_{i'}$ holds. The proof is completed.

□

From Lemma 4.3, it is proved that if $\theta_1 < \ldots < \theta_i < \ldots < \theta_N$, then $U^P(\theta_1, q_{1j}) < \ldots < U^P(\theta_i, q_{ij}) < \ldots < U^P(\theta_N, q_{Nj})$, which means a higher-type node can achieve greater utility compared to a lower-type node.

4.2.2.5 Optimal Contract Design Under Information Asymmetry

In this part, we employ the feasibility conditions of the contract obtained in Sect. 4.2.2.4 to simplify the IR and IC constraints, respectively.

IR Constraints Reduction

Firstly, we simplify the IR constraints described in (4.34b), which require that the utility for each type of content producer must be greater than 0. Besides, according to Lemma 4.3, the content producer with the lowest type will obtain the minimum utility. Therefore, as long as the utility of θ_1-type content producer satisfies the IR constraint, the other types of content producers must be able to meet the IR constraint [33]. Thus, (4.34b) can be simplified to:

$$\theta_1 ln(1 + \pi_{1j}) - cq_{1j} \geq 0. \tag{4.50}$$

In order to maximize the utility of content consumer, the optimal contract (q_{1j}^*, π_{1j}^*) for θ_1-type content producer must satisfy:

$$\theta_1 ln(1 + \pi_{1j}^*) - cq_{1j}^* = 0. \tag{4.51}$$

Proof We use proof by contradiction. Assuming $\theta_1 ln(1 + \pi_{1j}^*) - cq_{1j}^* > 0$, then content consumer j, in order to increase its expected utility while guaranteeing the feasibility of the contract, can either choose to decrease π_{1j}^* or increase q_{1j}^* until $\theta_1 ln(1 + \hat{\pi}_{1j}) - c\hat{q}_{1j} = 0$ is satisfied. In this case, we have:

$$\theta_1 ln(1 + \hat{\pi}_{1j}) - c\hat{q}_{1j} > \theta_1 ln(1 + \pi_{1j}^*) - cq_{1j}^*, \tag{4.52}$$

which contradicts the assumption that (q_{1j}^*, π_{1j}^*) is the optimal contract. Therefore, optimal contract must satisfy $\theta_1 ln(1 + \pi_{1j}^*) - cq_{1j}^* = 0$. The proof is finished. □

IC Constraints Reduction

Then, the IC constraints given in (4.34c) are reduced. Firstly, four concepts similar to those in [29], [34] are defined.

- downward incentive constraints (DICs): the IC between type-i content producer and type-i', where $i' \in \{1, \dots, i-1\}$ and $i \neq i'$.
- local downward incentive constraints (LDICs): the IC between type-i content producer and type-$i-1$, where $i > 1$.
- upward incentive constraints (UICs): the IC between type-i content producer and type-i', where $i' \in \{i+1, \dots, N\}$ and $i \neq i'$.

- local upward incentive constraints (LUICs): the IC between type-i content producer and type-$i+1$, where $i < N$.

Then, we prove that DICs can be substituted by LDICs.

Proof For type-i, type-$i-1$ and type-$i-2$ content producer, there exits two LDICs as follows:

$$\theta_i ln(1+\pi_{ij}) - cq_{ij} \geq \theta_i ln(1+\pi_{(i-1)j}) - cq_{(i-1)j}, \tag{4.53}$$

$$\theta_{i-1} ln(1+\pi_{(i-1)j}) - cq_{(i-1)j} \geq \theta_{i-1} ln(1+\pi_{(i-2)j}) - cq_{(i-2)j}. \tag{4.54}$$

Due to $\theta_i < \theta_{i-1}$, from (4.53) and (4.54) we can obtain:

$$\begin{aligned} &\theta_i[ln(1+\pi_{(i-1)j}) - ln(1+\pi_{(i-2)j})] \\ &\geq \theta_{(i-1)}[ln(1+\pi_{(i-1)j}) - ln(1+\pi_{(i-2)j})] \\ &\geq cq_{(i-1)j} - cq_{(i-2)j}. \end{aligned} \tag{4.55}$$

Therefore, (4.53) can be restated as:

$$\begin{aligned} \theta_i ln(1+\pi_{ij}) - cq_{ij} &\geq \theta_i ln(1+\pi_{(i-1)j}) - cq_{(i-1)j} \\ &\geq \theta_i ln(1+\pi_{(i-2)j}) - cq_{(i-2)j}, \end{aligned} \tag{4.56}$$

which can be summarized as

$$\begin{aligned} \theta_i ln(1+\pi_{ij}) - cq_{ij} &\geq \theta_i ln(1+\pi_{(i-1)j}) - cq_{(i-1)j} \\ &\geq \ldots \geq \theta_i ln(1+\pi_{1j}) - cq_{1j}. \end{aligned} \tag{4.57}$$

Thus, DICs can be reduced to LDICs. The proof is finished. □

Similarly, UICs can be substituted by LUICs:

$$\begin{aligned} \theta_i ln(1+\pi_{ij}) - cq_{ij} &\geq \theta_i ln(1+\pi_{(i+1)j}) - cq_{(i+1)j} \\ &\geq \ldots \geq \theta_i ln(1+\pi_{Nj}) - cq_{Nj}. \end{aligned} \tag{4.58}$$

Due to space constraints, the proof process is omitted.

To sum up, the IC constraints can be reduced as:

$$\begin{aligned} \theta_i ln(1+\pi_{ij}) - cq_{ij} \geq \theta_i ln(1+\pi_{(i-1)j}) - cq_{(i-1)j}, \\ i = 2, \ldots, N, j \in \mathcal{M}. \end{aligned} \tag{4.59}$$

After transforming, we have:

$$q_{ij} \leq q_{(i-1)j} + \frac{\theta_i}{c} ln(\frac{1+\pi_{ij}}{1+\pi_{(i-1)j}}),\ i = 2, \dots, N,\ j \in \mathcal{M}. \tag{4.60}$$

In order to increase the expected utility and maintain the feasibility of the contract, content consumer j must either decrease π_{ij} or increase q_{ij} until the equality holds. Therefore, for type-θ_i content producer, the optimal contract (q^*_{ij}, π^*_{ij}) should meet:

$$q^*_{ij} = q^*_{(i-1)j} + \frac{\theta_i}{c} ln(\frac{1+\pi^*_{ij}}{1+\pi^*_{(i-1)j}}),\ i = 2, \dots, N,\ j \in \mathcal{M}. \tag{4.61}$$

Therefore, based on the above constraint simplification, our initial optimization problem **P1** can be reformulated as:

$$\textbf{P2}: \quad \max_{q_{ij},\pi_{ij}} \sum_{i=1}^{N} p_i(\sqrt{aq_{ij}} - \pi_{ij}), \tag{4.62}$$

$$\text{s.t.} \quad q_j^{\min} \leq q_{ij} \leq q^{\max}, \tag{4.62a}$$

$$IR: \theta_1 ln(1+\pi_{1j}) - cq_{1j} = 0, \tag{4.62b}$$

$$IC: \theta_i ln(1+\pi_{ij}) - cq_{ij} = \theta_i ln(1+\pi_{(i-1)j}) - cq_{(i-1)j}, \tag{4.62c}$$

$$0 \leq \pi_{1j} \leq \dots \leq \pi_{ij} \leq \dots \leq \pi_{Nj}. \tag{4.62d}$$

Further simplify (4.62b) and (4.62c), we can obtain:

$$q_{1j} = \frac{\theta_1 ln(1+\pi_{1j})}{c}, \tag{4.63}$$

$$q_{ij} = q_{1j} + \sum_{i=2}^{N} \frac{\theta_i}{c} ln(\frac{1+\pi_{ij}}{1+\pi_{(i-1)j}}),\ \forall i = 2, \dots, N. \tag{4.64}$$

We can transform **P2** into an equivalent problem:

$$\textbf{P3}: \quad \min_{q_{ij},\pi_{ij}} \sum_{i=1}^{N} -p_i(\sqrt{aq_{ij}} - \pi_{ij}), \tag{4.65}$$

$$\text{s.t.} \quad q_j^{\min} \leq q_{ij} \leq q^{\max}, \tag{4.65a}$$

$$IR : cq_{1j} - \theta_1 ln(1 + \pi_{1j}) = 0, \tag{4.65b}$$

$$IC : cq_{ij} - \theta_i ln(1 + \pi_{ij}) = cq_{(i-1)j} - \theta_i ln(1 + \pi_{(i-1)j}), \tag{4.65c}$$

$$0 \leq \pi_{1j} \leq \ldots \leq \pi_{ij} \leq \ldots \leq \pi_{Nj}. \tag{4.65d}$$

We set the optimization problem (4.65) as F, and can easily obtain the Hessian matrix of the optimization problem as follows:

$$\mathbf{H} = \begin{bmatrix} \frac{\partial^2 F}{\partial q^2} & \frac{\partial^2 F}{\partial q \partial \pi} \\ \frac{\partial^2 F}{\partial \pi \partial q} & \frac{\partial^2 F}{\partial \pi^2} \end{bmatrix} = \begin{bmatrix} \frac{\partial^2 F}{\partial q^2} & 0 \\ 0 & \frac{\partial^2 F}{\partial \pi^2} \end{bmatrix} = \begin{bmatrix} \frac{1}{4} paq^{-\frac{3}{2}} & 0 \\ 0 & 0 \end{bmatrix}. \tag{4.66}$$

Given that the eigenvalues of the Hessian matrix are all greater than or equal to 0, it follows that the objective function is convex. Furthermore, since (4.65a) and (4.65d) are linear constraints, they form convex sets. Similarly, it is easy to prove that (4.65b) and (4.65c) are convex constraints, therefore the problem is a convex optimization problem. Thus, we can solve it by the Lagrange multiplier method. The corresponding Lagrangian function $\mathcal{L}$ can be written as:

$$\begin{aligned} &\mathcal{L}\Big(\{q_{ij}\}, \{\pi_{ij}\}, \{\alpha_i\}, \{\beta_i\}, \{\gamma_i\}, \{\eta_i\}\Big) \\ &= \sum_{i=1}^{N} p_i(\sqrt{aq_{ij}} - \pi_{ij}) + \sum_{i=1}^{N} \alpha_i (q_{ij} - q_j^{\min}) \\ &- \sum_{i=1}^{N} \beta_i (q_{ij} - q^{\max}) + \gamma_1[\theta_1 ln(1 + \pi_{1j}) - cq_{1j}] \\ &+ \sum_{i=2}^{N} \gamma_i \left[\frac{\theta_i}{c} \big(ln(1 + \pi_{ij}) - ln(1 + \pi_{(i-1)j})\big) + q_{(i-1)j} - q_{ij}\right] \\ &+ \eta_1 \pi_{1j} + \sum_{i=2}^{N} \eta_i (\pi_{ij} - \pi_{(i-1)j}), \end{aligned} \tag{4.67}$$

where $\{\alpha_i, i = 1, \ldots, N\}$, $\{\beta_i, i = 1, \ldots, N\}$, $\{\gamma_i, i = 1, \ldots, N\}$, $\{\eta_i, i = 1, \ldots, N\}$ are Lagrange multipliers. The KKT conditions are demonstrated as follows:

- Primal constraints:

$$\begin{cases} q_{ij} \geq q_j^{min}, \forall i \in \mathcal{N}, \forall j \in \mathcal{M}; \\ q_{ij} \leq q^{max}, \forall i \in \mathcal{N}, \forall j \in \mathcal{M}; \\ \pi_{1j} \geq 0, \forall j \in \mathcal{M}; \\ \pi_{ij} \geq \pi_{(i-1)j}, \forall i \in \mathcal{N}, i \neq 1, \forall j \in \mathcal{M}; \\ \theta_1 ln(1+\pi_{1j}) - cq_{1j} = 0, \forall j \in \mathcal{M}; \\ \theta_i ln(1+\pi_{ij}) - cq_{ij} = \theta_i ln(1+\pi_{(i-1)j}) - cq_{(i-1)j}, \\ \qquad \forall i \in \mathcal{N}, i \neq 1, \forall j \in \mathcal{M}. \end{cases} \tag{4.68}$$

- Dual constraints:

$$\alpha_i \geq 0, \beta_i \geq 0, \gamma_j \geq 0, \eta_i \geq 0, \forall i \in \mathcal{N}, \forall j \in \mathcal{M}. \tag{4.69}$$

- Complementary slackness:

$$\begin{cases} \alpha_i (q_{ij} - q_j^{\min}) = 0, \forall i \in \mathcal{N}, \forall j \in \mathcal{M}; \\ \beta_i (q_{ij} - q^{\max}) = 0, \forall i \in \mathcal{N}, \forall j \in \mathcal{M}; \\ \eta_1 \pi_{1j} = 0, \forall j \in \mathcal{M}; \\ \eta_i (\pi_{ij} - \pi_{(i-1)j}) = 0, \forall i \in \mathcal{N}, i \neq 1, \forall j \in \mathcal{M}; \\ \gamma_1 [\theta_1 ln(1+\pi_{1j}) - cq_{1j}] = 0, \forall i \in \mathcal{N}, \forall j \in \mathcal{M}; \\ \gamma_i \left(\frac{\theta_i}{c} \cdot ln \frac{1+\pi_{ij}}{1+\pi_{(i-1)j}} + q_{(i-1)j} - q_{ij} \right) = 0, \\ \qquad \forall i \in \mathcal{N}, i \neq 1, \forall j \in \mathcal{M}. \end{cases} \tag{4.70}$$

- The first-order conditions of the Lagrangian:

$$\begin{cases} \frac{\partial L}{\partial q_{ij}} = \frac{p_i}{2\sqrt{aq_{ij}}} + \alpha_i - \beta_i - \gamma_i + \gamma_{i+1} = 0, \\ \qquad \forall i \in \mathcal{N}, i \neq N, \forall j \in \mathcal{M}; \\ \frac{\partial L}{\partial q_{Nj}} = \frac{p_N}{2\sqrt{aq_{Nj}}} + \alpha_N - \beta_N - \gamma_N = 0, \\ \qquad \forall j \in \mathcal{M}; \\ \frac{\partial L}{\partial \pi_{ij}} = -p_i + \frac{1}{1+\pi_{ij}}(\frac{\gamma_i\theta_i}{c} - \frac{\gamma_{i+1}\theta_{i+1}}{c}) \\ + \eta_i - \eta_{i+1} = 0, \forall i \in \mathcal{N}, i \neq N, \forall j \in \mathcal{M}; \\ \frac{\partial L}{\partial \pi_{Nj}} = -p_N + \frac{1}{1+\pi_{Nj}} \cdot \frac{\gamma_N\theta_N}{c} + \eta_N = 0, \\ \qquad \forall j \in \mathcal{M}. \end{cases} \tag{4.71}$$

According to the KKT conditions above, optimal contract under information asymmetry must satisfy:

$$q_{ij} = \begin{cases} \frac{p_i^2}{4a(\gamma_i - \gamma_{i+1} + \beta_i - \alpha_i)^2}, \ i \neq N, \\ \frac{p_N^2}{4a(\gamma_N + \beta_N - \alpha_N)^2}, \ i = N. \end{cases} \tag{4.72}$$

$$\pi_{ij} = \begin{cases} \frac{\gamma_i\theta_i - \gamma_{i+1}\theta_{i+1}}{c(p_i + \eta_{i+1} - \eta_i)} - 1, \ i \neq N, \\ \frac{\gamma_N\theta_N}{c(p_N - \eta_N)} - 1, \ i = N. \end{cases} \tag{4.73}$$

4.2.2.6 Optimal Contract Design Without Information Asymmetry

In the scenario with complete information, we assume that there exists a selfish content consumer who greatly knows the type of each content producer, and will increase its profit by utilizing the complete information. In this case, content producers are unable to misrepresent their types, thus only the IR constraints need to be satisfied.

Specifically, the content consumer will decrease π_{ij} while making the contract meet IR constraints until the equality holds. Therefore, the optimal contract (q_{ij}^*, π_{ij}^*) with complete information must satisfy $\theta_i ln(1 + \pi_{ij}^*) - cq_{ij}^* = 0, \forall i \in \mathcal{N}, \forall j \in \mathcal{M}$.

The optimization problem without information asymmetry can be written as

$$\mathbf{P4}: \quad \max_{q_{ij},\pi_{ij}} \sum_{i=1}^{N} p_i(\sqrt{aq_{ij}} - \pi_{ij}), \tag{4.74}$$

$$\text{s.t.} \quad q_j^{\min} \leq q_{ij} \leq q^{\max}, \tag{4.74a}$$

$$IR: \theta_i ln(1+\pi_{ij}) - cq_{ij} = 0, \tag{4.74b}$$

$$0 \leq \pi_{1j} \leq \ldots \leq \pi_{ij} \leq \ldots \leq \pi_{Nj}. \tag{4.74c}$$

which can also be addressed using the method of Lagrange multipliers.

4.2.3 Consumer-Producer Matching Algorithm

The optimal contracts derived from the previous section provide rewards and data volumes for each consumer-producer pair, and the types of content producers are also exposed to content consumers. However, in practice, there may be multiple content consumers requesting the same content at the same time. The optimal content producer for each content consumer is the one with the highest type, also known as type N, whom they are most eager to sign contracts with. Thus, it is necessary to make a reasonable and precise match between UAVs on both sides.

4.2.3.1 Matching Definition

In this subsection, we first provide an introduction to some key components of the matching game, and subsequently, we present the definition of the many-to-one matching game in UNDN.

We define the consumer-producer matching game by the following components:

- The set of content consumers $\mathcal{M}$;
- The set of content producers $\mathcal{N}$;
- The preference lists $\mathcal{P}^{C}, \mathcal{P}^{P}$ for content consumers and content producers, which is generated by sorting the utility values.

Definition 4.5 We define the matching function Φ as a mapping from $\mathcal{M} \cup \mathcal{N} \rightarrow \mathcal{M} \cup \mathcal{N}$, where each content consumer $j \in \mathcal{M}$ and content producer $i \in \mathcal{N}$ satisfy the following conditions:

(1) $\Phi(j) \subset \mathcal{N}$ and $|\Phi(j)| \leq 1$;
(2) $\Phi(i) \subset \mathcal{M}$ and $|\Phi(i)| \leq K_i$;
(3) $\Phi(j) = i$ if and only if $\Phi(i) = j$,

Algorithm 10 Consumer-producer matching algorithm

Input: Utility matrix of Content Consumer $\boldsymbol{U}_m^C$, utility matrix of content producer $\boldsymbol{U}_m^P$, matching capacity matrix $\boldsymbol{K}$
Output: Matching results $\boldsymbol{\Phi}$

```
1: Initialization:
2: Construct the preference lists of content consumers P^C
3: Construct the preference lists of content producers P^P
4: Content consumers propose to content producers:
5: for all content consumers who are not matched do
6:     Propose to the content producer at the top of preference list
7: end for
8: Content producers make decisions:
9: for all content producers i ∈ N do
10:     if |Φ(i)| ≤ K_i then
11:         content producer i accepts all the proposed content consumers
12:     else
13:         content producer i accepts the most preferred K_i content consumers and rejects the rest
            ones
14:     end if
15: end for
16: if there is still a consumer has not been matched then
17:     Go to step 4
18: else
19:     Go to step 21
20: end if
21: End of algorithm
```

where K_i denotes the quota of content producer i and is determined by the residual battery B_i:

$$K_i = B_i K_{\max}. \tag{4.75}$$

K_{max} is the maximum number of matches in the scene.

Conditions (1) and (2) imply that each content consumer can match with at most one content producer, while each content producer can match with multiple content consumers. This defines a many-to-one matching scenario. Condition (3) illustrates that if content consumer j is matched with content producer i, then i is also matched with j.

4.2.3.2 Preference List Construction

Given that real content transmission occurs after matching, it is crucial for content producers to factor in the transmission cost and update the utility functions of both parties throughout the matching process.

For content consumer, after completing the contract design phase, the type of content producer and the selected contract are fully exposed. Each content consumer hopes to choose a content producer with a higher type, as such a content producer can provide more content and has higher reputation value and residual battery, which enables better completion of transmission tasks. Therefore, this type of content producer deserves more rewards. The utility function of content consumers in the matching stage are given as follows:

$$U_m^C(\theta_i, q_{ij}) = \varepsilon \ln(1 + q_{ij}) - \delta_i, \tag{4.76}$$

For a fixed type of content producer, the rewards they can receive from different content consumers are the same. Therefore, it is necessary to consider the issue of their own transmission costs. They tend to choose content consumers who can reduce their transmission costs to save energy. Thus, the utility function of content producers in the matching stage are given as follows:

$$U_m^P(\theta_i, q_{ij}) = \delta_i - \lambda E_{ij}^T, \tag{4.77}$$

where $\delta_i = m\theta_i$ represents the reward provided by content consumer j to θ_i-type content producer. λ denotes the unit transmission energy cost, and ε is a positive weighting factor.

Based on the outcomes of the contract, content consumers and content producers calculate the utilities according to (4.76) and (4.77), then establish their preference lists $\mathcal{P}^C, \mathcal{P}^P$ by sorting utility values in descending order. Specifically, we define a symbol $\succeq$ to express the preference relation between UAVs. For example, $i \succeq_j i'$ means that j prefers i to i' or considers them equally preferable.

The content producer seeks increased rewards and reduced transmission energy consumption to maximize utility. Hence, it favors content consumers who can contribute to the higher utility. Therefore, for $\forall i \in \mathcal{N}, \forall j, j' \in \mathcal{M}$, we have:

$$j \succeq_i j' \Longleftrightarrow U_m^P(\theta_i, q_{ij}) > U_m^P(\theta_i, q_{ij'}). \tag{4.78}$$

Similarly, for a content consumer, it intends to obtain more content and provide less reward to achieve greater utility, so it prefers the content producer who can make it achieve greater utility. Therefore, for $\forall i, i' \in \mathcal{N}, \forall j \in \mathcal{M}$, we have:

$$i \succeq_j i' \Longleftrightarrow U_m^C(\theta_i, q_{ij}) > U_m^C(\theta_{i'}, q_{i'j}). \tag{4.79}$$

4.2.3.3 GS Matching Algorithm

In this subsection, we will provide specific steps for the GS matching algorithm.

(1) **Construct preference lists:** Content consumers and content producers calculate their utilities separately and then rank the utility values in descending order to establish their own preference lists.

(2) **Both sides make choices:** Each content consumer proposes a matching request to the top nodes in its preference list, while each content producer accepts the top K items and rejects the remaining content consumers.
(3) **Both sides reselect:** The rejected content consumer removes the rejecting content producer from its preference list and repeats step (2) until either all content consumers are matched with content producers or all content producers reach their maximum capacity for matching.

The matching process is summarized in Algorithm 10.

4.2.3.4 Stability and Pareto Optimality Analysis

In this subsection, we will give the stability and Pareto optimality analysis of the consumer-producer matching algorithm. A stable matching means that none of the participants have motivations to change their partners. The definition of stability is shown as follows.

Definition 4.6 (Stability) A matching Φ is stable if there exits no blocking pairs [35, 36]. A matching is blocked by a pair (i, j) if:

(1) blocking occurs simultaneously in i and j;
(2) i or j chooses each other will result in a higher utility compared to the currently matched partners, i.e.,

$$\begin{aligned} &\exists i \in \mathcal{N},\ j \in \mathcal{M}, \Phi(j) = i^*, \Phi(i) = j^*, i^* \neq i, j^* \neq j, \\ &U_m^P(\theta_i, q_{ij}) > U_m^P(\theta_i, q_{ij^*}) \\ &and\ \ U_m^C(\theta_i, q_{ij}) > U_m^C(\theta_{i^*}, q_{i^*j}) \end{aligned} \tag{4.80}$$

Condition (1) illustrates that pair blocking occurs in pairs, not in a single party, and condition (2) explains that the blocking pair must have a higher utility than the current matching pair.

Given Definition 4.6 and Algorithm 10, we present Lemma 4.4 below and prove that our matching algorithm is stable.

Lemma 4.4 *Our proposed consumer-producer matching algorithm is a stable matching.*

Proof As we discussed before, a stable matching is essentially characterized by the absence of blocking pairs. Therefore, we verify that there are no blocking pairs in the matching algorithm. The preference lists of content consumers and content producers are constructed according to (4.78) and (4.79), thus are strictly monotone increasing. Furthermore, we can observe from line 6 and lines 10 ∼13 in Algorithm 10, the content consumer proposes to the preferred content producer, and the content producer accepts the top K content consumers among the proposed content consumers based on its preference list. This ensures the absence of any blocking pairs, and therefore the matching is stable. The proof is finished. □

Based on the properties of stable matching, we can obtain Lemma 4.5 as follows.

Lemma 4.5 *The obtained matching result set* $\boldsymbol{\Phi}$ *is weak Pareto optimal for each content consumer* $j \in \mathcal{M}$ *and content producer* $i \in \mathcal{N}$.

Proof This proof process largely follows [37] and [38]. In our proposed consumer-producer matching algorithm, each participant selects matching objects based on the ranking of the preference list, which indicates that the current choices of both parties are the positions available in the preference lists with the highest priority. That is to say, they cannot achieve higher utility by selecting other matching objects. This is consistent with the matching stability we demonstrated in Lemma 4.4. Therefore, there is no Pareto improvement throughout the matching process, indicating that the matching algorithm is weak Pareto optimal. The proof is finished. □

4.2.3.5 Complexity Analysis

This section provides a complexity analysis of the proposed mechanism. In the first stage, i.e., the contract-based incentive mechanism, solving the optimal contract (4.65) is a convex optimization problem. Since there are N contract items (q, π), the number of equality constraints is N and the number of inequality constraints is $3N$. Therefore, the computation complexity is $O(N)$. As we discussed before, there are M content consumers in the system, and all of them need to design contracts, thus the total computation complexity is $O(MN)$.

In the second stage, i.e., the matching mechanism, the matching results can be derived after the GS algorithm. For a content consumer, the complexity of constructing a preference list of N content producers with an efficient sorting algorithm is $O(NlogN)$ [29]. Therefore, the computing complexity can be $O(MNlogN)$ for M content consumers. Similarly, N content producers can build the preference lists with a computing complexity of $O(NMlogM)$. Then, the input size of GS algorithm is $\sum_{i \in \mathcal{N}} |P_i^C| + \sum_{j \in \mathcal{M}} |P_j^P| = 2MN$, where $|P|$ represents the length of the preference list, and the computing complexity of the loop in Algorithm 10 is $O(MN)$. Therefore, the overall computing complexity of the algorithm is $O(MN(logMN + 1)) = O(MNlogMN)$.

4.2.4 *Simulation Results and Discussions*

In this section, we evaluate the proposed mechanism through multiple sets of simulation experiments. We use *Python* to address the convex optimization problem. The parameter values used for simulation can be found in Table 4.1.

Table 4.1 Parameters

Parameter	Value
Number of content producers N	$10, 20 \sim 100$
Energy consumption factor of free space model η_{fs}	$10\,\text{pJ}/(\text{bit} * \text{m}^2)$ [39]
Number of content consumers M	$1, 5, 15, 20 \sim 100$
Energy consumption factor of multi-path fading model η_{mp}	$0.0013\,\text{pJ}/(\text{bit} * \text{m}^4)$ [39]
Probability of type-θ_i producer p_i	$\frac{1}{N}$
Maximum number of matching K_{max}	5
Weight factors a, ε, m	$50, 1, 2$
Content size q	$[0, 20]$ Mb
Caching cost per Mb c	0.1
Reputation value R	$[0, 1]$
Transmission cost per bit λ	0.5
Residual battery B	$[0, 1]$
Energy consumption per bit E_{elec}	$50\,\text{nJ/bit}$ [39]
Weight factors α_1, α_2	$0.6, 0.4$

4.2.4.1 Contract Feasibility

We consider one content consumer requests a kind of content while 10 content producers can offer the content. As mentioned before, the number of content producers is assumed to be equal to the number of contract types, and the type distribution of content producers follows a uniform distribution, i.e., $p_i = \frac{1}{N}, \forall i \in \mathcal{N}$. We conduct experiments under the situations with and without information asymmetry and compare them with two other algorithms: *linear pricing* [34] and *take-it-or-leave contract* [40]. *Linear pricing* is a mechanism employed under information asymmetry. In this method, the content consumer sets the reward proportional to content amounts, which obeys $\pi = \rho q$ and ρ is the unit reward. In this part, inspired by Su et al. [29], Zhopu et al. [41], Kazmi et al. [42], we set ρ as 0.3 and 1.5. In *take-it-or-leave contract*, there exits a type threshold θ_{th}. Content consumer designs a set of same contract items (q_{th}, π_{th}) based on θ_{th}. And a content producer with type greater than this threshold, i.e., $\forall \theta_i \in \Theta,\ \theta_i \geq \theta_{th}$ will accept and sign the contracts, while a content producer with type lower than this threshold, i.e., $\forall \theta_i \in \Theta,\ \theta_i < \theta_{th}$ will reject and discard the contracts. In this part, θ_{th} is set as θ_5 followed by [33]. Figure 4.9 shows the verification results of contract feasibility. The shared content amounts and corresponding rewards of the four algorithms are shown in Fig. 4.9a and Fig. 4.9b, respectively. As we can see from these two figures, the content amounts and the corresponding rewards all increase with the type of content producer, which has been proven in Lemmas 4.1 and 4.2. Moreover, the numerical results indicate that in the situation of complete information, content producers provide the most content and get the most reward, while the case of information asymmetry shows the middle content contributions and reward. In the case of *linear pricing*, the contributed content amounts and

Fig. 4.9 Contract feasibility. (**a**) Content amounts. (**b**) Rewards. (**c**) The utility of content producers

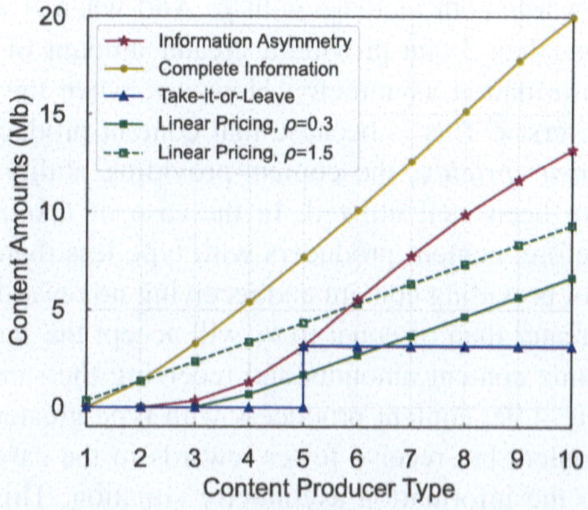

(a) Content Amounts

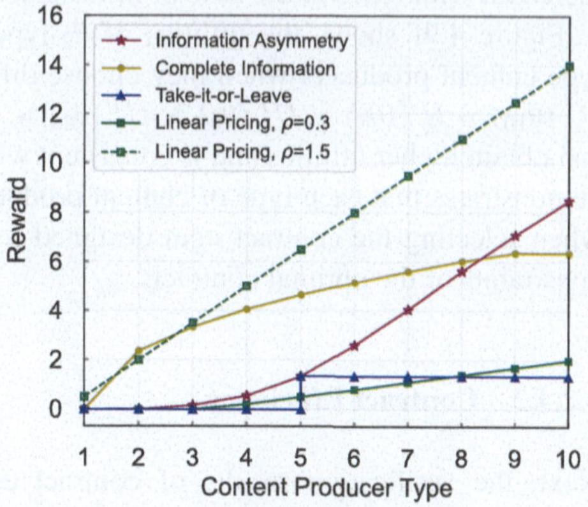

(b) Rewards

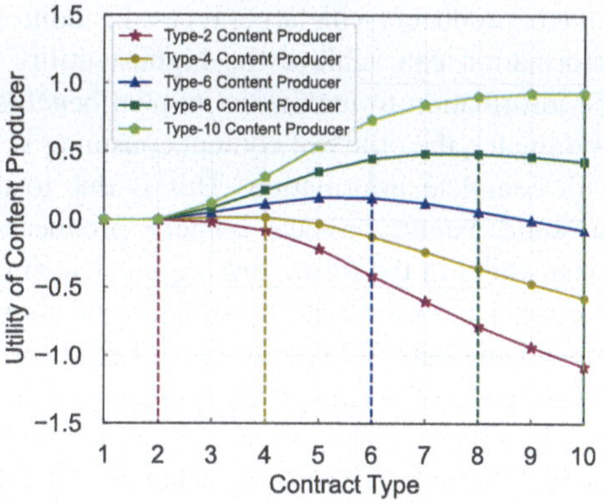

(c) The Utility of Content Producers

rewards both increase with ρ. And when $\rho = 1.5$, content producers with a type less than 5 can provide a greater amount of content compared to situations with information asymmetry. However, when the type is greater than 5, the situation reverses. This is because that content producers are regarded as the same type in *linear pricing*, the content-providing ability of high-type content producers has not been well utilized. In the case of *take-it-or-leave contract*, as we discussed before, content producers with type less than θ_5 will refuse the contracts, thereby not providing content and receiving no rewards, while content producers with types greater than or equal to θ_5 will accept the same contract items, thus providing the same content amounts and receiving the same rewards. It is worth noting that in Fig. 4.9e, content producers with type greater than or equal to θ_8 contribute more content but receive fewer rewards in the case of complete information, compared to the information asymmetry situation. This is because in the case of complete information, the utilities of content producers are always 0, while the utilities are increased with types in the case of information asymmetry.

Figure 4.9f shows the utilities of θ_2-type, θ_4-type, θ_6-type, θ_8-type and θ_{10}-type content producers when they choose different contract items. As we can see, $U^P(\theta_{10}) > U^P(\theta_8) > U^P(\theta_6) > U^P(\theta_4) > U^P(\theta_2)$, which means the higher type can obtain higher utilities and is consistent with Lemma 4.3. Furthermore, Fig. 4.9f demonstrates that each type of content producer can achieve maximum utility only when selecting the contract item designed for its type, which conforms to the IC constraints of the optimal contracts.

4.2.4.2 Contract Efficiency

Next, the verification results of contract efficiency are presented in Fig. 4.10. Figure 4.9d shows how the utility of the content consumer varies with types. We can observe that the content consumer can achieve higher utility by choosing content producers with larger types. In addition, the content consumer with complete information can achieve the highest utility values, since it can perfectly utilize the information to maximize its own benefits. Besides, In the case of information asymmetry, the utility of content consumer is lower compared to that in the scenario with complete information. This is due to the fact that asymmetric information, to some extent, protects content producers from being excessively exploited. Meanwhile, in the *linear pricing* mechanism, when $\rho = 0.3$, the utility of content consumer is lower than the situation of asymmetric information due to the lowest content amounts. While when $\rho = 1.5$, the utility of content consumer is higher than the situation of asymmetric information when the chosen type of content producer θ is less than 4. This is because more content amounts dominates when $\theta \leq 4$, while more reward dominates when $\theta > 4$. Furthermore, according to the principles of *take-it-or-leave contract*, content producers with type less than θ_5 refuse the contracts and therefore do not offer content to the content consumer, resulting in a utility of 0. Conversely, content producers with types greater than or equal to

Fig. 4.10 Contract efficiency. (**a**) The utility of content consumer. (**b**) The utility of content producers. (**c**) Comparison of utility

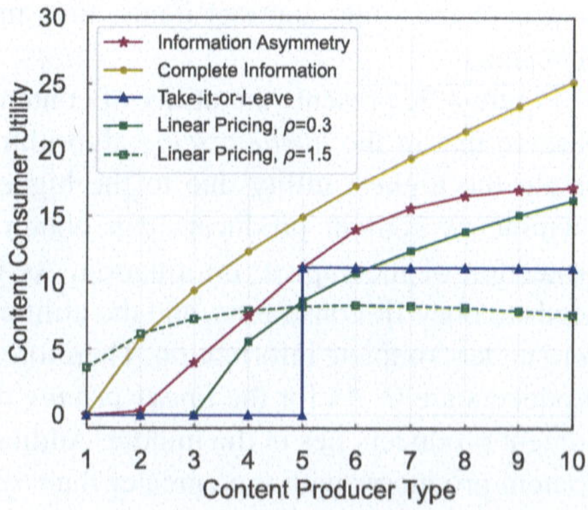

(a) The Utility of Content Consumer

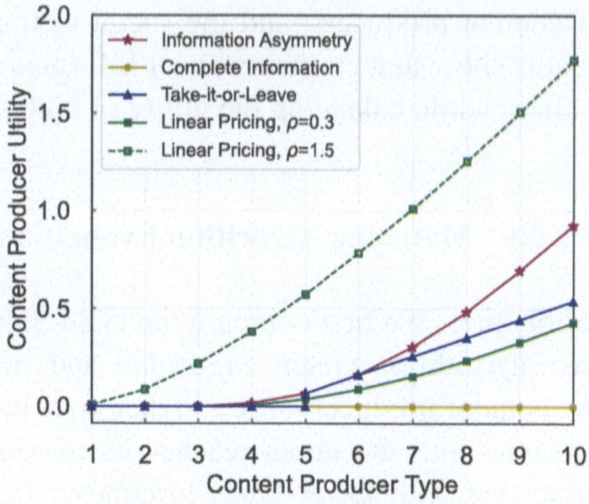

(b) The Utility of Content Producers

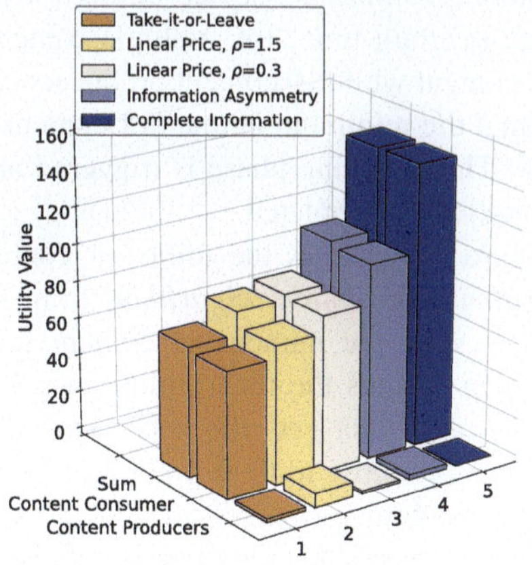

(c) Comparison of Utility

θ_5 accept the same contract items, thus making the utility of content consumers unchanged.

Figure 4.9e presents the utility of content producers changes with types. We can observe that in the *linear pricing* algorithm when $\rho = 1.5$, content producers can obtain the highest utility due to the highest rewards. In the case of information asymmetry, content producers can obtain the second highest utility due to the protection of incomplete information. As we elaborated earlier, the contract only needs to obey IR constraints, and the utilities of content producers meet no less than zero under complete information. Therefore, the utilities of different types of content producers are 0. As for the *linear pricing* mechanism when $\rho = 0.3$, the utility of content producers lies in the middle. Additionally, in *take-it-or-leave contract*, only content producers with type greater than θ_5 will accept the contract, and thus these content producers can achieve non-zero utilities.

Figure 4.9f demonstrates the utility of the content consumer, the entire utility of content producers, and the social welfare of these four algorithms. Simulation results show that contracts under information asymmetry can achieve higher social welfare while balancing the utility of both sides.

4.2.4.3 Matching Algorithm Evaluation

In this part, We first compare the Gale-Shapley matching algorithm with the other two algorithms: *greedy algorithm* and *random algorithm*. In *greedy algorithm*, the content producer with a higher type has priority to choose content consumers to serve until the quota reaches its maximum K. Next, it is the turn of content producers with lower types to choose. In *random algorithm*, content consumers and content producers are randomly matched to each other. The performance of *random algorithm* is assessed through averaging over 500 iterations. We consider that at a certain time slot, multiple content consumers simultaneously request the same content while 10 content producers can provide such content. It's important to note that the minimum amount of content q^{min} each content consumer can accept varies. The matching phase is triggered after the contract phase under asymmetry information is completed.

Figure 4.11 shows the utility of content consumers of GS algorithm, *greedy algorithm* and *random algorithm*, from which we can observe that the utilities increase with the number of content consumers. Besides, the case of the GS algorithm obtains the most utility values among the three, while the case of the *random algorithm* gets the least. This is because, in the GS algorithm, content consumers propose to their preferred content producers based on utility functions, and the matching process is led by content consumers. Therefore, it is more possible to arrive at results that are more beneficial to content consumers.

Figure 4.12 presents the utility of content producers of the three algorithms mentioned above. Firstly, we can observe that as the number of content consumers increases, the utility of content producers increases. Besides, it is evident that the *greedy algorithm* can reach the most utility of content producers, and the utility in

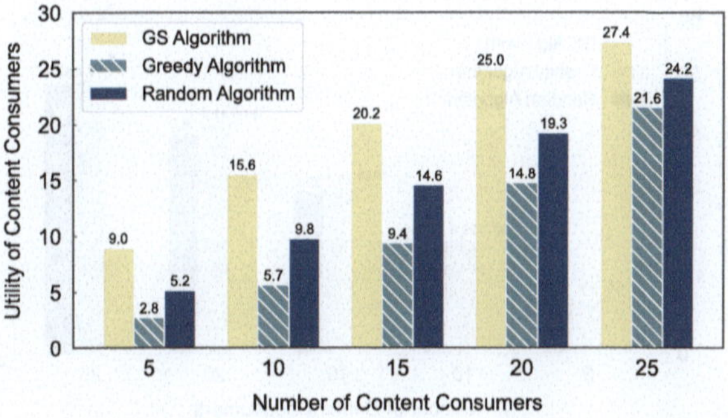

Fig. 4.11 Utility of content consumers

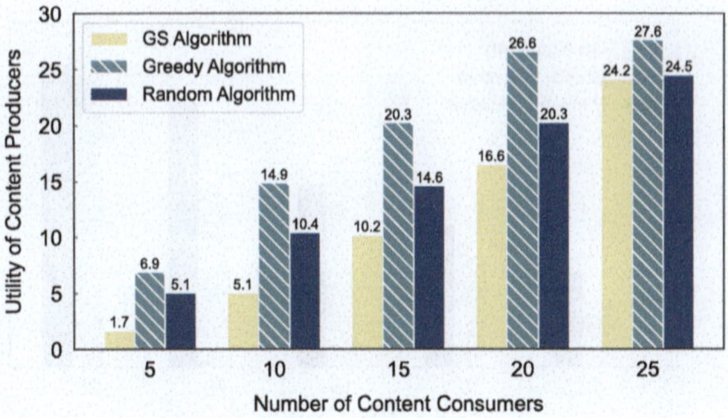

Fig. 4.12 Utility of content producers

GS algorithm is higher than in *random algorithm*. As we explained earlier, in the *greedy algorithm*, the high-type content producer is fully loaded before moving to the next one. Each content producer selects the top few content consumers in the preference list constructed based on its utility function, which means the matching process is led by content producers. Therefore, the *greedy algorithm* can yield results that are profitable for content producers.

Then, we compare the social welfare of these three algorithms in Fig. 4.13. The social welfare of the matching process increases with the number of content consumers. Furthermore, it is demonstrated that the GS algorithm has the best performance among the three.

The running time of different algorithms is also simulated as depicted in Fig. 4.14. The experimental results indicate that with an increase in the number of content consumers, the running time becomes longer, and overall, the GS algorithm has the shortest time consumption. In summary, the GS algorithm can achieve outstanding performance among the three algorithms. Finally, we increase the network size and evaluate the performance of the GS algorithm in Fig. 4.15. Two scenarios are considered: the number of content producers is fixed to 20 while the number of content consumers increases from 20 to 100, as shown in Fig. 4.15c, and

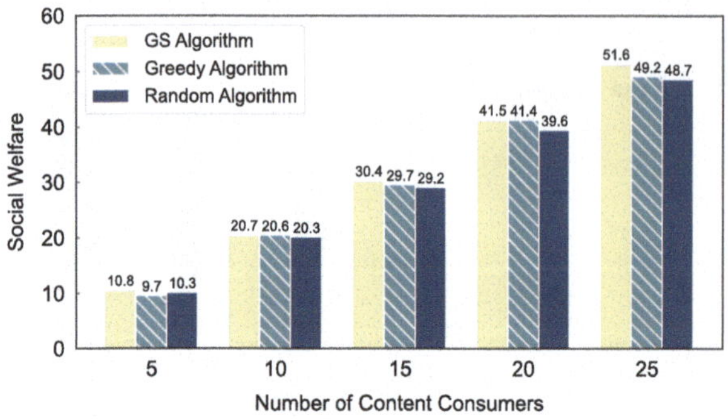

Fig. 4.13 Social welfare

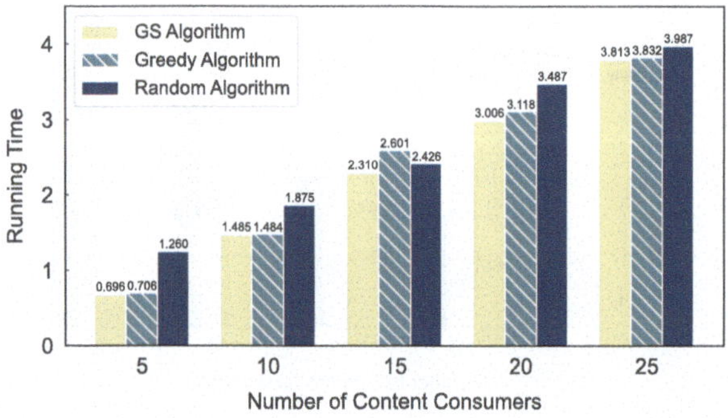

Fig. 4.14 Running time

the number of content consumers is fixed to 20, the number of content producers increases from 20 to 100, as shown in Fig. 4.15d. In Fig. 4.15c, we can observe that as the number of content consumers grows, the utility of both sides and the social welfare also increase until the number of content consumers reaches around 60. This is because the maximum matching number of all content producers is around 60. When the number of content consumers exceeds this limit, they start competing with each other for content, which causes their utilities to decline. In Fig. 4.15d, we can obtain that as the number of content producers increases, their utility gradually decreases, but the social welfare remains basically stable. This is because some content producers can already meet the content needs of all content consumers, and an increase in the number of content producers intensifies competition among them, leading to a decrease in their utility.

4.2.4.4 The Energy Efficiency of the Proposed Mechanism

In this part, we evaluate the energy efficiency of our proposed mechanism by comparing it with other algorithms and the original content-sharing mechanism of

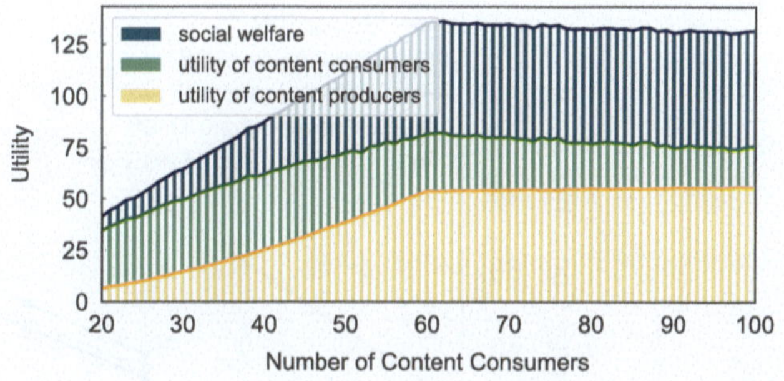

(a) 20 content producers with the number of content consumers changes

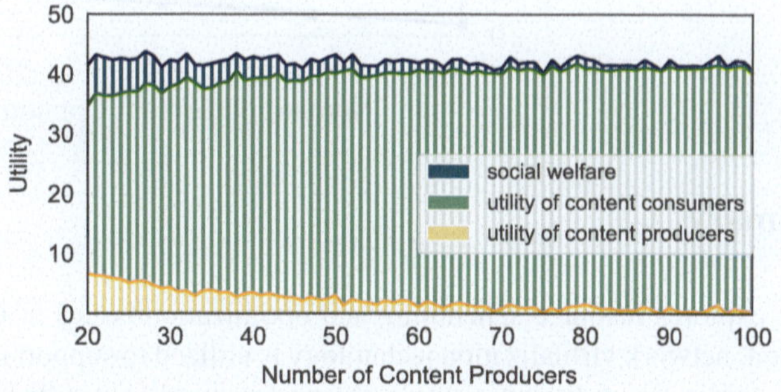

(b) 20 content consumers with the number of content producers changes

Fig. 4.15 Changes in the utility of content consumers and content producers. (**a**) 20 content producers with the number of content consumers changes. (**b**) 20 content consumers with the number of content producers changes

NDN. We fix the number of content producers at 10, vary the number of content consumers, and calculate the energy consumption of content producers when all content consumers request content once. As shown in Fig. 4.16, we observe that after conducting many-to-one matching, the transmission energy consumed by content producers is significantly reduced due to the absence of redundant data packet transmission. From the comparative experiments, it can be seen that our proposed mechanism demonstrates good energy efficiency performance. Additionally, the *take-it-or-leave* algorithm with GS matching shows the lowest energy consumption due to its low contributed content amounts and the content producers who signed the contracts being unable to meet the content provision needs of all content consumers.

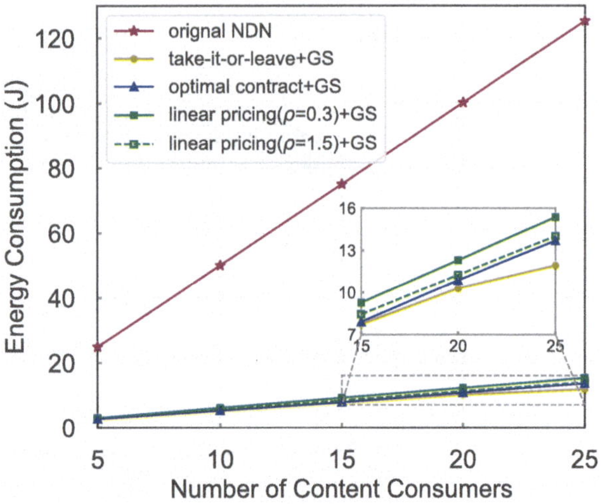

Fig. 4.16 Energy consumption

4.3 Summary

This chapter explores resource scheduling and optimization issues in UAV swarm networks. First, network virtualization technology is utilized to support multiple virtual network applications sharing a single physical network, providing customized services. To optimize performance, a Graph Transformer-aided VNE algorithm is proposed, which extracts higher-order relationships between nodes in the underlying network and partitions the network based on service capabilities. Secondly, a novel resource-efficient content-sharing approach in UNDN, combining contract theory and matching theory, is proposed to address the issue of content redundancy during the content Backhaul process. Redundant content packets in the network are reduced, leading to decreased energy consumption through many-to-one matching between content producers and content consumers. To establish preference lists for both sides, content consumers design optimal contract sets to identify content producers and incentivize them to share content.

References

1. C. Zhang, H. Yao, T. Mai, Graph transformer aided resource virtualization embedding in UAV swarm networks, in *Proceedings of the International Conference on Computing, Machine Learning and Data Science* (2024), pp. 1–6
2. C. Jin, H. Yao, T. Mai, J. Xu, Q. Zhang, F. R. Yu, A resource-efficient content sharing mechanism in large-scale uav named data networking. IEEE Trans. Netw. **33**(1), 114–129 (2025)
3. P. Yang, X. Xi, K. Guo, T.Q. Quek, J. Chen, X. Cao, Proactive UAV network slicing for URLLC and mobile broadband service multiplexing. IEEE J. Sel. Areas Commun. **39**(10), 3225–3244 (2021)

4. Q. Wu, J. Xu, Y. Zeng, D.W.K. Ng, N. Al-Dhahir, R. Schober, A.L. Swindlehurst, A comprehensive overview on 5G-and-beyond networks with UAVs: From communications to sensing and intelligence. IEEE J. Sel. Areas Commun. **39**(10), 2912–2945 (2021)
5. Z. Yuan, G.-M. Muntean, Airslice: A network slicing framework for UAV communications. IEEE Commun. Mag. **58**(11), 62–68 (2020)
6. G. Cappello, G. Colajanni, P. Daniele, L. Galluccio, C. Grasso, G. Schembra, L. Scrimali, Optimizing FANET lifetime for 5G softwarized network provisioning. IEEE Trans. Netw. Serv. Manag. **19**(4), 4629–4649 (2022)
7. N.M.K. Chowdhury, M.R. Rahman, R. Boutaba, Virtual network embedding with coordinated node and link mapping, in *IEEE INFOCOM 2009* (IEEE, Piscataway, 2009), pp. 783–791
8. X. Cheng, S. Su, Z. Zhang, H. Wang, F. Yang, Y. Luo, J. Wang, Virtual network embedding through topology-aware node ranking. ACM SIGCOMM Comput. Commun. Rev. **41**(2), 38–47 (2011)
9. N. Shahriar, S.R. Chowdhury, R. Ahmed, A. Khan, S. Fathi, R. Boutaba, J. Mitra, L. Liu, Virtual network survivability through joint spare capacity allocation and embedding. IEEE J. Sel. Areas Commun. **36**(3), 502–518 (2018)
10. H. Yao, X. Chen, M. Li, P. Zhang, L. Wang, A novel reinforcement learning algorithm for virtual network embedding. Neurocomputing **284**, 1–9 (2018)
11. Z. Yan, J. Ge, Y. Wu, L. Li, T. Li, Automatic virtual network embedding: A deep reinforcement learning approach with graph convolutional networks. IEEE J. Sel. Areas Commun. **38**(6), 1040–1057 (2020)
12. X. Li, T. Zhang, J. Li, A particle swarm mobility model for flying ad hoc networks, in *GLOBECOM 2017-2017 IEEE Global Communications Conference* (IEEE, Piscataway, 2017), pp. 1–6
13. M.A.M. De Oca, T. Stutzle, M. Birattari, M. Dorigo, Frankenstein's PSO: a composite particle swarm optimization algorithm. IEEE Trans. Evol. Comput. **13**(5), 1120–1132 (2009)
14. A. Vaswani, N. Shazeer, N. Parmar, J. Uszkoreit, L. Jones, A.N. Gomez, Ł. Kaiser, I. Polosukhin, Attention is all you need, in *Advances in Neural Information Processing Systems*, vol. 30 (2017)
15. A.E. Mirino et al., Best routes selection using Dijkstra and Floyd-Warshall algorithm, in *2017 11th International Conference on Information & Communication Technology and System (ICTS)* (IEEE, Piscataway, 2017), pp. 155–158
16. A. Hagberg, P. Swart, D.S. Chult, Exploring network structure, dynamics, and function using networkx. Los Alamos National Lab.(LANL), Los Alamos, NM (United States), Tech. Rep. (2008)
17. T. Megan, Z. Ellen, Generation and analysis of random graphs to model internetworks. Technical Report GIT-CC-94-46, Georgia Institute of Technology (1994)
18. S. Ma, H. Yao, T. Mai, J. Yang, W. He, K. Xue, M. Guizani, Graph convolutional network aided virtual network embedding for internet of thing. IEEE Trans. Netw. Sci. Eng. **10**(1), 265–274 (2022)
19. L. Gupta, R. Jain, G. Vaszkun, Survey of important issues in UAV communication networks. IEEE Commun. Surv. Tutorials **18**(2), 1123–1152 (2015)
20. I. Bekmezci, O.K. Sahingoz, Ş. Temel, Flying ad-hoc networks (FANETs): A survey. Ad Hoc Netw. **11**(3), 1254–1270 (2013)
21. Z. Wang, H. Yao, T. Mai, Z. Xiong, X. Wu, D. Wu, S. Guo, Learning to routing in UAV swarm network: A multi-agent reinforcement learning approach. IEEE Trans. Veh. Technol. **72**(5), 6611–6624 (2023)
22. A. Miglani, N. Kumar, Blockchain-based co-operative caching for secure content delivery in CCN-enabled V2G networks. IEEE Trans. Veh. Technol. **72**(4), 5274–5289 (2023)
23. L. Zhang, A. Afanasyev, J. Burke, V. Jacobson, K. Claffy, P. Crowley, C. Papadopoulos, L. Wang, B. Zhang, Named data networking. ACM SIGCOMM Comput. Commun. Rev. **44**(3), 66–73 (2014)
24. P. Bolton, M. Dewatripont, *Contract Theory* (MIT Press, 2004)

25. L. Lovász, M.D. Plummer, *Matching Theory*, vol. 367 (American Mathematical Society, Providence, 2009)
26. L.E. Dubins, D.A. Freedman, Machiavelli and the Gale-Shapley algorithm. Am. Math. Mon. **88**(7), 485–494 (1981)
27. B. Zhu, E. Bedeer, H.H. Nguyen, R. Barton, J. Henry, UAV trajectory planning in wireless sensor networks for energy consumption minimization by deep reinforcement learning. IEEE Trans. Veh. Technol. **70**(9), 9540–9554 (2021)
28. N. Gupta, J. Singh, S.K. Dhurandher, Z. Han, Contract-theory-based incentive design mechanism for opportunistic IoT networks. IEEE Internet Things J. **10**(4), 2881–2892 (2021)
29. C. Su, F. Ye, T. Liu, Y. Tian, Z. Han, Computation offloading in hierarchical multi-access edge computing based on contract theory and Bayesian matching game. IEEE Trans. Veh. Technol. **69**(11), 13686–13701 (2020)
30. S. Wang, X. Zhang, L. Wang, J. Yang, W. Wang, Joint design of device to device caching strategy and incentive scheme in mobile edge networks. IET Commun. **12**(14), 1728–1736 (2018)
31. L. Xu, C. Jiang, Y. Chen, Y. Ren, K.R. Liu, Privacy or utility in data collection? A contract theoretic approach. IEEE J. Sel. Top. Signal Process. **9**(7), 1256–1269 (2015)
32. A. Asheralieva, D. Niyato, Combining contract theory and Lyapunov optimization for content sharing with edge caching and device-to-device communications. IEEE/ACM Trans. Netw. **28**(3), 1213–1226 (2020)
33. Z. Zhou, P. Liu, J. Feng, Y. Zhang, S. Mumtaz, J. Rodriguez, Computation resource allocation and task assignment optimization in vehicular fog computing: A contract-matching approach. IEEE Trans. Veh. Technol. **68**(4), 3113–3125 (2019)
34. Y. Zhang, L. Song, W. Saad, Z. Dawy, Z. Han, Contract-based incentive mechanisms for device-to-device communications in cellular networks. IEEE J. Sel. Areas Commun. **33**(10), 2144–2155 (2015)
35. S. Bayat, Y. Li, L. Song, Z. Han, Matching theory: Applications in wireless communications. IEEE Signal Process. Mag. **33**(6), 103–122 (2016)
36. D. Gusfield, R.W. Irving, *The Stable Marriage Problem: Structure and Algorithms* (MIT Press, 1989)
37. Z. Zhou, K. Ota, M. Dong, C. Xu, Energy-efficient matching for resource allocation in D2D enabled cellular networks. IEEE Trans. Veh. Technol. **66**(6), 5256–5268 (2016)
38. Y. Du, J. Li, L. Shi, T. Liu, F. Shu, Z. Han, Two-tier matching game in small cell networks for mobile edge computing. IEEE Trans. Serv. Comput. **15**(1), 254–265 (2019)
39. W.B. Heinzelman, A.P. Chandrakasan, H. Balakrishnan, An application-specific protocol architecture for wireless microsensor networks. IEEE Trans. Wirel. Commun. **1**(4), 660–670 (2002)
40. L. Duan, L. Gao, J. Huang, Cooperative spectrum sharing: A contract-based approach. IEEE Trans. Mob. Comput. **13**(1), 174–187 (2012)
41. Z. Zhou, H. Liao, B. Gu, S. Mumtaz, J. Rodriguez, Resource sharing and task offloading in IoT fog computing: A contract-learning approach. IEEE Trans. Emerg. Top. Comput. Intell. **4**(3), 227–240 (2019)
42. S.A. Kazmi, T.N. Dang, I. Yaqoob, A. Manzoor, R. Hussain, A. Khan, C.S. Hong, K. Salah, A novel contract theory-based incentive mechanism for cooperative task-offloading in electrical vehicular networks. IEEE Trans. Intell. Transp. Syst. **23**(7), 8380–8395 (2021)

Chapter 5
Distributed Learning Empowered UAV Swarm Networks Cooperation

Traditional intelligent UAV swarm systems that rely on conventional machine learning solutions are often constrained by cloud-centric architectures, necessitating the transmission of large volumes of data from UAVs to centralized servers for training. This approach not only incurs significant communication overhead–thereby reducing system response times–but also raises privacy concerns, as entities are reluctant to send raw data to third-party edge servers. Recently, distributed learning, particularly Federated Learning (FL), has emerged as a promising solution to address these challenges. FL facilitates collaborative model training across multiple UAV devices while preserving data privacy. However, the independence and self-interest of distributed devices require the design of effective incentive mechanisms. To address this, in this chapter, we propose a double auction mechanism to facilitate model trading, coupled with an Iterative Dual Auction (IDA) algorithm and reinforcement learning-based Experience Weighted Attraction Learning-Double Auction (EWA-DA) algorithm to optimize participants' strategies [1]. Moreover, conventional FL models typically assume uniform data structures across devices–a presumption that does not hold in the heterogeneous environments of UAV swarms. To overcome this limitation, a Cluster-based Federated Learning (CFL) architecture is introduced, wherein the UAV swarm is partitioned into clusters based on the similarity of their optimization objectives, and FL is applied within each cluster for model updates. To further optimize resource allocation within this architecture, a three-stage Stackelberg game model is proposed, involving model owners, cluster heads, and UAV workers in the model trading process. Additionally, a hierarchical reinforcement learning algorithm is introduced to search for the game equilibrium within the CFL framework, thereby enhancing both resource utilization and overall system efficiency [2].

H. Yao, T. Mai, *UAV Swarm Cooperation*, Wireless Networks,
https://doi.org/10.1007/978-3-031-96444-2_5

5.1 Federated Learning Service Market for UAV Swarm Networks

The past decade has witnessed an exponential growth of diverse UAV applications ranging from industrial production to medical care [3]. Billions of UAV devices (e.g., smart cameras, smartwatches) around the world are now connected to the Internet, all generating, collecting, and sharing massive amounts of data. The significant increase in the number of UAV devices and the data volume presents huge challenges for the current AI training architecture (i.e., centralized edge/cloud server-based architecture) [4]. All local data have to be transmitted to one central server, and then used for training corresponding machine learning models. In many cases, sending data to the server may raise data security issues, privacy issues, and even regulatory and judicial issues.

Recently, FL offers significant opportunities to address this challenges [5]. FL enables distributed devices to collaboratively learn a shared AI model without the need to transfer the local data to cloud servers. In federated learning, distributed FL participants (i.e., UAV devices) can compute the updates to the current global model over their local data. Then, the updates to the model parameters (e.g., the weights and biases of neural network) will be aggregated by the central server (i.e., FL platform), and broadcasted it back [6]. This iterative process will repeat until achieving the target model accuracy level.

However, there are still some critical challenges when implementing the federated learning framework over massive UAV devices. Compared to centralized learning, the FL participants are independent, uncontrollable, and self-interested. Because of the energy costs suffered from model training, UAV devices are unwilling to participate in FL tasks or provide the redundant or low-quality local model. Therefore, designing an effective incentive scheme to simulate the data owner to actively participate in FL tasks is thus a priority problem.

To tackle this problem, we establish a double auction-based FL service market, where the training models can be traded between UAV devices and FL tasks. In this market, the FL tasks announce their service demand and how much they are willing to pay (i.e., payment price). Meanwhile, data owners (i.e., UAV devices) can announce their service value (i.e., model accuracy) and asking price. A centralized broker is introduced to determines how much service value of each data owner will be provided to each task and at what deal price.

However, due to the asymmetric information between two sides, the auction market is difficult to converge to the global and social efficiency. In this part, to elicit hidden information of both sides to the broker, we propose an iterative double auction (IDA) algorithm [7]. In the IDA mechanism, a controller is implemented to guide participants to adjust their policies for the sake of maximizing individual participates utilities as well as social welfare. The simulation result shows that the system can quickly converge to social welfare maximization. While the IDA algorithm can quickly converge to the optimal solution, it relies on a centralized controller and some strong assumptions, thus reducing the system's robustness. On this basis, we propose a reinforcement learning-based scheme, termed the Experience-Weighted Attraction Learning-Double Auction (EWA-DA) mechanism [8]. The EWA-DA combines reinforcement learning and belief learning algorithms. It enables the agent to improve their strategies directly by participating in the auction continuously, without the controller participating. The extensive and in-depth simulations show that the EWA-DA algorithm enable the market can quickly converge to the competitive equilibrium.

5.1.1 System Model and Problem Formulation

In this section, we first discuss the federated learning framework. Then, we present the system model and design the cost function and the utility function respectively.

5.1.1.1 Federated Learning Framework

As shown in Fig. 5.1, the FL framework consists of a central FL platform and a set of data owners. These devices can collaboratively train a shared AI model through the FL platform coordinating. During the initialization phase, the FL platform sets training tasks, including the corresponding data requirements, the algorithm structure, and a round of global iteration time, etc., and broadcasts them to all data owners. After receiving information, data owners decide whether they will participate in tasks according to their available resources (e.g., computing resources, available data) and the local strategy. Then, the FL platform initializes model parameters and transmits them to all FL participants to begin the training. During the learning phase, the data owners train the model over their local training data and send the updated parameters to the FL platform. The FL platform then aggregates the received model updates into a single global model and feeds it back to all participants. The learning phase will repeat until the target model accuracy is achieved.

Within the framework, we define the $\theta \in [1, \infty)$ as the local model accuracy and $\epsilon \in [1, \infty)$ as the global accuracy to evaluate the model quality. Note that the model accuracy is commonly defined at [0, 1]. In this part, to meet the monotonicity demand in utility function design, we adopt its reciprocal as our accuracy value. Here, the larger values of θ show better model accuracy and vice versa, where the $\theta = \infty$ means the local training tasks are solved to optimality, and $\theta = 1$ means that it doesn't make any progress whatsoever [9].

According to [10], the number of local iterations to attain the accuracy θ can be given by:

$$I_l(\theta) = \gamma\theta, \tag{5.1}$$

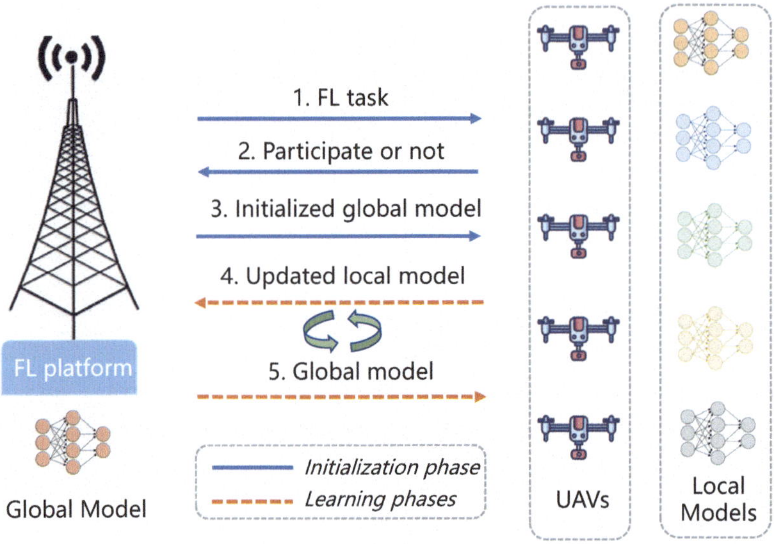

Fig. 5.1 Federated learning framework

where γ is the difficulty coefficient, which relies on the local sub-problems and data set on the data owner. In addition, according to [11], the number of global iterations is dependent on the local accuracy and the global accuracy, which can be denoted as:

$$I_g(\epsilon, \theta) = \frac{\alpha \log(\epsilon)}{1 - \frac{1}{\theta}}. \tag{5.2}$$

From this equation, we can find that the local accuracy θ is inversely proportional to global iterations times I_g. The higher local accuracy are, and the fewer global iterations are.

5.1.1.2 System Model

Consider an FL market consisting of N FL tasks and M data owners. The FL task is indexed as i where $i\epsilon\{1, 2, \ldots, N\}$, and the data owner is indexed as j where $j\epsilon\{1, 2, \ldots, M\}$. We assume that the jth data owner are willing to participate in the ith FL task and supply local trained model with accuracy $\theta_{j,i}^{wo} \geq 1$. And we denote $\boldsymbol{\theta}_{\mathbf{j}}^{\mathbf{wo}}$ as the $1 \times N$ supply vector of jth data owner for all FL tasks. Since data owners are assumed to be resource-constrained UAV devices, the training model will incur the cost of computing and energy resources. Therefore, we denote a cost function $C_j(\boldsymbol{\theta}_{\mathbf{j}}^{\mathbf{wo}})$ to reflect the drop in utility of jth devices caused by supplying trained model with the supply vector $\boldsymbol{\theta}_{\mathbf{j}}^{\mathbf{wo}}$.

On the other hand, we assume that ith FL task requests a trained model with accuracy $\theta_{i,j}^{ag} \geq 1$ from the jth data owner. Then, we can define a $1 \times M$ demand vector of ith task as $\boldsymbol{\theta}_{\mathbf{i}}^{\mathbf{ag}}$. Also, we define an utility function of the ith task with demand vector $\boldsymbol{\theta}_{\mathbf{i}}^{\mathbf{ag}}$ as $U_i(\boldsymbol{\theta}_{\mathbf{i}}^{\mathbf{ag}})$. In the following, we will design the cost function $C_j(\cdot)$ and the utility function $U_i(\cdot)$ in detail.

Computation Cost

This cost is incurred by the local model training. For attaining a relative accuracy $\theta_{j,i}^{wo}$, the data owner has to perform a large number of training iterations on its local data. It can be denoted as:

$$E_{j,i}^{com} = I_l(\theta_{j,i}^{wo})(\rho\zeta D_{j,i} p_{j,i}^2), \tag{5.3}$$

where $D_{j,i}$ is the size of local data, $p_{j,i} \in [0, P^{max}]$ is the contributed computing resource (e.g., CPU frequency), ζ is the required CPU cycles for processing a single data point in local model training, and ρ is the CPU effective parameter.

Besides, the time duration of one local iteration can be formulated as $\frac{D_{j,i} \times \zeta}{p_{j,i}}$. Considering that the training process should be finished in one round of global iteration time, the minimum of the contributed computing resource can be described as:

$$p_{j,i}^{min} = \frac{D_{j,i} \times \zeta I_l(\theta_{j,i})}{t_g}, \tag{5.4}$$

where t_g is the one round of global iteration time.

Communication Cost

This cost is incurred when the data owner interacts with the FL platform for parameters updating. During one round of communication with the FL platform, we denote v as the size (in bits) of the model parameters. Note that v is the same for all participants in the same task. We formulate the transmission time as:

$$T_{j,i} = \frac{v_i}{R_j}, \tag{5.5}$$

where R_j is the transmission rate.

For simplicity, we consider that data owners are allocated with orthogonal sub-channels. According to Shannon's formula, the achievable transmission rate for data owner j can be expressed as:

$$R_j = B\log_2(1+\frac{tp_j|G_j|^2}{N_j}), \tag{5.6}$$

where tp_j is the transmission power, B is the allocated bandwidth, N_j is the Gaussian noise power and G_j is the average channel gain [12]. Then, we can formulate the transmission time $T_{j,i}$ for each round as:

$$T_{j,i} = \frac{v_i}{B\log_2(1+\frac{tp_j|G_j|^2}{N_j})}. \tag{5.7}$$

Therefore, the energy consumed during transmission using transmit power tp_j can be given by:

$$E_{j,i}^{trans} = \frac{v_i tp_j}{B\log_2(1+\frac{tp_j|G_j|^2}{N_j})}. \tag{5.8}$$

5.1.1.3 Cost Model

As discussed above, we discuss computation cost and communication cost. Then, the total cost of the jth data owner in one round global iteration with supply vector $\boldsymbol{\theta}_{\mathbf{j}}^{\mathbf{wo}}$ can be formulated as:

$$E_j(\boldsymbol{\theta}_{\mathbf{j}}^{\mathbf{wo}}) = \sum_{i=0}^{N}(\chi_j \cdot E_{j,i}^{trans} + (1-\chi_j)\cdot E_{j,i}^{com}), \tag{5.9}$$

where $0 \le \chi_j \le 1$ is the normalized weight to balance the computation and communication costs. Thus, the total cost can be formulated as:

$$C_j(\boldsymbol{\theta}_{\mathbf{j}}^{\mathbf{wo}}) = I_g(\epsilon, \boldsymbol{\theta}_{\mathbf{j}}^{\mathbf{wo}}) \times E_j(\boldsymbol{\theta}_{\mathbf{j}}^{\mathbf{wo}}), \tag{5.10}$$

where the $I_g(\epsilon, \boldsymbol{\theta}_{\mathbf{j}}^{\mathbf{wo}})$ is the global iteration times.

As discuss above, after the FL system initialized (i.e., the global accuracy is fixed), the global iteration times is only determined by the local accuracy vector. Thus, the $\alpha\log(\frac{1}{\epsilon})$ in Eq. (5.2) can be regarded as a constant. In this part, we normalize it to 1. In addition, for simplifying the derivation, we use the first-order Taylor's approximation to $f(\theta) = \frac{1}{1-\frac{1}{\theta}}$, where for large θ, the approximation results $f(\theta)|_{\theta\to\infty} = 1 + \frac{1}{\theta}$. Then, we can characterize the total cost as:

$$C_j(\boldsymbol{\theta}_{\mathbf{j}}^{\mathbf{wo}}) = \sum_{i=0}^{N}(1+\frac{1}{\theta_{j,i}^{wo}})(\chi_j \cdot E_{j,i}^{trans} + (1-\chi_j)\cdot E_{j,i}^{com}). \tag{5.11}$$

From Eq. (5.11), we notice that the computation and communication costs have contradictory relations with each other. An increase in computation cost decreases the communication cost and vice versa. The data owner can adjust its preference by the weight metric χ_j. The higher χ_j indicates that the data owner incline to improve the local accuracy, thus avoiding frequent interaction with the FL platform (e.g., under an insecure wireless network environment). Also, the smaller value of χ_j indicates that the data owner is more willing to increase the iteration number to alleviate local computation pressure (e.g., under resource-constraint IoT devices scenario).

5.1.1.4 Utility Model

In this part, we design a utility function $U(\cdot)$ to measure the earning with respect to the demand vector $\boldsymbol{\theta}_{\mathbf{i}}^{\mathbf{ag}}$. In the FL system, the FL tasks prefer to seek a higher quality model (i.e., higher local accuracy) from the data owners. Thus, the utility function $U(\cdot)$ should be positive, increasing and concave with the increase of $\boldsymbol{\theta}_{\mathbf{i}}^{\mathbf{ag}}$. The utility value should have an upper bound, and the increment ratio of $U(\cdot)$ should decrease with the increase of θ_i^{ag}. Therefore, we denote the average utility of ith FL task as:

$$U(\boldsymbol{\theta}_{\mathbf{i}}^{\mathbf{ag}}) = \beta(\frac{1}{M}(\sum_{j=1}^{M} a \cdot \theta_{i,j}^{ag} + b))^{0.5}, \tag{5.12}$$

where β is system parameters, a and b are dynamical parameters.

5.1.1.5 Monotonicity and Concave-Convex Analysis

We will discuss the monotonicity and concave-convex of cost function and utility function in this part. For cost function, we take the first and second-order derivatives with respect to θ, which can be formulated as:

$$\begin{aligned} \frac{\mathrm{d}C}{\mathrm{d}\theta} &= \gamma\rho\zeta Dp^2 - \frac{\chi_j \cdot E_{j,i}^{trans}}{\theta^2}, \\ \frac{\mathrm{d}^2C}{\mathrm{d}\theta^2} &= 2 \times \frac{\chi_j \cdot E_{j,i}^{trans}}{\theta^3} > 0. \end{aligned} \tag{5.13}$$

We notice that while the cost value firstly decrease and then increase, $C_j(\theta)$ is still an increasing, positive and strictly convex function of θ when $\theta > 1$, due to $\chi_j \cdot E_{j,i}^{trans} \ll \gamma\rho\zeta Dp^2$.

Also, for utility function, we take the first and second-order derivatives with respect to θ, which can be formulated as:

Table 5.1 Notations

Parameter	Definition
I_g	The global iterations
I_l	The local iterations
N	The number of FL task
M	The number of data owners
ϵ	The global accuracy
$\theta^{wo}_{j,i}$	The supply model accuracy of the jth data owner for the ith task
$\theta^{ag}_{i,j}$	The demand model accuracy of ith task from the jth data owner
$E^{com}_{j,i}$	The computation cost in one round iteration
$E^{trans}_{j,i}$	The communication cost in one round iteration
$U_i(\cdot)$	The utility function
$C_j(\cdot)$	The total cost function
$P_i(\cdot)$	The payment rule
$E_j(\cdot)$	The pricing rule

$$\begin{aligned}\frac{\mathrm{d}U}{\mathrm{d}\theta} &= 0.5\beta \cdot a(a\theta + b)^{-0.5} > 0,\\ \frac{\mathrm{d}^2U}{\mathrm{d}\theta^2} &= -0.25\beta \cdot a^2(a\theta + b)^{-1.5} < 0.\end{aligned} \tag{5.14}$$

Obviously, the utility function $U(\cdot)$ is a positive, increasing and strictly concave function of demand $\boldsymbol{\theta}_{\mathbf{i}}^{\mathbf{ag}}$. As shown in Table 5.1, we list the notations of this part.

5.1.1.6 Social Welfare Maximization Problem

As discussed above, the objectives of FL tasks and data owners are conflicting, where FL tasks try to maximize their utility whereas data owners try to minimize their cost. Specifically, FL tasks are willing to obtain the local model with higher accuracy to improve utility. Meanwhile, data owners tend to reduce the supply accuracy to avoid high energy costs. Therefore, there is a need for a centralized market broker to determine the accuracy supply vector and accuracy demand vector [13]. From a social perspective, the optimal solution is when trading maximizes social welfare. Mathematically speaking, the broker can find the optimal $\boldsymbol{\theta}_{\mathbf{i}}^{\mathbf{ag}}$ and $\boldsymbol{\theta}_{\mathbf{j}}^{\mathbf{wo}}$ by solving a social welfare optimization problem ($Object1$):

$$\max_{\boldsymbol{\theta}_{\mathbf{i}}^{\mathbf{ag}},\boldsymbol{\theta}_{\mathbf{j}}^{\mathbf{wo}}} \{\sum_{i=1}^{N} U_i(\boldsymbol{\theta}_{\mathbf{i}}^{\mathbf{ag}}) - \sum_{j=1}^{M} C_j(\boldsymbol{\theta}_{\mathbf{j}}^{\mathbf{wo}})\}, \tag{5.15}$$

subject to:

$$\sum_{j=1}^{M} \theta_{j,i}^{ag} = \boldsymbol{\theta}_{\mathbf{i}}^{\mathbf{ag}}\mathbf{1} \leq \theta_{i}^{max}; \qquad i \in \{1, 2, \ldots, N\}, \tag{5.16a}$$

$$\sum_{i=1}^{N} p_{j,i}^{min} = \mathbf{p}_{\mathbf{j}}\mathbf{1} \leq p_{j}^{max}; \qquad j \in \{1, 2, \ldots, M\}, \tag{5.16b}$$

$$\theta_{i,j}^{ag} = \theta_{j,i}^{wo}; \qquad i \in \{1, 2, \ldots, N\}, \; j \in \{1, 2, \ldots M\}. \tag{5.16c}$$

Constraints (5.16a) indicate that the total demand of the task cannot exceed the maximum demand. Constraints (5.16b) indicate that the contributed computing resource cannot exceed the maximum resource capacity. And constraints (5.16c) indicate that the supply accuracy should equal to the demand accuracy when the deal is completed.

As discussed above, the Eq. (5.4) indicate that the $p_{j,i}^{min}$ has a positive relation with $\theta_{j,i}^{wo}$. Meanwhile, considering the difference of data size, a round of global iterations time between different tasks are small. Thus, the Constraints (5.16b) can be approximately reformulated as:

$$\sum_{i=1}^{N} \theta_{j,i}^{wo} = \boldsymbol{\theta}_{\mathbf{j}}^{\mathbf{wo}}\mathbf{1} \leq \theta_{j}^{max}; \; j \in \{1, 2, \ldots, M\}, \tag{5.17}$$

where $\theta_{j}^{max} = \frac{\overline{t_g}}{\zeta \gamma D_j}$.

Based on the concave-convex analysis in Sect. 5.1.1.5, the objective function is strictly concave with respect to $\boldsymbol{\theta}_{\mathbf{i}}^{\mathbf{ag}}$ and $\boldsymbol{\theta}_{\mathbf{j}}^{\mathbf{wo}}$, and the constraints is compact and convex. Therefore, the $Objection1$ possesses a unique optimal solution which can be described by applying Karush-Kuhn-Tucker (KKT) conditions. Specifically, we relax constraints and yield the following Lagrangian L_1:

$$\begin{aligned} L_1(\boldsymbol{\theta}_{\mathbf{i}}^{\mathbf{ag}}, \boldsymbol{\theta}_{\mathbf{j}}^{\mathbf{wo}}, \beta_i, \sigma_j, \psi_{i,j}) = &\sum_{i=1}^{N} U_i(\boldsymbol{\theta}_{\mathbf{i}}^{\mathbf{ag}}) - \sum_{j=1}^{M} C_j(\boldsymbol{\theta}_{\mathbf{j}}^{\mathbf{wo}}) - \\ &\sum_{i=1}^{N} \beta_i(\boldsymbol{\theta}_{\mathbf{i}}^{\mathbf{ag}}\mathbf{1} - \theta_i^{max}) - \sum_{j=1}^{M} \sigma_j(\boldsymbol{\theta}_{\mathbf{j}}^{\mathbf{wo}}\mathbf{1} - \theta_j^{max}) + \\ &\sum_{i=1}^{N}\sum_{j=1}^{M} \psi_{i,j}(\theta_{j,i}^{wo} - \theta_{i,j}^{ag}), \end{aligned} \tag{5.18}$$

where $\boldsymbol{\beta} \triangleq (\beta_1, \beta_2, \ldots, \beta_N)$, $\boldsymbol{\sigma} \triangleq (\sigma_1, \sigma_2, \ldots, \sigma_M)$, and $\boldsymbol{\psi} \triangleq (\psi_{1,j} \ldots, \psi_{i,j}, \ldots, \psi_{N,M})$ are Lagrange multipliers corresponding to the inequality and equality constraints respectively. Then, the KKT conditions that yield optimal primal variables and optimal dual variables for $Objection1$ can be ar given by:

$$\nabla_{\boldsymbol{\theta}_{\mathbf{i}}^{\mathbf{ag}}} U_i - \beta_i \mathbf{1} - \boldsymbol{\psi}_{\mathbf{i}} = 0, \tag{5.19a}$$

$$\nabla_{\boldsymbol{\theta}_{\mathbf{j}}^{\mathbf{wo}}} C_j + \sigma_j \mathbf{1} - \boldsymbol{\psi}_{\mathbf{j}} = 0, \tag{5.19b}$$

$$\boldsymbol{\theta}_{\mathbf{i}}^{\mathbf{ag}} \mathbf{1} - \theta_i^{max} = 0, \tag{5.19c}$$

$$\boldsymbol{\theta}_{\mathbf{j}}^{\mathbf{wo}} \mathbf{1} - \theta_j^{max} = 0, \tag{5.19d}$$

$$\theta_{i,j}^{ag} - \theta_{j,i}^{wo} = 0. \tag{5.19e}$$

Obtaining the optimal solution of $Objection1$ requires the complete information of all data owners' cost functions and FL tasks' utility functions. However, since the non-cooperation relationship among FL tasks and data owners, the local information cannot be aware to the broker. The broker needs to design an incentive mechanism to encourage the market participants to report their hidden information. Such a mechanism should processes the following four economic properties, including economic efficiency (i.e., maximizes the social welfare), individually rationality (i.e., participants never get worse or get negative payoffs by participating than those obtained by not participating), incentive compatibility (i.e., the participants are willing to reveal their truthful requirements), and budget balance (i.e., the broker never need to pay additional money for conducting the mechanism). However, the current double auction (DA) mechanism can not satisfy all these four properties at the same time [14]. To tackle this challenge, we design an iterative double auction (IDA) mechanism. We will detail this in the following section.

5.1.2 *Iterative Double Auction-Based AI Training Service Market*

5.1.2.1 Two-Stage IDA Mechanism

In this section, we propose an iterative double auction (IDA) algorithm to elicit hidden information of FL tasks and data owners to the broker. In the IDA mechanism, the broker facilitates multiple buyers (i.e., FL tasks) and multiple sellers (i.e., data owners) to iteratively interact and adjust their bidding and pricing strategy [15]. As shown in Fig. 5.2, there are two stages in the IDA algorithm. In the first stage, each FL task will submit a $1 \times N$ bidding vector $\boldsymbol{\omega}$ to the broker. We denote the $\omega_{i,j}$ as the ith FL task's bidding for the jth data owner. Meanwhile, each data owner will submit a $1 \times M$ pricing vector $\mathbf{g}$ to the broker. Similarly, $g_{j,i}$ represents the jth data owner's pricing to the ith FL task. These prices will indicate the buyer's and seller's preferences respectively, i.e., model accuracy demand and supply.

In the second stage, based on the submitted price vectors, the broker will determine the supply model accuracy level by solving the following optimization problem ($Objective2$).

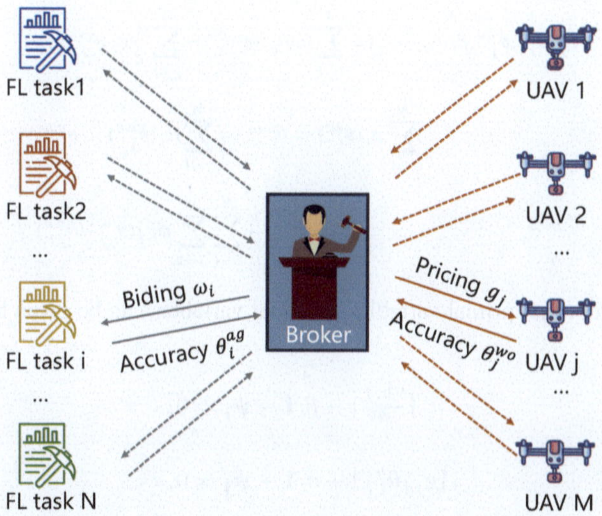

Fig. 5.2 Double auction market

$$\max_{\boldsymbol{\theta}_{\mathbf{i}}^{\mathbf{ag}},\boldsymbol{\theta}_{\mathbf{j}}^{\mathbf{wo}}} \{\sum_{i=1}^{N}\sum_{j=1}^{M} \omega_{i,j} \log \theta_{i,j}^{ag} - \frac{1}{2} g_{j,i} \theta_{j,i}^{wo2}\}, \tag{5.20}$$

subject to:

$$\sum_{j=1}^{M} \theta_{j,i}^{ag} = \boldsymbol{\theta}_{\mathbf{i}}^{\mathbf{ag}} \mathbf{1} \le \theta_i^{max}; \qquad i \in \{1, 2, \ldots, N\}, \tag{5.21a}$$

$$\sum_{i=1}^{N} \theta_{j,i}^{wo} = \boldsymbol{\theta}_{\mathbf{j}}^{\mathbf{wo}} \mathbf{1} \le \theta_j^{max}; \qquad j \in \{1, 2, \ldots, M\},, \tag{5.21b}$$

$$\theta_{i,j}^{ag} = \theta_{j,i}^{wo}; \qquad i \in \{1, 2, \ldots, N\}, \quad j \in \{1, 2, \ldots M\}. \tag{5.21c}$$

where the objective function design is motivated by the auction rule in [13] and [7]. In function, the $\log \theta_{i,j}^{ag}$ and $\theta_{j,i}^{wo2}$ component are used to capture the concave utility function of FL tasks and convex cost function of data owners, respectively. Then, we can define the Lagrangian function of the $Objective2$ by relaxing constraints (5.21) as:

$$L_2(\boldsymbol{\theta}_{\mathbf{i}}^{\mathbf{ag}},\boldsymbol{\theta}_{\mathbf{j}}^{\mathbf{wo}},\beta_i,\sigma_j,\psi_{i,j})=\sum_{i=1}^{N}(\omega_{i,j}\log\theta_{i,j}^{ag}-\sum_{j=1}^{M}\frac{1}{2}g_{j,i}\theta_{j,i}^{wo2})-$$
$$\sum_{i=1}^{N}\beta_i(\boldsymbol{\theta}_{\mathbf{i}}^{\mathbf{ag}}\mathbf{1}-\theta_i^{max})-\sum_{j=1}^{M}\sigma_j(\boldsymbol{\theta}_{\mathbf{j}}^{\mathbf{wo}}\mathbf{1}-\theta_j^{max}) \tag{5.22}$$
$$+\sum_{i=1}^{N}\sum_{j=1}^{M}\psi_{i,j}(\theta_{i,j}^{ag}-\theta_{j,i}^{wo}).$$

Then, the optimal primal variables and dual variables can be given by following functions:

$$[\frac{\omega_{i,j}}{\theta_{i,j}^{ag}}]-\beta_i\mathbf{1}-\boldsymbol{\psi}_{\mathbf{i}}=0, \tag{5.23a}$$
$$[g_{j,i}\theta_{j,i}^{wo}]+\sigma_i\mathbf{1}-\boldsymbol{\psi}_{\mathbf{j}}=0, \tag{5.23b}$$
$$\boldsymbol{\theta}_{\mathbf{i}}^{\mathbf{ag}}\mathbf{1}-\theta_i^{max}=0, \tag{5.23c}$$
$$\boldsymbol{\theta}_{\mathbf{j}}^{\mathbf{wo}}\mathbf{1}-\theta_j^{max}=0, \tag{5.23d}$$
$$\theta_{i,j}^{ag}-\theta_{j,i}^{wo}=0. \tag{5.23e}$$

And the optimal accuracy supply can be described as:

$$\theta_{i,j}^{ag}=\frac{\omega_{i,j}}{\psi_i+\beta_i}, \tag{5.24}$$

$$\theta_{j,i}^{wo}=\frac{\psi_j-\sigma_j}{g_{j,i}}. \tag{5.25}$$

Comparing two KKT conditions of $Objective1$ and $Objective2$, we notice that Eqs. (5.19c), (5.19d), and (5.19e) are identical to Eqs. (5.23c), (5.23d), and (5.23e), but Eqs. (5.19a) and (5.19b) are different from Eqs. (5.23a) and (5.23b). Therefore, if the FL task and data owner submit their prices as:

$$\nabla_{\boldsymbol{\theta}_{\mathbf{i}}^{\mathbf{ag}}}U_i(\boldsymbol{\theta}_{\mathbf{i}}^{\mathbf{ag}})=[\frac{\omega_{i,j}}{\theta_{i,j}^{ag}}]\Rightarrow\omega_{i,j},=\theta_{i,j}^{ag}\cdot\frac{\partial}{\partial\theta_{i,j}^{ag}}U_i(\boldsymbol{\theta}_{\mathbf{i}}^{\mathbf{ag}}), \tag{5.26}$$

$$\nabla_{\boldsymbol{\theta}_{\mathbf{j}}^{\mathbf{wo}}}U_i(\boldsymbol{\theta}_{\mathbf{j}}^{\mathbf{wo}})=[g_{j,i}\theta_{j,i}^{wo}]\Rightarrow g_{j,i}=\frac{1}{\theta_{j,i}^{wo}}\cdot\frac{\partial}{\partial\theta_{j,i}^{wo}}C_j(\boldsymbol{\theta}_{\mathbf{j}}^{\mathbf{wo}}), \tag{5.27}$$

the optimal solutions of $Objective2$ is equal to the solution of the $Objective1$. In other words, if market participants submit their price according to Eqs. (5.26) and (5.27) respectively. Then the IDA algorithm yields a solution identical to the optimal solution of the social welfare maximization problem, and therefore elicit the hidden information from the market.

5.1.2.2 Settlement and Earning Rules Design

Since participants are selfish, they only concern about maximizing their profit, i.e., solving its local utility optimization problem. Therefore, we design a settlement rule for FL tasks and an earning rule for data owners to revise the participants' local utility problem, therefore inducing participants to submit their price as (5.26) and (5.27).

For FL task, we denote $P_i(\boldsymbol{\omega}_\mathbf{i})$ as the settlement rule of ith FL task given by the broker when it received the bidding vector $\boldsymbol{\omega}_\mathbf{i}$. After receiving the settlement rule, the local optimization problem of ith FL platform can be revised as:

$$\max_{\boldsymbol{\omega}_\mathbf{i}} \{U_i(\boldsymbol{\theta}_\mathbf{i}^\mathbf{ag}) - P_i(\boldsymbol{\omega}_\mathbf{i})\}. \tag{5.28}$$

By solving the above payoff maximization problem individual, each FL task can solve its optimal bidding vector. The optimal solution satisfies the following optimality conditions:

$$\frac{\partial P_i(\boldsymbol{\omega}_\mathbf{i})}{\partial \boldsymbol{\omega}_\mathbf{i}} = \frac{\partial U_i(\boldsymbol{\theta}_\mathbf{i}^\mathbf{ag})}{\partial \boldsymbol{\theta}_\mathbf{i}^\mathbf{ag}} \cdot \frac{\partial \boldsymbol{\theta}_\mathbf{i}^\mathbf{ag}}{\partial \boldsymbol{\omega}_\mathbf{i}} = \frac{1}{\psi_i + \beta_i} \cdot \frac{\partial U_i(\boldsymbol{\theta}_\mathbf{i}^\mathbf{ag})}{\partial \boldsymbol{\theta}_\mathbf{i}^\mathbf{ag}}. \tag{5.29}$$

To induce each FL task submits their bidding according to (5.26), we have

$$\frac{\partial P_i(\boldsymbol{\omega}_\mathbf{i})}{\partial \boldsymbol{\omega}_\mathbf{i}} = \frac{1}{\psi_i + \beta_i} \cdot \frac{\omega_{i,j}}{\theta_{i,j}^{ag}} = 1. \tag{5.30}$$

Then, the settlement rule can be described as:

$$P_i(\boldsymbol{\omega}_\mathbf{i}) = \sum_{i=0}^{N} \boldsymbol{\omega}_\mathbf{i}. \tag{5.31}$$

Similarly, for data owner, $E_j(\mathbf{g_j})$ denote the earning rule of the jth data owner given by the broker when it received the pricing vector θ_j^{wo}. Then, the local optimization problem of data owner j can be formulated as:

$$\max_{\mathbf{g_j}} \{E_j(\mathbf{g_j}) - C_j(\boldsymbol{\theta}_\mathbf{j}^\mathbf{wo})\}, \qquad s.t. g_{j,i} \geq 0, \forall i \in N. \tag{5.32}$$

Taking the first derivative of objective function (5.32) with respect to g_i, we can derive the optimality conditions as:

$$\frac{\partial E_j(\mathbf{g_j})}{\partial \mathbf{g_j}} = \frac{\partial C_j(\boldsymbol{\theta}_{\mathbf{j}}^{\mathbf{wo}})}{\partial \boldsymbol{\theta}_{\mathbf{j}}^{\mathbf{wo}}} \cdot \frac{\partial \boldsymbol{\theta}_{\mathbf{j}}^{\mathbf{wo}}}{\partial \mathbf{g_j}} = \frac{\psi_j - \sigma_j}{g_{j,i}^2} \cdot \frac{\partial C_j(\boldsymbol{\theta}_{\mathbf{j}}^{\mathbf{wo}})}{\partial \boldsymbol{\theta}_{\mathbf{j}}^{\mathbf{wo}}}. \tag{5.33}$$

To induce each data owner submits their pricing according to (5.27), we have

$$\frac{\partial E_j(\mathbf{g_j})}{\partial \mathbf{g_j}} = -\frac{\psi_j - \sigma_j}{g_{j,i}^2} \cdot \theta_{j,i}^{wo} g_{j,i} = -\frac{(\psi_j - \sigma_j)^2}{g_{j,i}^2}. \tag{5.34}$$

Then, the earning rule can be given by:

$$E_j(\mathbf{g_j}) = \frac{(\psi_j - \sigma_j)^2}{g_{j,i}}. \tag{5.35}$$

The proposed IDA algorithm is shown in Algorithm 11.

Algorithm 11 The IDA algorithm for federated learning task

Initialize bidding vectors $\boldsymbol{\omega}$ and pricing vector $\mathbf{g}$
Initialize index $conv = 0$, $t = 0$, Lagrange multipliers $\boldsymbol{\beta}$, $\boldsymbol{\sigma}$, and $\boldsymbol{\psi}$
while $conv = 0$ **do**
  Broker compute $\boldsymbol{\theta}_{\mathbf{i}}^{\mathbf{ag}}$, $\boldsymbol{\theta}_{\mathbf{j}}^{\mathbf{wo}}$ and Lagrange multiplies $\beta_i, \sigma_j, \psi_{i,j}$ by solving $Objective2$
  Broker announces $\boldsymbol{\theta}_{\mathbf{i}}^{\mathbf{ag}}$ and $\boldsymbol{\theta}_{\mathbf{j}}^{\mathbf{wo}}$ to market
  Compute pricing rules $P_i(\boldsymbol{\omega}_\mathbf{i})$ and $E_j(\mathbf{g_j})$ according to Eqs. (5.31) and (5.35)
  for all FL platforms **do**
    Derives optimal bidding vector $\boldsymbol{\omega}_\mathbf{i}$ by solving:
    $\max_{\boldsymbol{\omega}_\mathbf{i}} \{U_i(\boldsymbol{\theta}_{\mathbf{i}}^{\mathbf{ag}}) - P_i(\boldsymbol{\omega}_\mathbf{i})\}$
    Submits the $\boldsymbol{\omega}_\mathbf{i}$ to the broker
  end for
  for all data owners **do**
    Derives optimal pricing vector $\mathbf{g_j}$ by solving:
    $\max_{\mathbf{g_j}} \{E_j(\mathbf{g_j}) - C_j(\boldsymbol{\theta}_{\mathbf{j}}^{\mathbf{wo}})\}$
    Submits the $\mathbf{g_j}$ to the broker
  end for
  Broker announce bid vector $\boldsymbol{\omega}_\mathbf{i}$ and $\mathbf{g_j}$, $t \doteq t + 1$
  if $|\omega_{i,j}^{(t)} - \omega_{i,j}^{(t-1)}| < \varepsilon$ and $|g_{j,i}^{(t)} - g_{j,i}^{(t-1)}| < \varepsilon$ **then**
    $conv = 1$
  end if
end while

5.1.2.3 Economic Properties Analysis of IDA Algorithm

In this part, we will analyze the four economic properties of the IDA algorithm separately.

Economic Efficiency

As discussed above, in the IDA algorithm, the settlement and earning rules design will induce participants to submit their prices as the (5.26) and (5.27). These prices will enable the broker to yield the $Objective1$'s solution. Therefore, the system can converge to the optimal solution of the social welfare maximization problem.

Individually Rationality

Since U_i is a strictly concave function of $\boldsymbol{\theta}_{\mathbf{i}}^{\mathbf{ag}}$ and $U_i(0) = 0$, then we have:

$$U_i(\theta_i^*) > U_i(0) = 0 + \theta_i^* \frac{\partial U_i(\theta_i^*)}{\partial \theta_i^{ag}} = \theta_i^* \frac{\partial U_i(\theta_i^*)}{\partial \theta_i^{ag}}, \tag{5.36}$$

where θ_i^* is the optimal solution point. According to the (5.29) and (5.30), we have:

$$\theta_i^* \frac{\partial U_i(\theta_i^*)}{\partial \theta_i^{ag}} = \theta_i^* \cdot \frac{\omega_i}{\theta_i^*} = \omega_i^* = P_i(\omega_i^*). \tag{5.37}$$

Therefore, it can implies that:

$$U_i(\theta_i^*) > P_i(\omega_i^*). \tag{5.38}$$

Similarly, for data owners, we can easily proof that:

$$E_j(g_j^*) > C_j(\theta_j^*). \tag{5.39}$$

So in conclusion, each FL task and data owner will always obtain a positive utility (i.e., satisfying Individually Rationality conditions).

Incentive Compatibility

According to the settlement and earning rules design, the participants will submit their bids as (5.26) and (5.27). Through the iteratively updated optimal bids, the broker can yield a solution identical to the $Objective1$ by solving the $Objective2$, i.e., revealing data owners' hidden cost and tasks' hidden utility in $Objective1$.

Therefore, we can conclude that the IDA processes the characteristics of Incentive Compatibility.

Budget Balance

When optimal bids from both sides of participants in double-auction are submitted (i.e., $\theta_{i,j}^{ag} = \theta_{j,i}^{wo}$), the budget of the broker can be calculated by:

$$\begin{aligned} P_i(\omega_{i,j}^*) - E_j(g_{j,i}^*) &= \omega_{i,j}^* - \frac{(\psi_j - \sigma_j)^2}{g_{j,i}^*} \\ &= (\psi_i + \beta_i)\theta_{i,j}^{ag} - (\psi_{j,i} - \sigma_j)\theta_{j,i}^{wo} \geq 0, \end{aligned} \tag{5.40}$$

when implementing earning and settlement rules formulated in (5.35) and (5.31). Therefore, the broker will always get a nonnegative budget, i.e., satisfying the budget balance.

According to the above analysis, we can conclude that the FL system with the IDA algorithm can satisfy the four economic properties at the same time ideally. The maximum social welfare will always be achieved in the FL market. And both the FL tasks and data owners are encouraged to participate in the trading process.

5.1.3 *Reinforcement-Learning-Based Double Auction Algorithm*

As discussed above, we design an iterative double auction algorithm. While the system can quickly converge to social welfare maximization, it relies on some strong assumptions (e.g., monotonicity and concave-convex requirement). Besides, in IDA, a centralized controller have to be implemented to guide all participants' bid strategy updating, and therefore reducing the system robustness. In this part, inspired by the recent success of distributed reinforcement learning, we design a more general method, termed the EWA-DA algorithm [16]. The participants can learn to improve their strategies directly by participating in the auction continuously. We will detail the algorithm design in the following.

5.1.3.1 The Multi-Agent System

The market participants constitute a multi-agent system, which can be formulated as a decentralized partially observable Markov decision process (Dec-POMDP) [17]. Formally, the Dec-POMDP can be molded as a 4-tuple $< A, O, \Pi, R >$, where A is the action space, O is the observations space, Π is the policy space, and R is the immediate rewards [18]. In this part, our system exists two kinds of agents (i.e., the FL task and the data owner). We denote $\Pi^{wo} = \{\pi_1^{wo}, \pi_2^{wo}, \ldots, \pi_M^{wo}\}$ as the set of the strategies of data owners, and $\Pi^{ag} = \{\pi_1^{ag}, \pi_2^{ag}, \ldots, \pi_N^{ag}\}$ as the set of the strategies of FL tasks. The strategy space of the game can be represented by $\Pi = \Pi^{wo} \times \Pi^{ag}$. In addition, we use the $\pi_{-j}^{wo} = (\pi_1^{wo}, \ldots \pi_{j-1}^{wo}, \pi_{j+1}^{wo}, \ldots, \pi_M^{wo}, \pi_1^{ag}, \ldots, \pi_N^{ag})$ to denote a strategy combination of all agents except π_j^{wo}.

At each double auction round, data owners and FL tasks take action simultaneously according to current policy. For simplicity, we assume that one FL task sets the same demand accuracy and biding price for all data owners. Thus, we define $a_i^{ag} = (\theta_i^{ag}, p_i^{ag})$ as the action space of ith FL task. Similarly, for data owners, we define $a_j^{wo} = (\theta_j^{wo}, p_j^{wo})$ as its action space. Then, the broker determines if jth data owner accepts ith FL task accuracy demand and bidding price (i.e., $\theta_j^{wo} \geq \theta_i^{ag}$ and $p_j^{wo} \leq p_i^{ag}$). In this part, we use $\kappa_{i,j}$ to denote the accuracy that the ith FL task offer to the jth data owner, and the $\rho_{i,j}$ to denote the price that the ith FL task pays to the jth data owner. If there is no deal between the ith task and the jth data owner, $\kappa_{i,j}$ and $\rho_{i,j}$ both equals to 0.

By the end of the round, an immediate reward R will be generated from the underlying environment. Note that the reward not only relies on the local policy π_i but also other agents' policy π_{-i}. In our scenario, we define the reward function for the data owner as:

$$r_j^{wo}(\pi_j^{wo}, \pi_{-j}^{wo}) = \sum_{k=0}^{N} \rho_{k,j} - C_j(\kappa_{\mathbf{j}}^{\mathbf{wo}}). \tag{5.41}$$

Similarly, the reward function of FL task is:

$$r_i^{ag}(\pi_i^{ag}, \pi_{-i}^{ag}) = U(\kappa_{\mathbf{i}}^{\mathbf{ag}}) - \sum_{k=0}^{M} \rho_{i,k}. \tag{5.42}$$

5.1.3.2 Experience-Weighted Attraction Learning Algorithm

In this section, we apply the Experience-Weighted Attraction Learning (EWA) algorithm to double auction markets. The EWA combines reinforcement learning and belief learning model. Strategies use attractions value to reflect choice predispositions and players update the attractions based on payoff experience during the game [19]. We denote the $A_{\pi_a}^{ag_i}(t)$ and $A_{\pi_a}^{wo_j}(t)$ as the attraction values of the FL task i's and data owner j's current strategy π_i^{ag} and π_j^{wo} (i.e., choice action a_i^{ag}, a_j^{wo}) in round t. The updating rules can be formulated as:

$$A_{\pi_a}^{ag_i}(t) = \frac{\phi N(t-1)A_{\pi_a}^{ag_i}(t-1) + \Delta \times r_i^{ag}(\pi_{ag}^{i}, \pi_{ag}^{-i})}{N(t)}$$
$$\Delta = \delta + (1-\delta)I(\pi_a^{ag_i}, \pi^{ag_i}(t)). \tag{5.43}$$

Similarly,

$$A_{\pi_a}^{wo_j}(t) = \frac{\phi N(t-1)A_{\pi_a}^{wo_j}(t-1) + \Delta \times r_j^{wo}(\pi_{wo}^{j}, \pi_{wo}^{-j})}{N(t)}$$
$$\Delta = \delta + (1-\delta)I(\pi_a^{wo_j}, \pi^{wo_j}(t)). \tag{5.44}$$

The updating rule for A is consists of 3 part.

1. **Keeping old attraction:** The first entry $\phi N(t-1)A_{\pi_a}^{ag_i}(t-1)$ in the numerator is used to keep old attraction, where ϕ is the decay rate. This demonstrates that the previous attraction will decay with the learning environment changing constantly.
2. **Adding value of the current round outcome:** The second entry $(\delta + (1-\delta)I(\pi_a^{ag_i}, \pi^{ag_i}(t))) \times r_i^{ag}(\pi_{ag}^{i}, \pi_{ag}^{-i})$ is the attraction value of the current round. I is the indicator function, which equals to 1 if the strategy was used this round, and equals to 0 if the strategy was not used. And δ is the imagining factor, which determines weight added for used and unused strategies, and therefore encourages the agent to explore unknown strategy [20]. For example, when $\delta = 1$, the used and unused strategies are treated as the same. And when $\delta = 0$, only used strategy is updated (as in reinforcement learning). The agent can adjust δ to balance the exploration and exploitation.
3. **Normalization with experience:** $N(t)$ is the number of "experience-equivalent" of past experiences, which is roughly the same as previous rounds of experience. The updating rule is:

$$N(t) = \rho N(t-1) + 1, t \geq 1, \tag{5.45}$$

where previous experience is depreciated with a discount factor ρ.

Then, attractions are mapped into a behavior choice probabilities function according to logit form [21]. For ith FL task:

$$P_{\pi_a}^{ag_i}(t+1) = \frac{e^{\lambda A_{\pi_a}^{ag_i}(t)}}{\sum\limits_{\pi \in \pi_A} e^{\lambda A_{\pi}^{ag_i}(t)}}. \tag{5.46}$$

For jth worker:

$$P_{\pi_a}^{wo_j}(t+1) = \frac{e^{\lambda A_{\pi_a}^{wo_j}(t)}}{\sum\limits_{\pi \in \pi_A} e^{\lambda A_{\pi}^{wo_j}(t)}}. \tag{5.47}$$

where $P_{\pi_a}^{ag_i}(t+1)$ is the probability of agent i to play strategy π_a in round $t+1$ and λ is a payoff sensitivity parameter. The proposed EWA algorithm is shown in Algorithm. 12.

Algorithm 12 Experience-weighted attraction learning

Initialize marketing environment
Initialize system parameters and learning parameters
for $t = 1$ to $max - episode - length$ do **do**
  All agents choice their strategy according to Eqs. (5.46) and (5.47)
  Broker broker determines if worker j accepts agent i's accuracy requirement and bidding price
  for $j \in M$ and $i \in N$ **do**
    if $\theta_j^{wo} \geq \theta_i^{ag}$ and $p_j^{wo} \leq p_i^{ag}$ **then**
      $\kappa_{i,j} = \theta_i^{ag}$ and $\rho_{i,j} = p_j^{wo}$
    else
      $\kappa_{i,j} = 0$ and $\rho_{i,j} = 0$
    end if
  end for
  Update $A(t)$ in all agents according to Eqs. (5.43) and (5.44)
end for

5.1.4 *Simulation Results and Discussions*

In this section, we present the experiment results to evaluate the validity and effectiveness of the proposed algorithm. Also, we demonstrate the convergence and economic properties of the IDA-based FL mechanism.

5.1.4.1 Simulation Settings

We simulate an FL market with multiple FL tasks and multiple data owners. For each data owner, we assume that the computing resource f are uniformly distributed on 1–2 Ghz, training size D are uniformly distributed on 5–10 MB, ζ are uniformly distributed on 10–30 cycles/bit, and α is set as $1 * 10^{-28}$. Besides, for the wireless communication model, we assume that the channel gains G follow the exponential distribution with the mean $-40\,\mathrm{dB}(1/\mathrm{d})^4$, where d is the distance between these FL platforms and data owners. We assume that the distance d is uniformly distributed between 2 to 100 m. In addition, we set the bandwidth as $B = 1$ Mhz, the Gaussian noise power as $N = 10^{-20}$ W, the size of local parameters as $v = 5000$ bit, and the transmission power P are limited from 0.1 to 2 W. The detail parameters setting can be found in Table 5.2.

Table 5.2 Detailed system parameters

Parameter	Value
System parameter β	4
Dynamical parameter a	1
Dynamical parameter b	0
Constant of global iterations α	100
Global FL model accuracy ϵ	100
Constant of local iterations γ	10
CPU parameter of computing ρ	2×10^{-28} J/W
Number of CPU cycles ζ	25 cycles/bit
Local data samples D_n	10 MB
Computation resource p	1 GB
Size of local parameters υ_i	5000 bit
Transmission rate R	4000 KB
Total bandwidth B	1 Mhz
Transmission power P_j	1 W
Average channel gain G_j	$-40\,\mathrm{dB}(1/\mathrm{d})^4$
Distance between worker and agent d	48 m
Gaussian noise power N_j	10^{-20} W

5.1.4.2 Convergence Analysis

Firstly, we evaluate the convergence and economic properties of the designed IDA-based FL learning mechanism. We simulate a FL market with $N = 3$ and $M = 3$. The three FL tasks perform different utility preferences function, i.e., adopting different β's value. We set the β_1 as 2.1172, the β_2 as 1.4659, and β_3 as 1.0001. Then, we implement the IDA algorithm to the FL market to search for social welfare maximization solution.

As shown in Figs. 5.3 and 5.4, we plot the price strategy and accuracy strategy of both FL task and data owner for each iteration of the IDA algorithm. We notice that both the FL task and data owner can quickly converge to the social welfare-maximizing point after 10 times iterations. Also, comparing the Figs. 5.3 and 5.4, we notice that the supply vector $\boldsymbol{\theta}_{\mathbf{j}}^{\mathbf{wo}}$ of each data owner is always equal to the demand vector $\boldsymbol{\theta}_{\mathbf{i}}^{\mathbf{ag}}$. This means that the whole market converges to equilibrium.

Besides, in Fig. 5.3, we find that $FL\ task1$'s bidding and accuracy demand are always larger than the other two tasks. This is because that the utility preferences parameter β_1 is larger than β_2 and β_3. The higher return prompts the participants to bid for a higher quality commodity (i.e., higher accuracy level).

As shown in Fig. 5.4, we notice that the pricing from the data owner decreases with the increase of the iteration times. In contrast, the supply vectors $\boldsymbol{\theta}_{\mathbf{j}}^{\mathbf{wo}}$ of data owners present a different tendency to $\mathbf{g_j}$. This is caused by the settlement rule design, where the settlement price is inversely proportional to the pricing

Fig. 5.3 FL task with bidding ω and θ_i^{ag}

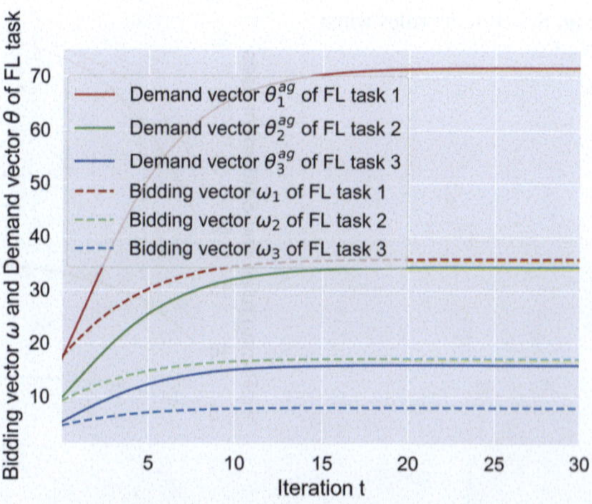

Fig. 5.4 Data owner with pricing g and supply vector θ_j^{wo}

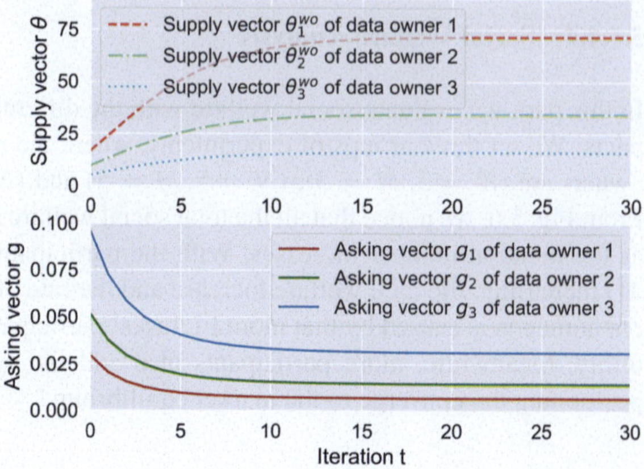

g. Therefore, the actual payment from the broker increase along with the supply accuracy level increasing.

5.1.4.3 Price Rules Analysis

Then, we analyze the price strategy as the iteration number increases. As shown in Fig. 5.5, we plot the earning price and settlement price of the FL task and data owner. We notice that $P_i(\boldsymbol{\omega}_\mathbf{i})$ and $E_j(\mathbf{g}_\mathbf{j})$ hold similar tendencies to $\boldsymbol{\theta}_\mathbf{i}^\mathbf{ag}$ and $\boldsymbol{\theta}_\mathbf{j}^\mathbf{wo}$, respectively. This is caused by settlement rules and earning rules design in Eqs. (5.31) and (5.35). Meanwhile, this phenomenon also demonstrates that the market holds the budget balance properties, i.e., the broker never needs to pay additional money for conducting the mechanism.

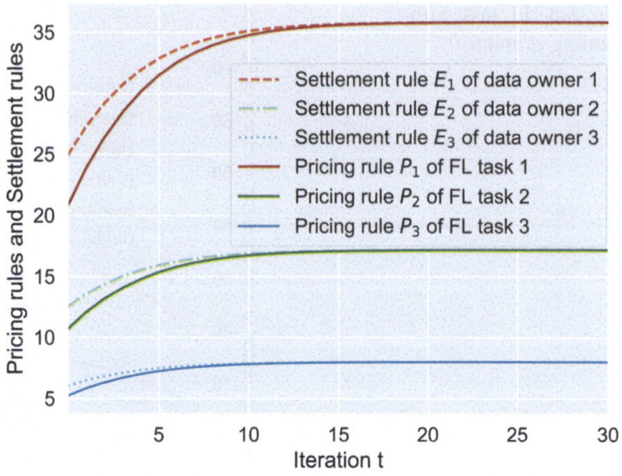

Fig. 5.5 Pricing rules when N=3

5.1.4.4 Social Welfare Analysis

In this part, we evaluate social welfare with the different number of market participants. We set three groups of experiments, where the number of FL tasks and data owners are ($N = 3, M = 3$), ($N = 5, M = 5$) and ($N = 7, M = 7$) respectively. From Fig. 5.6, we notice that all the total social welfare is gradually increasing along as the iteration number increases. With the participant (i.e., the number of N and M) increasing, the total welfare increase and the rate of convergence decrease. This phenomenon is caused by that more FL tasks' participation will bring more external utility. Meanwhile, more participants also add the system complexity, therefore postponing the converge to the market equilibrium.

5.1.4.5 Economic Properties Analysis

Then, we evaluate the economic properties of the designed IDA-based FL learning market. As shown in Table 5.3, we present the numerical results after the market reaching the equilibrium. We list the optimal demand vectors $\boldsymbol{\theta}_\mathbf{i}^\mathbf{ag}$, submitted bidding price vector $\boldsymbol{\omega}_\mathbf{i}$, settlement price $P_i(\boldsymbol{\omega}_\mathbf{i})$, utility value $U_i(\boldsymbol{\theta}_\mathbf{i}^\mathbf{ag})$ of each FL task, and optimal supply vectors $\boldsymbol{\theta}_\mathbf{j}^\mathbf{wo}$, submitted pricing vector $\mathbf{g_j}$, earning price $E_j(\mathbf{g_j})$ and cost $C_j(\boldsymbol{\theta}_\mathbf{j}^\mathbf{wo})$ of each data owner. Also, we list optimal values of Lagrange multiplies β_i, σ_j, and $\psi_{i,j}$ at the equilibrium point.

From the Table 5.3, we notice that the demand accuracy $\boldsymbol{\theta}_\mathbf{i}^\mathbf{ag}$ is equal to the supply accuracy $\boldsymbol{\theta}_\mathbf{j}^\mathbf{wo}$. This equilibrium indicates that the FL tasks and data owners can make a deal by accuracy allocation approaches (5.24) and (5.25) in the broker. Bedsides, the settlement price $P_i(\boldsymbol{\omega}_\mathbf{i})$ is equal to the earning price $E_j(\mathbf{g_j})$. This equilibrium validates the settlement rule and earning rule design and demonstrates the IDA-based market holds the budget balance property.

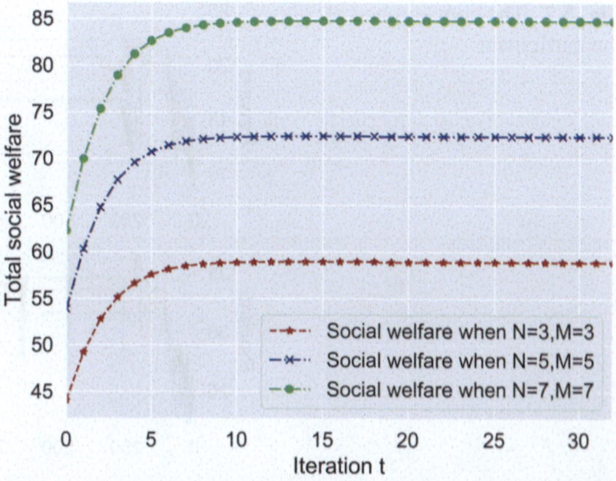

Fig. 5.6 Total social welfare when N=3,N=5 and N=7

Table 5.3 Simulation results

Definition	Parameter	Agent1	Agent2	Agent3
Model quality	$\boldsymbol{\theta}_{\mathbf{i}}^{\mathbf{ag}}$	71.7157	34.3890	16.0264
	$\boldsymbol{\theta}_{\mathbf{i}}^{\mathbf{wo}}$	71.7157	34.3890	16.0264
Bids	$\boldsymbol{\omega}_{\mathbf{i}}$	35.8590	17.1927	8.0074
	$\mathbf{g}_{\mathbf{j}}$	0.0069	0.0145	0.0311
Lagrange multiplies	β_i	0	0	0
	σ_j	0	0	0
	$\psi_{i,j}$	0.5000	0.4999	0.4996
Utility	$U_i(\boldsymbol{\theta}_{\mathbf{i}}^{\mathbf{ag}})$	59.7653	27.8766	13.6810
Cost	$C_j(\boldsymbol{\theta}_{\mathbf{i}}^{\mathbf{wo}})$	25.6035	12.0056	6.5565
Pricing rules	$P_i(\boldsymbol{\omega}_{\mathbf{i}})$	35.8590	17.1927	8.0074
	$E_j(\mathbf{g}_{\mathbf{j}})$	35.8590	17.1927	8.0074

In addition, we notice that the utility $U_i(\boldsymbol{\theta}_{\mathbf{i}}^{\mathbf{ag}})$ are greater than the cost $C_j(\boldsymbol{\theta}_{\mathbf{j}}^{\mathbf{wo}})$. Such results demonstrate the efficient properties of the IDA-based market. Meanwhile, results in Table 5.3 show the $E_j(\mathbf{g}_{\mathbf{j}})$ are greater than $C_j(\boldsymbol{\theta}_{\mathbf{j}}^{\mathbf{wo}})$, and $U_i(\boldsymbol{\theta}_{\mathbf{i}}^{\mathbf{ag}})$ are greater than $P_i(\boldsymbol{\omega}_{\mathbf{i}})$. These show that the return (utility) is greater than the cost (expenditure) of all participants, therefore validating the individual rationality property of the market.

5.1.4.6 Convergence Analysis

Then, in this section, we evaluate the performance of the EWA-DA algorithm in the federated learning market and demonstrate the dynamic behavior of the agent in the double auction game. In this part, we use C++ to develop our experiment environment. Firstly, we simulate an FL market with 3 FL tasks and 3 data owners. Each participant is acting as an EWA-DA algorithm agent. These agents can learn to improve their strategies by participating in the auction round

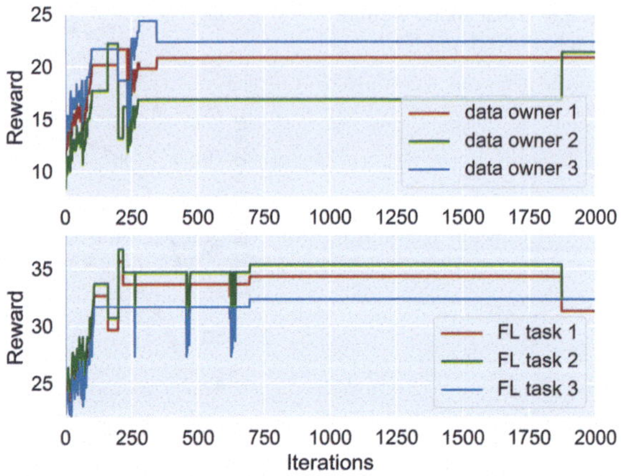

Fig. 5.7 The convergence of the participant

Table 5.4 EWA algorithm simulation results

Definition	Parameter	Agent1	Agent2	Agent3
Model quality	θ_i^{ag}	70	30	20
	θ_i^{wo}	70	30	20
Utility	$U_i(\theta_i^{ag})$	70	32	17
Cost	$C_j(\theta_i^{wo})$	35	15	10
Pricing rules	$\mathbf{p}_i^{ag}$	40	20	15
	$\mathbf{e}_i^{wo}$	40	20	15

constantly. In the EWA-DA agent, we set the discount factors ρ as 0.9, ϕ as 0.8, and the imagining factor δ as 0.4. To compress the search space of the EWA-DA algorithm, we discretize the action space. We denote the accuracy action space as $(10, 20, \ldots, 100)$, and the price space as $(1, 2, \ldots, 100)$.

Firstly, we evaluate the convergence of the EWA-DA algorithm. As shown in Fig. 5.7, the EWA-DA algorithm exhibits an excellent convergence performance. The two-side participants' reward increases as the iteration increase and converges after approximately 800 rounds. All supply accuracy and deal prices after convergence can be found in Table 5.4. Compared to Table 5.3, we notice the converge value of the EWA-DA algorithm is closed to the IDA algorithm. The slight difference is caused by the discretization of action space. This demonstrates that the convergence point of EWA-DA is near to the market equilibrium point.

5.1.4.7 Performance Analysis

In this part, we set the Q-Learning, Policy Gradient (PG), Deep Q-learning (DQN), and Deep Deterministic Policy Gradient (DDPG) to evaluate the performance of our algorithm. All parameters setting of FL market is same to Sect. 5.1.4. The learning rate α and the discount factor β of all baseline algorithms are set as 0.2 and 0.8. The other algorithm parameters setting of DQN and DDPG can be found in

Table 5.5 Convergence performance

Algorithm	Conv Acc	Conv steps
EWA	(70,70)	840
Q-Learning	(60, 60)	1630
Policy Gradient	(70,70)	1180
DQN	(70,70)	4450
DDPG	(60,60)	6890

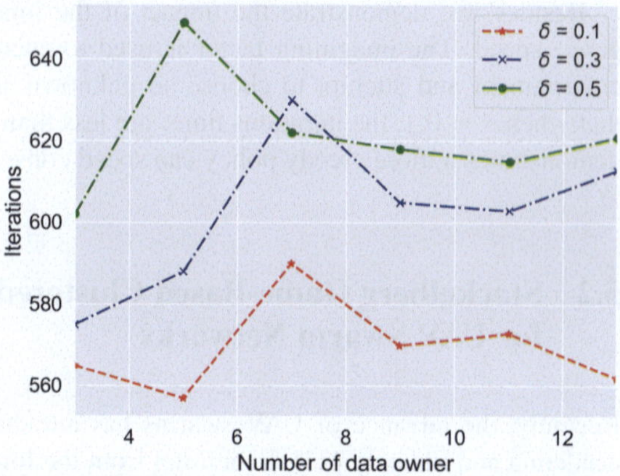

Fig. 5.8 The convergence speed with different parameter

[22]. In Table 5.5, we present the convergence steps and the convergence accuracy action of agent. As illustrated in Table 5.5, while all algorithms can converge to a small neighborhood of the maximum welfare point, they present different rates of convergence. The EWA algorithms converge in around 840 steps, which exhibit the best performance. This result demonstrates that the belief learning mechanism can effectively improve the learning rate of reinforcement learning. Moreover, we notice that the DQN and DDPG present the worst convergence performance. This is because that they have too many parameters that need to be updated. Therefore, we can take a lesson that while the complex neural network design enables them to solve complex tasks, simple learning algorithms may more efficient for simple tasks.

5.1.4.8 Convergence Speed

Then, we evaluate the convergence performance with the different number of participants and imagining factor δ. In our experiments, we set the $N = 3$, and increase M from 3 to 13 with 2 new data owner. In addition, we set three groups of parameter, where $\delta_1 = 0.1$, $\delta_1 = 0.3$, and $\delta_1 = 0.5$. As shown in Fig. 5.8, within the same parameter setting, the number of iterations to converge does not increase significantly as the number of participants increases. The market can converge to a stable around 550 iterations. This demonstrates that our EWA-DA-based market can still keep stabilizing with the new participants joining the market.

Note that the fluctuation with the number of data owners increasing is caused by the uncertainty in the learning process of EWA-DA. EWA-DA updates policy based on rewarding desired behaviors and/or punishing undesired ones according to Eqs. (5.43) and (5.44). Such a trial and error learning mechanism will make the policy optimization process uncontrollable. Besides, the behavior choice function (i.e., exploration mechanism) in EWA design will further exacerbate the randomness of the optimization process.

Besides, we demonstrate the impact of the imagining factor δ on the convergence speed. The imagining factor is used to encourage the agent to explore the environment and attempt to choose an unknown strategy. For Fig. 5.8, we notice that when $\delta = 0.1$, the iterations times are less than it in $\delta = 0.3$ and $\delta = 0.5$. That demonstrates a more greedy policy can speed convergence in our market.

5.2 Stackelberg Game-Based Clustered Federated Learning for UAV Swarm Networks

Recently, the advance of UAV swarms has attracted a lot of attention from both academia and industry [23]. Benefiting from the low cost, flexible maneuverability, and fast deployment, UAV swarms have breathed new life into military, healthcare, and industrial scenarios [24]. In particular, with the development of AI technologies, integrating intelligence at UAV swarms by fully utilizing the data resources of UAVs and applying AI methods to promote the expansion of distributed applications [25, 26].

Traditional UAV swarms burdened with traditional machine learning solutions rely on cloud-centric architecture, requiring UAVs to send their data and train on centralized servers [27]. Sending massive raw data to a centralized entity incurs a huge network communication overhead. Such long-distance transmissions reduce the response speed of the whole system. Additionally, entities are reluctant to transmit raw data to a third-party edge server due to the private problem. To overcome these challenges, FL, as a compelling privacy-preserving and distributed learning paradigm, enables UAVs to conjunctively train shared AI models [28]. FL extracts intelligence and hashrate from such geographically distributed mobile devices more efficiently and safely, while keeping all training data on each device. Such distributed learning paradigm frees up the data collection and computing power of UAV swarms.

However, on account for the diversified services and differentiated function, UAV swarms have the statistical heterogeneity of data, such as non-independent and identically distributed (non-IID).The current FL framework relays on homogeneous data paradigms, which requires distributed devices hold the same data structure. Heterogeneous UAVs collect local data in various data structures downgrade the accuracy of the original FL training models mainly in the following two aspects [29]. The updated parameters are aggregated in a linear simulation way each round, resulting in that the model parameters of each UAV are only suitable for its local data. Additionally, an unbalanced sample size leads to different local training times, and the iterations that are not completed within a fixed time will be dropped by the system, which seriously affects the convergence performance. Consequently, throughout the FL process, the inconsistency of training paces, low communication efficiency, unpredictable delay, and the failure of single points badly affect the system performance. Many methods are proposed to solve the above challenges, such as weighted aggregation, gradient compression, semi-asynchronous aggregation, clustered federated learning, and model split. Considering such a scenario of massive dynamic heterogeneous devices, the calculation of weighted values and compressing gradients can incur additional overhead; the semi-asynchronous aggregation and model split can drop some significant nodes if waiting too long. The CFL adopts similar performance, functionalities, and datasets to group massive heterogeneous UAVs, why we construct a cluster structure to first reduce the scale and release heterogeneity [30, 31].To overcome some struggles such as reducing the high transmission delay, the CFL increases train efficiency on heterogeneous devices [32].

Combining the characteristics of UAV swarms for federated learning, we propose a CFL architecture based on local gradient directions. In the view that numerous UAVs are selfish, designing an effective and reliable incentive mechanism for CFL systems is a priority and practical challenge [33, 34]. In this part, we propose a clustered method based on a local learning structure, efficiently improving the performance of heterogeneous FL and the efficiency of UAV interaction while maintaining data privacy through gaming processes. Aided by a feasible incentive mechanism, collaborative learning within each UAV swarm can improve the efficiency of interaction when they will not share their own data under the federated learning paradigm. Given the synchronization of global aggregation and update, the training stage is easy to be delayed by heterogeneous devices. To stimulate a more efficient uplink of model updates, the task publisher purchases the service provided by cluster heads and rewards the clusters. Considering the non-cooperative UAVs, which expect to maximize their utilities, we model the association of the UAV swarms, cluster heads, and the model owner as a three-stage Stackelberg Game.

5.2.1 Data-Driven Clustered Federated Learning Architecture

In this section, we propose a data-driven clustered FL architecture for UAV swarms. In the original FL (OFL), the model owner trains the global model cooperatively with the contribution of updated parameters calculated by data owners. Above process requires all data owners to transmit local model parameters to a fusion center, such as the Base Station (BS) and the edge server. However, in a dynamic network environment of UAV swarms, UAVs cannot be connected to the BS directly considering high energy consumption or sometimes high transmission delays. Considering the high convergence delay derived from heterogeneous data samples,

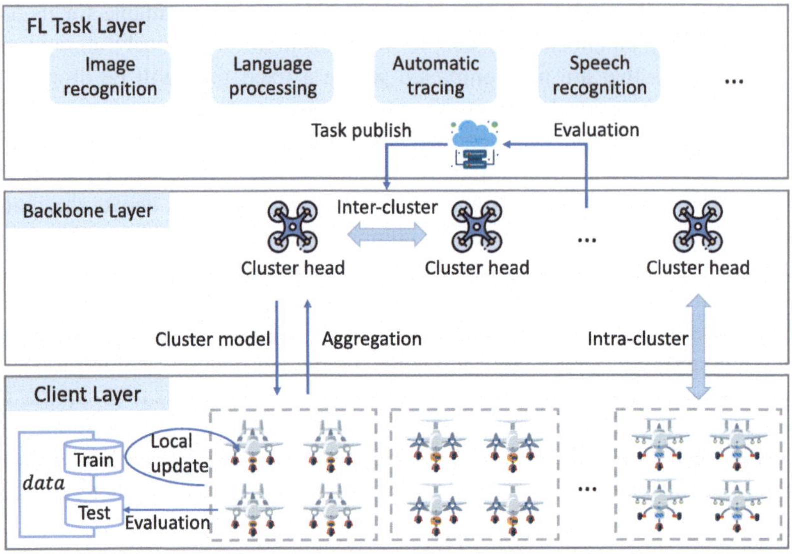

Fig. 5.9 Data driven clustered federated learning architecture

we draw a novel FL paradigm aided by a clustering structure. As shown in Fig. 5.9, the proposed federated learning architecture includes three layers: the client layer, the backbone layer, and the FL-task layer.

5.2.1.1 FL Task Layer

The FL task layer is composed of massive various FL tasks such as image recognition, language processing, etc.We consider that the model owner publishes a single task each time. In the process of the FL, the model is trained on UAVs, where raw data is generated. Then, the locally updated model parameters are transmitted to the model owner for aggregation without the disclosure of raw data. After the parameter aggregation in a centralized platform, the updated global model is responded to UAVs for the next round of training until converging to the target. Given a model $f_{\omega} : \mathscr{X} \rightarrow \mathscr{Y}$ is parameterized by $\boldsymbol{\omega} \in \boldsymbol{\Omega}$ and a loss function is L. The distributed objective of the FL system can be further described as:

$$\min_{\boldsymbol{\omega}}\{f(\boldsymbol{\omega}) \triangleq \sum_{n=1}^{N} p_n F_n(\boldsymbol{\omega})\}, \tag{5.48}$$

where $p_n = \frac{|D_n|}{D}$ denotes the proportion of worker n samples and $\sum_n p_n = 1$. Hereby, the local train goals $F_n(\boldsymbol{\omega})$ represent the local empirical risk over possibly differing data distribution Ψ_n.

$$F_n(\omega) \triangleq \mathbb{E}_{(x,y)\sim\psi_n} L(x, y; \omega). \tag{5.49}$$

5.2.1.2 Backbone Layer

The backbone layer consists of UAV swarm heads which usually have the highest node centrality within a cluster or own rich communication resources. When denoted as a backbone node, the UAV receives the global model from the task layer and calls on the devices at the client layer to join its cluster, where a new master-slave game is thus established. On the one hand, each cluster head acts as the central aggregation server for the cluster, aggregating and updating the training model intra-cluster. On the other hand, they execute inter-cluster interaction to aggregate intermediate training results. Our work focuses on the design of incentive mechanisms based on clustered federated learning structures. When generating cluster heads, we can consider two approaches. One approach is to pre-define all the cluster heads, usually choosing a UAV with a more central location and more resources available. The other is to elect the cluster head in the unmanned cluster, which can be selected by energy, trust, etc. However, these are not the key challenge of this part, where we choose nodes with more available resources as cluster heads. To achieve the global target and be more feasible for the cluster federated learning system, we clarify the process of CFL as follows. The UAVs in each sub-group G_h establish an OFL system, where the cluster head acts as the model owner to aggregate and update.

5.2.1.3 Client Layer

The client layer is composed of UAV devices potential to participate in the FL task. For each cluster, the data can be considered isomorphic, and we make the following assumptions. Accordingly, we can consider the global optimization problem $f(\omega)$ as a joint optimization problem for multiple clusters. To achieve the global target and be more feasible for a UAV-assisted federated learning system, in each iteration k, CFL contains the following three steps:

1. The workers first synchronize the task with their cluster head h by downloading the last round model $\omega_h{}^{(k)}$. Specifically, we denote the model aggregated by all cluster heads as the global model, whose performance decides the training time of the whole CFL system. The model aggregated by UAVs within the same cluster is denoted as an intermediate model and the model trained on local devices is denoted as the local model.
2. The UAV acts as a worker node and then improves the performance of the intermediate model by executing multiple iterations of stochastic gradient descent with mini-batches sampled from its local data D_n [35], leading a vector

$$\Delta\omega_{h,n}{}^{(k)} = \text{SGD}(\omega_h{}^{(k)}, D_n) - \omega_h{}^{(k)}. \tag{5.50}$$

3. The cluster head h aggregates local parameters from the participants in its cluster to derive an updated intermediate model ${\omega_h}^{(k+1)}$.

$$
{\omega_h}^{(k+1)} = {\omega_h}^{(k)} + \sum_{n=1}^{N} \frac{|D_n|}{D} \Delta {\omega_{h,n}}^{(k+1)}. \tag{5.51}
$$

At intervals $\widetilde{k}$, cluster heads transmit intermediate model parameters ${\omega_h}^{(k+\widetilde{k})}$ to execute global aggregation on the model owner, where obtains global model parameters $\omega^{(k+\widetilde{k})}$. If not meeting the training targets, the intermediate model ${\omega_h}^{(k+1)}$ is transmitted back to the previous clusters for the $(k+1)^{th}$ local training.

Heterogeneous data drives UAVs to self-organize into different clusters and execute OFL based on the clustering structure we state above. Furthermore, each cluster head broadcasts the intermediate model to the workers in their cluster and then aggregates the updated results from these workers using FedAvg in parallel, which is described as intra-group aggregation. When all training rounds of clusters are complete, the intermediate models are aggregated with an appropriate weight, denoted as inter-group aggregation. As a federated learning task publisher, the model owner provides reasonable compensation for participants in the CFL system to achieve a utility balance between itself and each cluster.

5.2.2 *System Model and Problem Formulation*

In this section, we explain the mathematical model of the economic problem under CFL. According to the CFL architecture stated above, we describe the model of cluster heads and UAVs in clusters in detail, followed by the cost functions for model owners.

We consider a CFL architecture for UAV swarms as shown in Fig. 5.10, which consists of N worker UAVs established by the set of $\mathcal{N} = \{1, \ldots, N\}$, with different dataset D_n of each client. A group of workers forms a cluster G_h training for the same task and the clusters are represented by the set of $\mathcal{G} = \{G_1, \ldots, G_h, \ldots, G_H\}$, where $h \in H$ and $\mathcal{H} = \{1, \ldots, H\}$ is used to represent the cluster heads. We denote $G_h = \{1, \ldots, n_h\}$ and there exists $\dot{U}_{h=1} G_h = \{1, \ldots, N\}$. As the task publisher, the goal of the model owner is to train the model iteratively through $K$to minimize the global loss function $L^K(\omega)$, where ω represents the global model parameter.

5.2.2.1 Worker Model

Each worker has local data samples D_n to participate in the clustered federated learning task. The payment from cluster head h depends on the data contribution of

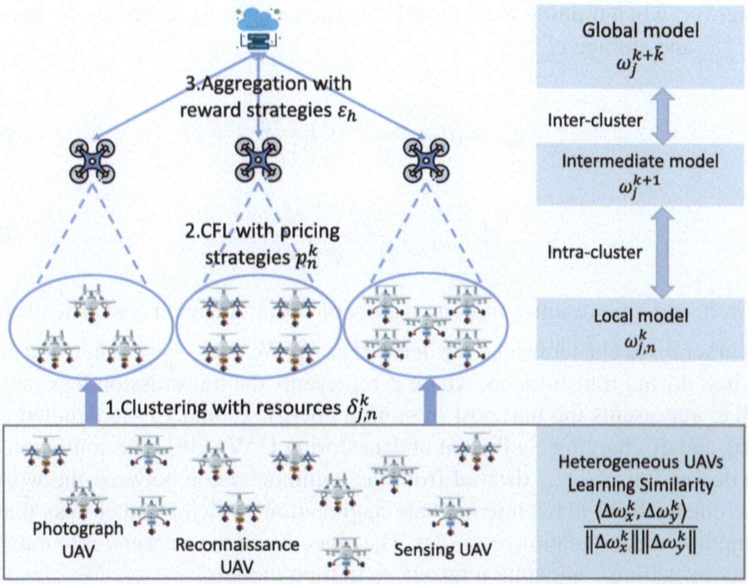

Fig. 5.10 The system model of clustered federated learning aided by an incentive mechanism

the corresponding worker. We solve the heterogeneous problem of UAV swarms by clustering. Based on the homogeneity of workers belonging to the same cluster, the data contribution can be defined as the number of training samples $p_n = \frac{D_n}{D}$. Considering the learning process within a cluster consumes computational resources, the computing cost can be depicted as

$$c_{n,comp}^k = \tau \frac{\alpha_n^k}{2} c_n D_n (\delta_n^k)^2, \tag{5.52}$$

where τ denotes the number of rounds during one iteration k and c_n represents the unit cost of energy consumptions. We set α_n^k to represent the effective capacitance parameter of the computing chipset for worker n, which has the computation ability decided by the CPU as δ_n^k. We can further obtain,

$$\sigma_{h,n}^k = \tau c_n D_n, \tag{5.53}$$

where $\sigma_{h,n}^k$ represents the CPU-cycle frequency of worker n and the time of each worker training the model is denoted as τ. Further, we can obtain the time consumption of computing in the iteration k,

$$t_{n,comp}^k = \frac{\tau c_n D_n}{\delta_n^k}. \tag{5.54}$$

Moreover, when training local models, workers consume its energy by transmission c_{h,n_h}^{comm} and storage c_{h,n_h}^{store}.

$$c_{n,comm}^{k} = a(\frac{\delta_n^k}{1-\phi})^2 + b\frac{\delta_n^k}{1-\phi} + z, \tag{5.55}$$

$$c_{n,store}^{k} = \frac{\delta_n^k}{1-\phi}\frac{c_s}{\kappa_c \kappa_d}, \tag{5.56}$$

where a, b, and z represent weighting factors of communication cost. The resources from worker n_h to cluster head h are denoted as δ_n^k. We use $\frac{\delta_n^k}{1-\phi}$ to denote the energy with a loss during transmission, where ϕ represents the transmission loss rate. The symbol c_s represents the unit cost of storing energy. κ_c and κ_d are denoted as the charging and discharging coefficient of data storing UAVs [36]. The communication cost is denoted as $c_{n,comm}^{k}$ derived from the communication between the worker n and the cluster head h for intermediate aggregation, which is much less than the consumption of computation resources. The objective of the worker n is to maximize its utility by selling computation resources to train models.

$$U_n^k(\delta_n^k) = p_n^k \delta_n^k - c_{comp}^{k,n} - c_{comm}^{k,n} - c_{store}^{k,n}. \tag{5.57}$$

5.2.2.2 Cluster Head Model

We focus on the heterogeneity of the UAV network, derived from various onboard equipment, different data structures, imbalanced sample sizes, etc. To avoid discarding important nodes and unstable global training performance caused by statistical heterogeneity, we divide the optimization objective into several subproblems following the Pluralistic Group architecture. Considering the data structure and UAVs' function in the same cluster not changing, we propose a static strategy to make clusters of UAV swarms based on the similarity between the UAVs' local optimization results. Specifically, we make use of Decomposed Cosine similarity to make cluster decisions of high dimensional data samples. It is an effective method to obtain the cosine similarity between UAV learning gradients by calculating backpropagation or model parameter update [37], which is defined as:

$$S(x, y) \triangleq \frac{< \Delta\omega_x^t, \Delta\omega_y^t >}{\| \Delta\omega_x^t \| \| \Delta\omega_y^t \|}. \tag{5.58}$$

Due to the high-dimension data, we further take advantage of the Euclidean distance of Decomposed Cosine (EDC) similarity to simplify the clustering process as follows,

$$EDC(x, y) = \frac{1}{J}\sqrt{\sum_{v \in V}(S(x, v) - S(y, v))^2},$$
$$V = SVD(\Delta\Omega^T, J), V \subset \mathbb{R}^{d_\omega \times J}, \tag{5.59}$$

where $d_\omega \times J$ represents the number of parameters. Based on the formula (5.59), it decomposes the updates of the local model into j directions by facilitating a truncated Singular Value Decomposition (SVD) algorithm, then the problem is turned into the similarity between the updates and directions. In other words, we cluster the worker with a more similar learning target so that the CFL obtains a better global performance.

Meanwhile, there is no central fusion center in our model where the FL system needs to self-organize to seek the cluster heads first. After clustering, we can obtain the cosine similarity matrix $\mathcal{M} \in \mathbb{R}^{N \times N}$ as follows,

$$M = K(\Delta\boldsymbol{\Omega}, \Delta\boldsymbol{\Omega}), M_{xy} = S(x, y), \tag{5.60}$$

where the function $K(\cdot)$ is the cosine similarity kernel. At the beginning of the training, the cluster head selects a set of workers in its group. Each cluster trains the model to obtain a temporary cluster model $\tilde{\boldsymbol{\omega}}_h^{k+1}$. Then, the cluster parameters are updated after inter-cluster aggregation. As the backbone UAV nodes, cluster heads purchase resources of workers with unit cost p_n^k and obtain the reward strategies $\boldsymbol{\varepsilon}_h$ from the model owner to maximize their utility. Thus, the reward that the cluster head gains increases as the workers' computation resources increase. However, the incremental speed of cluster heads' utility gradually diminishes when the number of CPU frequency gets larger on account of the generic economic law of "diminishing marginal return" [38]. Based on the information on energy resources in the above content, we define the utility of cluster head h in the k-th iteration as follows,

$$U_h^k = \frac{\varepsilon_h}{1 + \frac{1}{\sum_{n=1}^{n_h} p_n^k \delta_n^k}} - \sum_{n=1}^{n_h} p_n^k \delta_n^k - c_{h,comm}^k - c_{h,comp}^k, \tag{5.61}$$

where ε_h represents the payment for the cluster head h from the model owner. After inter-cluster aggregation, each cluster heads h and uploads the immediate model to the model owner. The transmission time is calculated by

$$t_h^{k,comm} = \frac{\xi}{B_h^k}. \tag{5.62}$$

We denote the learning model size as ξ and the mean transmission rate of cluster head h as B_h^k, where B_h^k can be expressed as

$$B_h^k = \frac{1}{t^{k+1} - \Delta t_h^k - t_h^{k,comm} - t^k} \int_{t^k + t_h^{k,comm}}^{t^{k+1} - \Delta t_h^k} B_t dt. \tag{5.63}$$

We set t^{k+1} as the starting of the k-th iteration and Δt_h^k is the free time of cluster head h. Therefore, the total time consumption of the FL for cluster G_h can be denoted as

$$T_h^k = \mathbb{E}[max\, t_{h,n_h}^{k,comp}] + t_h^{k,comm}. \tag{5.64}$$

In our system, the model owner evaluates the resources of each cluster and generates its reward allocation strategies $\boldsymbol{\varepsilon} = \{\varepsilon_1, \ldots, \varepsilon_h, \ldots, \varepsilon_H\}$ to minimize its cost value.

$$Cost_{MO} = \lambda T_{\text{global}}^k + \frac{\varepsilon_h}{1 + \frac{1}{\sum_{n=1}^{n_h} p_n^k \delta_n^k}}, \tag{5.65}$$

where $T_{\text{global}}^k = max T_h^k$.

5.2.2.3 Three-Stage Stackelberg Game

We use the three-stage Stackelberg game to model the problems modeled in our work. In this part, we construct problem formulations for each stage. Then, we apply backward induction methods to study the Stackelberg equilibrium and accordingly prove the existence and uniqueness of the Stackelberg equilibrium.

Stage I

At the beginning of each iteration k, the cluster head h generates optimal resource allocation $\boldsymbol{p}_n^{k,*} = \{p_1^{k,*}, \ldots, p_N^{k,*}\}$ to maximize the utility.

$$\begin{aligned} \max_{p_n^k} U_h^{k,(H)} &= \frac{\varepsilon_h}{1 + \frac{1}{\sum_{n=1}^{n_h} p_n^k \delta_n^k}} - \sum_{n=1}^{n_h} p_n^k \delta_n^k \\ &\quad - c_{h,comm}^k - c_{h,comp}^k. \\ (P1) \quad \text{s.t.} &\sum_{n=1}^{n_h} p_n^k \delta_n^k \le \varepsilon_h. \\ &0 \le \delta_n^k \le \delta_n^{max}. \end{aligned} \tag{5.66}$$

Stage II

The expected benefit of each worker is determined by its net utility, which is the difference between the payoff obtained for following the cluster head and the payoff for participating in FL training.

$$
\begin{aligned}
&\max_{\delta_n^k} U_n^{k(W)} = p_n^k \delta_n^k - c_{comp}^{k,n} - c_{comm}^{k,n} - c_{store}^{k,n}. \\
&\text{s.t.} \sum_{n=1}^{n_h} p_n^k \delta_n^k \le \varepsilon_h, \\
(P2)\quad & 0 \le \delta_n^k \le \delta_n^{max}, \\
& c_{store}^{k,n} \ll c_{comp}^{k,n}, \\
& c_{comm}^{k,n} \ll c_{comp}^{k,n}.
\end{aligned}
\tag{5.67}
$$

Stage III

The reward is the payoff for cluster heads which group a set of workers to train the model successfully. The total latency of the federated learning system depends on the total usage of the CPU power of all UAVs including cluster heads and workers. In the optimization problem we construct, the time and energy consumption of the CFL process are conflicting goals. If the UAV swarm performs the training task with full CPU ability, the time cost can be minimized, while the energy consumption would be high to an extreme degree, and the training will be quickly withdrawn. Considering the marginal benefit of finishing federated learning with resources, the model owner forms a subgame problem as follows,

$$
\begin{aligned}
&\min_{\varepsilon_h} Cost_{MO} = \lambda T_{\text{global}}^k + \frac{\varepsilon_h}{1 + \frac{1}{\sum_{n=1}^{n_h} p_n^k \delta_n^k}}. \\
(P3)\quad & \text{s.t.} \quad T_{\text{global}}^k = \max T_h^k, \\
& t^{k+1} = t^k + T^k, \forall 1 \le k \le K.
\end{aligned}
\tag{5.68}
$$

The problems to be solved at each stage are coupled, so it is not possible to optimize them in isolation considering:

1. The pricing strategy at Stage I affects the result of (P2) while the results in turn provide the information required to formulate the strategy.
2. The participation of workers at Stage II controlled by the strategy of (P2) affects the time consumption of (P3) and the result of (P2).

3. The timing cost of Stage III is decided by the Stage II results and affects the payment in (P1).

The above three sub-games construct a whole three-stage Stackelberg game.

5.2.2.4 The Analysis of Stackelberg Equilibrium

The best response function of the followers can be obtained by solving the problems we stated above. Then, we discuss the existence of a unique Stackelberg Equilibrium.

Theorem *There exists unique equilibrium of the stage* $\mathcal{G}_H = \{\mathcal{H}, \{\mathcal{U}_h^{(H)}\}_{h \in H}, [0, +\infty)^H\}$*, which is calculated by:*

$$(p_n^k)^* = \frac{\varepsilon_h}{\sum_{n=1}^{n_h} \delta_n^k}. \tag{5.69}$$

Proof: Based on the formula (5.66), we can calculate the first and second derivatives of utility $U_h^{k,(H)}$*. To simplify, we denote* $A = \sum_{n=1}^{n_h} p_n^k \delta_n^k$*. The contribution of different workers is decided by data size.*

$$\frac{\partial U_h^{k,(H)}}{\partial A} = \frac{\varepsilon_h}{(A+1)^2} - 1, \tag{5.70}$$

and

$$\frac{\partial^2 U_h^{k,(H)}}{\partial A^2} = \frac{-2\varepsilon_h}{(A+1)^3}. \tag{5.71}$$

As the description of payments, ε_h is non-negative and the second derivative is negative, we can know the utility function is concave. Then, we explain the existence of SE. We set $\boldsymbol{p}^k \triangleq (p_1^k, \ldots, p_{n_h}^k)$, and $\mathcal{U}^{(H)}(\boldsymbol{p}) \triangleq \mathcal{U}_1^{(H)}(\boldsymbol{p}), \ldots, \mathcal{U}_{n_h}^{(H)}(\boldsymbol{p})$, we can depict point-to-set mapping $\boldsymbol{W} = \boldsymbol{W}(\mathcal{U}^{(H)}) = [\Delta_{p_n} U_n^{(H)}(\boldsymbol{p})]_{n=1}^{n_h}$, where

$$\begin{aligned}
\Delta \boldsymbol{W} &= \Delta \boldsymbol{W}(\mathcal{U}^{(H)}(\boldsymbol{p})) \\
&= \begin{bmatrix} \Delta_{11}^2 U_1^{(H)}(\boldsymbol{p}) & \Delta_{12}^2 U_1^{(H)}(\boldsymbol{p}) & \cdots & \Delta_{1N}^2 U_1^{(H)}(\boldsymbol{p}) \\ \Delta_{21}^2 U_1^{(H)}(\boldsymbol{p}) & \Delta_{22}^2 U_1^{(H)}(\boldsymbol{p}) & \cdots & \Delta_{2N}^2 U_1^{(H)}(\boldsymbol{p}) \\ \vdots & \vdots & \vdots & \vdots \\ \Delta_{N1}^2 U_1^{(H)}(\boldsymbol{p}) & \Delta_{N2}^2 U_1^{(H)}(\boldsymbol{p}) & \cdots & \Delta_{NN}^2 U_1^{(H)}(\boldsymbol{p}) \end{bmatrix} \\
&= \begin{bmatrix} \frac{\varepsilon_h}{\sum_{n=1}^{n_1} \delta_n^k} - 1 & \cdots & 0 \\ \vdots & \vdots & \vdots \\ 0 & \cdots & \frac{\varepsilon_h}{\sum_{n=1}^{n_h} \delta_n^k} - 1 \end{bmatrix}.
\end{aligned} \tag{5.72}$$

Therefore, $\Delta \boldsymbol{W} + \Delta \boldsymbol{W}^T$ is negative definite. Additionally, $\Delta \boldsymbol{W}$ is diagonally strictly concave. Then the stage $\mathscr{G}_N$ has a sole equilibrium.

Theorem 5.1 *There exists unique equilibrium of the stage* $\mathscr{G}_N = \{\mathscr{N}, \{\mathscr{U}_n^{(W)}\}_{n\in N}, [0, +\infty)^N\}$.

$$(\delta_n^k)^* = \frac{p_n^k}{\tau\alpha_n^k c_n D_n}. \tag{5.73}$$

Proof: Based on the formula (5.67), we further calculate the first and second derivatives of utility U_n^k *with respect to* δ_n^k,

$$\frac{\partial U_n^k}{\partial \delta_n^k} = p_n^k - \tau\alpha_n^k c_n D_n \delta_n^k, \tag{5.74}$$

and

$$\frac{\partial^2 U_n^k}{\partial (\delta_n^k)^2} = -\tau\alpha_n^k c_n D_n. \tag{5.75}$$

Then we set $\frac{\partial U_n^k}{\partial \delta_n^k} = 0$ and obtain the formula (5.73). Considering the description of CPU fixed coefficient α_n^k is non-negative and the second derivative is negative, the function $U^{(W)}$ of the worker is concave.

Theorem 5.2 *With the optimal pricing strategies of cluster heads* $\boldsymbol{p}_n^{k,*}$ *and resources allocated by workers* $\boldsymbol{\delta}_n^{k,*}$, *we can further discuss the optimal cost of the model owner, the optimization problem of which has a unique solution. When the variable* τ *in the formula (5.68) is fixed, we can prove the constraint set of* $Cost_{MO}$ *is convex. Proof: We can easily derive that* $h(\boldsymbol{p}^k) = \sum_{n=1}^{n_h} \frac{p_n^k}{\tau\alpha_n^k c_n D_n}$ *is a strictly convex function. Denote that* $\iota \in [0, 1]$ *and* $\boldsymbol{x}, \boldsymbol{y} \in \boldsymbol{p}^k$, *and* $g(\boldsymbol{p}^k) = max\{\frac{(c_n D_n)^2 \tau\alpha_n^k}{p_n^k} + \epsilon_n^k\}$. *The upload time for the training parameter is denoted as* $\epsilon_n = \frac{\xi}{B_n^k}$.

Then,

$$\begin{aligned}
g(\iota\boldsymbol{x} + (1-\iota)\boldsymbol{y}) &= max\{\frac{(c_n D_n)^2 \tau\alpha_n^k}{\iota x_n + (1-\iota) y_n} + \epsilon_n^k\} \\
&\le max\{\frac{\iota (c_n D_n)^2 \tau\alpha_n^k}{x_n} + \frac{(1-\iota)(c_n D_n)^2 \tau\alpha_n^k}{y_n} + \epsilon_n^k\} \\
&= \iota\, max\{\frac{(c_n D_n)^2 \tau\alpha_n^k}{x_n}\} + (1-\iota)\, max\{\frac{(c_n D_n)^2 \tau\alpha_n^k}{y_n} + \epsilon_n^k\} \\
&= \iota g(\boldsymbol{x}) + (1-\iota) g(\boldsymbol{y}),
\end{aligned} \tag{5.76}$$

which indicates that $g(\boldsymbol{p}^k)$ is a convex function. Therefore, $Cost_{MO}$ is a strictly convex function, and there exists a unique solution to $\boldsymbol{\varepsilon}_h$.■

5.2.2.5 Backward Induction

The constructed three-stage Stackelberg game model in our work is a large-scale nonlinear problem with a hierarchical structure. Backward induction is used to search for the dynamic game equilibrium. The so-called dynamic process shows that a sequence of actions of the players exists in the iterations interactions, and the players who act later can observe the previous actions. Considering the Stackelberg game under incomplete information and a dynamically sequential process, we first design the backward induction algorithm shown in Fig. 5.11 to seek equilibrium. There exists a sequence of actions of different players in the three-stage Stackelberg game, where the latter actions can observe the previous. We state the process in Algorithm 13. In our proposed model, the leaders (the model owner) first move and the followers (cluster heads) then respond, followed by the workers taking action accordingly. If we consider the model as a combination of two-stage Stackelberg games, it is a single leader with a multi-follower and a multi-leader with a multi-follower structure. The model owner first publishes a federated learning task and initializes resource pricing strategies. The determination of the pricing strategies is affected by the behavior of cluster heads and the efficiency of workers. Therefore, we build the above policy adjustments through an iterative process. After each participant decides and takes action, others decide whether they desire to adjust their strategies to obtain a higher payment. Until all participants obtain the optimal solution, the whole system reaches equilibrium. In the three-stage Stackelberg game in Fig. 5.12, there is at least one NE at each stage, including the resource strategies δ_n^* of workers in stage I, the unit price of computing services p_n^* decided by cluster heads in stage II, and the reward strategies ε_h^* determined by the model owner in stage III.

5.2.3 Hierarchical Multi-Agent Reinforcement Learning-Based Training Algorithm

In this section, we propose a multi-agent reinforcement learning (MARL) algorithm to search for the hierarchical game equilibrium solution and explain how the proposed algorithm seeks equilibrium. Additionally, we explain the detail of the proposed algorithm design.

To search for the SE in the hierarchical game, currently popular algorithms, such as Backward Induction, ideally assume existing fusion platforms can aggregate parameter information collected from all participants and compute the appropriate strategies. Unfortunately, the data collected by UAVs and the parameters of the

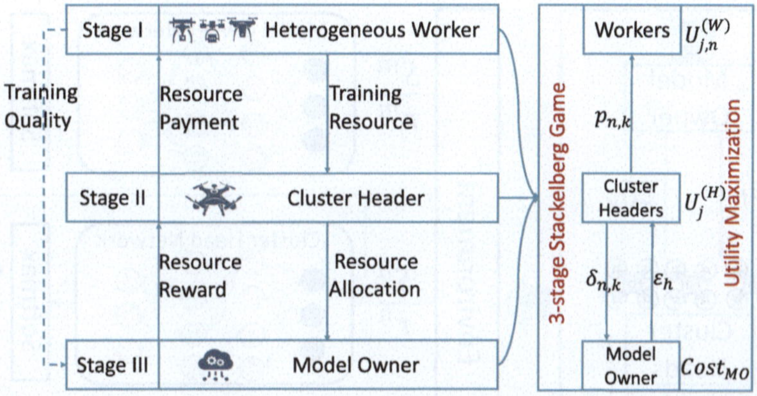

Fig. 5.11 Three-stage Stackelberg game with the backward induction algorithm

---

Algorithm 13 Backward induction algorithm to achieve Stackelberg equilibrium

---

Input: Model owner, cluster heads, worker attributes, system parameters, and the threshold ϵ
Output: p_n^*, δ_n^*, ε_h^*.

1: Repeat
2: Calculate the optimal solution for utility optimization of workers in cluster h (Internal Loop): Each worker UAV $n \in \mathbb{N}$ observes the announced purchase strategies p_n^k, $\forall k \in \mathbb{K}$, and decides on their individually expected resource amount δ_n^k by maximizing the derived utilities $U^{(M)}$ in P2.
3: Calculate the optimal solution for utility optimization of cluster heads (Internal Loop): According to the worker's actions δ_n^k and the payment ε_h decided by the model owner in the previous iteration, the cluster heads construct the optimal pricing strategies p_n^k by maximizing their utility $U^{(H)}$.
4: Calculate the optimal solution for cost optimization of the model owner (Eternal Loop): Based on the resources provided by cluster heads and the work ability of the workers in the previous iteration, the model owner obtains the optimal pricing strategies p_n^k by minimizing their cost $Cost_{MO}$.
5: $i = i + 1$;
6: Until $||U_{n,k,i}^{(H)} - U_{n,k,(i-1)}^{(H)}|| \leq \epsilon$

---

training model are locally private, which is also a feature that the federated learning architecture adheres to. As we know the deep reinforcement learning methods widely used in traffic management and energy scheduling under dynamic scenarios [39–41], we use the powerful data processing and inference capability of DRL to solve the equilibrium problem. We construct three parties in the CFL architecture as a multi-agent system to apply the RL method. Different from existing methods, MARL aims to learn a general action decision from historical experiences based on the current state and the given reward [42]. Considering the dynamic process of resource allocation by each participant, the decision not only is influenced by the reward from the buyer, but also by the actions of other participants.

To solve the proposed multi-stage master-slave game problem, we construct dynamic strategies of UAVs as a Markov Decision Process. Under incomplete infor-

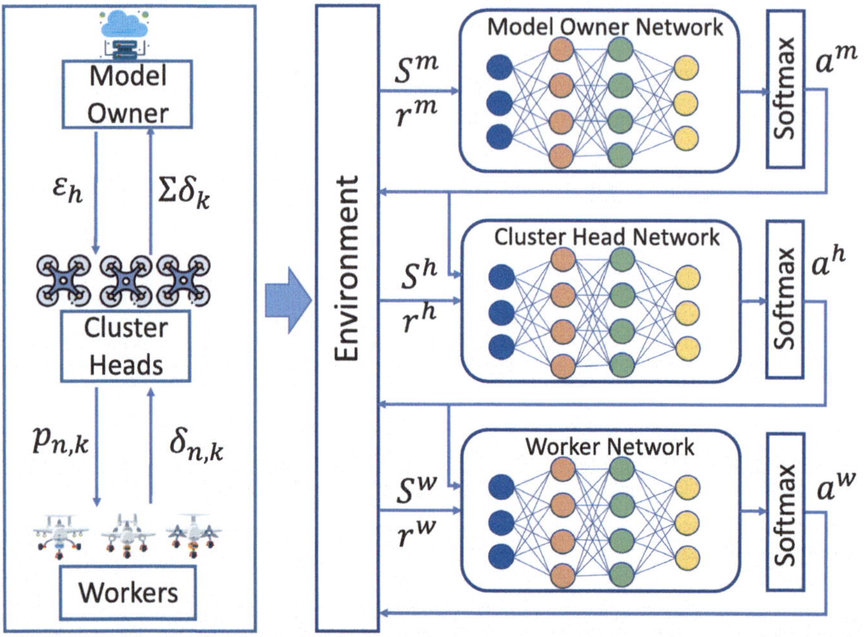

Fig. 5.12 Algorithm architecture

mation, we propose a hierarchical multi-agent reinforcement learning algorithm to solve the problem, as shown in Fig. 5.12. On the one hand, the existing heuristic algorithm has the threat of disclosing the private information of entities. On the other hand, multi-agents represent multiple entities in the game, which can better describe the process of a dynamic game. These agents can dynamically learn strategies when they interact with the environment. All agents simultaneously select and execute their respective actions according to the current environment state (or observation value), and the joint actions brought by the respective actions affect the transition and update of the environment state and determine the reward feedback obtained by the agent. Specifically, as illustrated in Fig. 5.12, when the k-th iteration starts, the entity can observe the state of the inter-clusters and decides how to act.

(1) State: The state information is the foundation for the agent to strategize and consider earnings from a long-term perspective. Therefore, the quality of the state space design directly affects whether the MARL algorithm can converge, the speed of convergence, and the final performance. We construct action spaces $\mathcal{S}^m$, $\mathcal{S}^h$, and $\mathcal{S}^w$ of the model owner, cluster heads, and workers, respectively. According to the process of environment, at round k of CFL, the state of workers at iteration t consists of the last round price strategies of the cluster head. We set the unit resource price p_h in a single cluster and unit reward ε_0 of a single model owner to be consistent, then the state space of the worker node can be expressed as $s_w^t = p_h^{t-1}$. As for the cluster head, the state observed

is $s_h^t = \{\varepsilon_0^{t-1}, [\delta_j^{t-1}]_{j\in n_h}\}$. To ensure the MARL model learns the trend of time consumption, the model owner first makes its payment decision based on $s_m^k = \{\sum \delta_j^{t-1}, Time_h^{t-1}\}$, where $Time_h^{t-1}$ represents the time consumption.

(2) Action: The design of the action space must first consider the possibility of achieving the desired goal and avoiding the unreachable state blind area in the task solution space and especially ensure the sufficient accessibility of the high-performance area. In order to improve the algorithm performance, the best action space as far as possible is simple and efficient, effectively reducing the training difficulty. We construct action spaces $\mathcal{A}^m$, $\mathcal{A}^h$, and $\mathcal{A}^w$ of the model owner, cluster heads, and workers, respectively. At round t, the cluster head first sets the unit resource price $p_n^t \in \mathcal{A}^h$ based on s_h^t. The action of the workers and the model owner is respectively signified by $\delta_n^t \in \mathcal{A}^w$ and $\varepsilon_0 \in \mathcal{A}^m$. After all, clusters complete this training round and upload the parameters, the model owner will record $Time_t$ and its state s_m^t can transit into s_m^{t+1}.

(3) Reward: The reward in the MARL algorithm is responsible for guiding the neural network to mine the decision-related factors in the state information, and after refining it, it is used for the calculation and generation of the action. When determing an action a^t under the state s^t, the agent receives a reward r^t from the CFL in the t-th round. The reward function of three tiers in our proposed hierarchical Stackelberg model is defined as formula (5.66)–(5.68), where the reward function of the model owner is denoted as the inverse function of cost.

5.2.3.1 Policy and Learning Objective

We describe the multi-agent interaction process for solving equilibrium in Algorithm 14. Our proposed algorithm is designed to search the Stackelberg Equilibrium under incomplete information built by three entities. We set α_m, α_h, and $\alpha_w \in (0, 1]$ of the policy network of the model owner, cluster heads, and workers, respectively. In Algorithm 14, $\gamma \in (0, 1]$ represents the future reward discount factor, E represents the total episode, and T represents the maximum time-step in each episode. When the agent completes decided actions, the clustered federated learning framework transfer to the next state. Accordingly, the agent obtains a reward defined as the utilities of each party. In each round, the process is repeated to calculate the cumulative reward. Considering the entity of each worker, the {state, action} are under the continuous vector δ_n^k, which can not be stored in a tabular with limited space. To enable this process, the adjustable parameters of the policy network π are denoted as $\boldsymbol{\theta}$. The proposed hierarchical multi-agent reinforcement learning algorithm consists of the interaction of three-party with the environment. The agent in the MARL algorithm finds the best policy π to map states to actions by continuous learning to maximize the cumulative reward. In the federated learning system composed of high-dynamic UAV swarms, the algorithm avoids the heavy oscillation situations caused by the huge state space and action space if using a single agent algorithm, so that the whole system can reach the equilibrium state

Algorithm 14 The hierarchical multi-agent reinforcement learning algorithm

Input: $\alpha_m, \alpha_h, \alpha_w, \gamma, E, T, M, \mathcal{A}^m, \mathcal{A}^h, \mathcal{A}^w, \mathcal{S}^m, \mathcal{S}^h, \mathcal{S}^w$

Initialize: $\theta_m(s_m^t, a_m^t), \theta_h(s_h^t, a_h^t), \theta_w(s_w^t, a_w^t)$
for $e = 1, 2, \ldots, E$ **do**
  Initialize action space of all agents.
  for $t = 1, 2, \ldots, T$ **do**
    1. Model owner stage.
      Observe the state s_m^t: $Time^{t-1}, \varepsilon_1^{t-1}, \ldots, \varepsilon_H^{t-1}$.
      $\theta_m \leftarrow \theta_m + \alpha_m \nabla_{\theta_m} \log \pi_m(s_m^t, \delta^t; \theta_m) R_m^t$.
    2. Cluster head stage.
      Observe the state s_h^t: $\delta_1^t, \ldots, \delta_{n_h}^t, \varepsilon_h^t$.
      $\theta_h \leftarrow \theta_h + \alpha_h \nabla_{\theta_h} \log \pi_h(s_h^t, p^t; \theta_h) R_h^t$.
    3. Worker stage.
    for $j = 1, 2, \ldots, N$ **do**
      Obtain the state $s_{w_n}^t : \delta_n^t$.
      $\theta_{w_n} \leftarrow \theta_{w_n} + \alpha_w \nabla_\theta \log \pi_{w_n}(s_{w_n}^t, q_j^t; \theta_w) R_{w_n}^t$.
    end for
  end for
end for

Output: π_m, π_h, π_w

in the economic game. In essence, the proposed hierarchical MARL is a distributed policy gradient algorithm. It trains a critic that needs global information and an actor that needs local information for each agent and allows each agent to have its reward function.

The policy gradient is a reinforcement learning algorithm that can process continuous or discrete action space, which adjusts policy parameters $\boldsymbol{\theta}$ along the direction of performance gradient $\nabla J(\pi_\theta)$. In the point of the stochastic policy gradient method, the policy gradient is defined as:

$$\begin{aligned} \nabla \bar{R}_\theta &= \sum_\tau R(\tau) \nabla p_\theta(\tau) \\ &= \mathbb{E}_{\tau \sim p_\theta(\tau)}[R(\tau) \nabla \log p_\theta(\tau)] \\ &\approx \frac{1}{M} \sum_{m=1}^{M} \sum_{t=1}^{T_m} R(\tau^m) \nabla \log p_\theta(a_t^m | s_t^m), \end{aligned} \tag{5.77}$$

In the policy network, the loss of each action is defined by cross-entropy, which is to update the parameters by adding up all the losses for the action in each turn. We adopt a distributed framework with multi-agents to deploy the algorithm.

As we discuss above, we combine the CFL system with a hierarchical MARL algorithm to search for the equilibrium of the three-stage Stackelberg Game. Furthermore, we integrate intra-cluster subtask aggregation with MARL-based inter-cluster resource allocation, as shown in Algorithm 15.

Algorithm 15 The clustered federated learning with game architecture

Input: Workers set $N \leftarrow \{n_1, n_2, \ldots, n_h\}$, clusters set $G \leftarrow \{g_1, g_2, \ldots, g_h\}$, initial intermediate parameters set $W_0^{(G)} \leftarrow \{\boldsymbol{w}_0^{g_1}, \boldsymbol{w}_0^{g_2}, \ldots, \boldsymbol{w}_0^{g_h}\}$, $g_j.clients \leftarrow \{n_i | n_i \text{ is in cluster } g_j, \forall n_i \in N\}$, initial global parameters $\boldsymbol{w}_0$, K communication rounds, M selected workers per round, inter-cluster learning rate η_g, proximal hyperparameter μ.

Output: Updated cluster model parameters $W_K^{(G)}$

procedure: Running Algorithm 14 to search SE and obtain stable cluster structure.
$W_1^{(G)} \leftarrow \{\boldsymbol{w}_1^{g_1}, \boldsymbol{w}_1^{g_2}, \ldots, \boldsymbol{w}_1^{g_h}\} \leftarrow$ initialized $W_0^{(G)}$ by $\boldsymbol{w}_0$.
for each communication round $k \in K$ **do**
  for each cluster g_h in G **do**
    Obtain local cluster $S_k^{(g_h)} \leftarrow \{n_i \in g_h.clients, \forall n_i \in S_k\}$.
    $\tilde{\boldsymbol{w}}_{k+1}^{(g_h)} \leftarrow$ IntraClusterUpdate($S^{(g_h)}, \boldsymbol{w}_k^{(g_h)}$)
    $\tilde{WG}_{k+1} \leftarrow \{\tilde{\boldsymbol{w}}_{k+1}^{(g_1)}, \tilde{\boldsymbol{w}}_{k+1}^{(g_2)}, \ldots, \tilde{\boldsymbol{w}}_{k+1}^{(g_h)}\}$
  end for
  for each cluster g_h in G **do**
    $W^G{}_{k+1} \leftarrow$ InterClusterAggregation($\tilde{W^G}_{k+1}$).
  end for
end for
function IntraClusterUpdate($S^{(g_h)}, \boldsymbol{w}_k^{(g_h)}$)
if S_k is $\emptyset$ **then**
  $\boldsymbol{w}_{k+1} \leftarrow FedAvg(S_k, \boldsymbol{w}_k)$
end if
return $\boldsymbol{w}_{k+1}$
function InterClusterAggregation($\tilde{W}_{k+1}^{G}$)
for each cluster parameter $\tilde{\boldsymbol{w}}_{k+1}^{(g_j)}$ in $\tilde{W}_{k+1}$ **do**
  $\Delta\tilde{\boldsymbol{w}}_{k+1}^{g_j} \leftarrow \eta_g \sum_{l \neq h} \frac{\tilde{\boldsymbol{w}}_{k+1}^{(g_l)}}{||\tilde{\boldsymbol{w}}_{k+1}^{(g_j)}||}$, then $\boldsymbol{w}_{k+1}^{(g_j)} \leftarrow \tilde{\boldsymbol{w}}_{k+1}^{(g_j)} + \Delta\tilde{\boldsymbol{w}}_{k+1}^{g_h}$.
end for
return $W_{k+1} \leftarrow \{\boldsymbol{w}_{k+1}^{(g_1)}, \boldsymbol{w}_{k+1}^{(g_2)}, \ldots, \boldsymbol{w}_{k+1}^{(g_h)}\}$

5.2.4 *Simulation Results and Discussions*

5.2.4.1 Experiment Setup

We evaluate our proposed algorithm on the Federated Extended MNSIT (FEMNIST) [43]. In the process of CFL, the local epoch is $E = 10$, the number of iterations is $K = 20$ at each round, and the local solver is a mini-batch SGD with 10. Moreover, the learning rate η in our experiments is 0.003. We present that the number of CPU cycles c_n for training a single data sample follows a uniform distribution, and the value is in the range of 20–50 cycles/bit. We set $\tau = 3$, the data size D_n as 1000–2000 bits, and α_n follows a Gaussian distribution with a mean value of 1 and variance of 0.01. We construct 30 UAVs into no more than 5 clusters. In the initial clustering phase, the number of local training rounds is set to 10.

Algorithm 16 FedAvg algorithm

procedure: Cluster head training
  Initialize intermediate model w_0, then $w_1 \leftarrow w_0$.
for each communication round $k \in K$ **do**
  $CH_k \leftarrow$ Cluster head selects a subset of A clients.
  Cluster head broadcasts w_{k+1} to all selected clients.
  for each UAV in the cluster $j \in CH_k$ **do**
    $\Delta w_{k+1} \leftarrow$ InterClusterClientUpdate(i,w_k).
  end for
  $w_{k+1} \leftarrow w_k + \sum_{i \in CH_k} \frac{n_i}{n} \Delta w^i_{k+1}$
end for
function InterClusterClientUpdate(i,w)
  $w^{k+1} \leftarrow w$
for each local epoch $e = 1, 2, \ldots, E$ **do**
  $\hat{w} \leftarrow w - \eta \nabla L(b; w)$ for local batch $b \in B_i$.
end for
return $\Delta w_{k+1} \leftarrow w - \hat{w}$

5.2.4.2 Simulation Results of the Clustered Federated Learning Architecture

In this section, we present the experimental results to evaluate the validity and effectiveness of the proposed incentive mechanism under clustered FL architecture. Furthermore, we demonstrate the performance of convergence and economic properties under the MARL-based CFL framework.

The Effectiveness of the CFL Structure

We evaluate the accuracy of heterogeneous data with original federated learning methods such as FedAvg and the proposed CFL, as shown in Fig. 5.13. The dispersion of the dataset increases in turn in our experiment, representing the heterogeneity. The blue line in Fig. 5.13 represents the clustered federated learning based on the cosines similarity between gradients, which can relatively improve the overall performance of federated learning on heterogeneous UAVs. The worst result of CFL is to converge to the FL structure without clustering. Theoretically, if we take the homogeneous dataset as the ideal case to observe the performance of OFL and CFL, where OFL has better convergence performance due to the highly consistent data. In the experiment, we rotate and disturb the data set, which is more consistent with the heterogeneity of the real situation. First, we gradually increased the heterogeneity of the data in four experiments by adding random twiddle factors to the data. We can observe that CFL shows obvious advantages in the distributed training of heterogeneous UAV swarms, where the divergence of datasets can lead to worse performances of federated learning systems such as lower accuracy, weaker convergence, and longer delay. When the data collected by the device has a higher disparity, there will be more instability and anomalies in the results of OFL training.

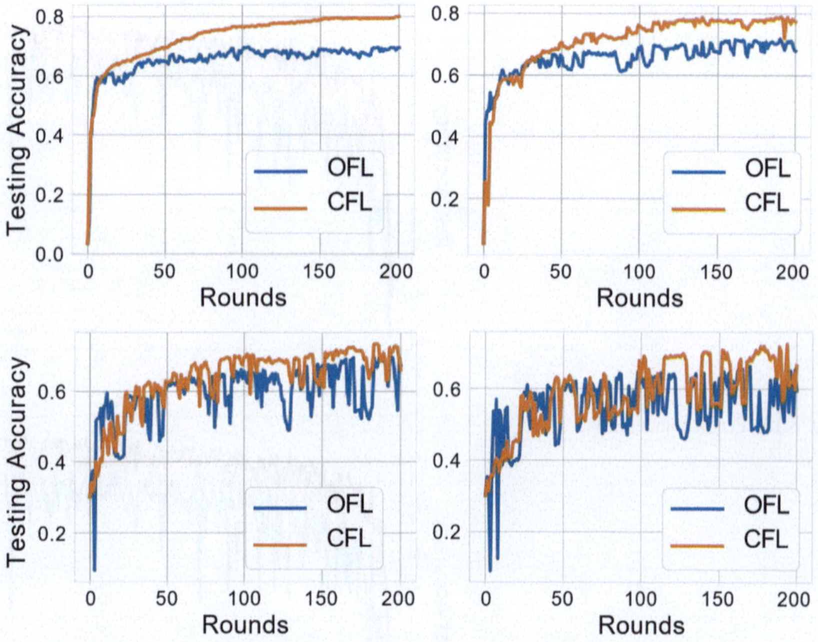

Fig. 5.13 Original federated learning vs. clustered federated learning with heterogeneous data

The Effectiveness of a Different Number of Clusters

Furthermore, we consider the effect of a different number of clusters on the performance of the CFL system and compare the performance of the system under a different preset number of clusters. Theoretically, the more grain-fined the clusters are, the better the training performances will be, while the higher depth of the cluster will also bring about longer training time and an unbalanced cluster structure. Clustering needs to take into account the balance of the size and structure of the cluster. It is not that the more the number of clusters, the better the system performance, considering extreme clustering returns to the OFL system. We set 30 clients with a different number of clusters as 1 (without clustering), 3, and 5. As shown in Fig. 5.14, the accuracy of the test dataset fluctuates greatly without clustering. Clustering based on the heterogeneity of data can effectively enhance the stability of the whole FL system. Given the CFL-based system have various accuracies of clusters with different sizes, we take advantage of the weighted precision to measure performance, which is proportional to the size of each set of data. As the number of clusters increases, the accuracy of the overall system will be higher.

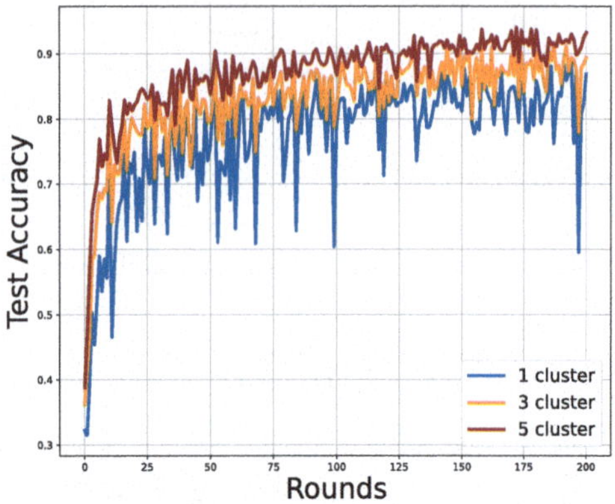

Fig. 5.14 Accuracy with a different number of clusters

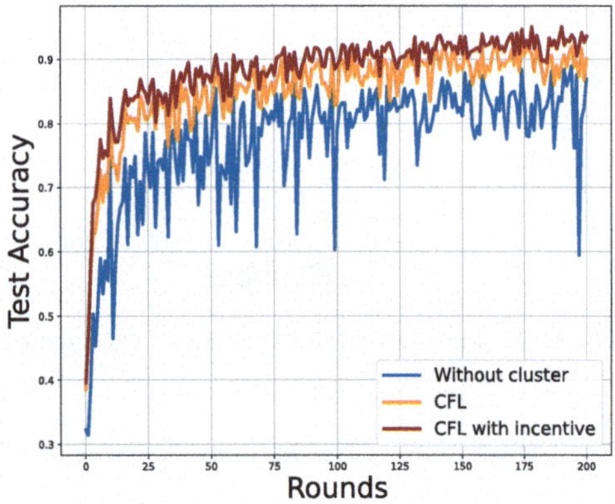

Fig. 5.15 Performance of 3 methods

The Effectiveness of the Incentive Mechanism

Based on the above analysis, we set the number of clusters as 5 controlling 30 clients. To evaluate the performance of the proposed incentive mechanism, we construct the model owner, cluster heads, and clients as the entities of the three-stage Stackelberg Game. As shown in Fig. 5.15, we can observe the test accuracy of original federated learning without clustering, CFL with 5 clusters, and CFL assisted with the proposed incentive mechanism. According to the analysis of the performance of the clustered federated learning architecture, we observe that the clustering method can effectively improve the accuracy of the whole federated learning system under a heterogeneous environment. Assisted with the incentive mechanism, we model the entities in the CFL as a three-stage Stackelberg game, a hierarchical master-slave game framework. Given the economic activities, appropriate game methods can stimulate more clients to contribute their computation resources so that the whole CFL system can achieve a higher utility.

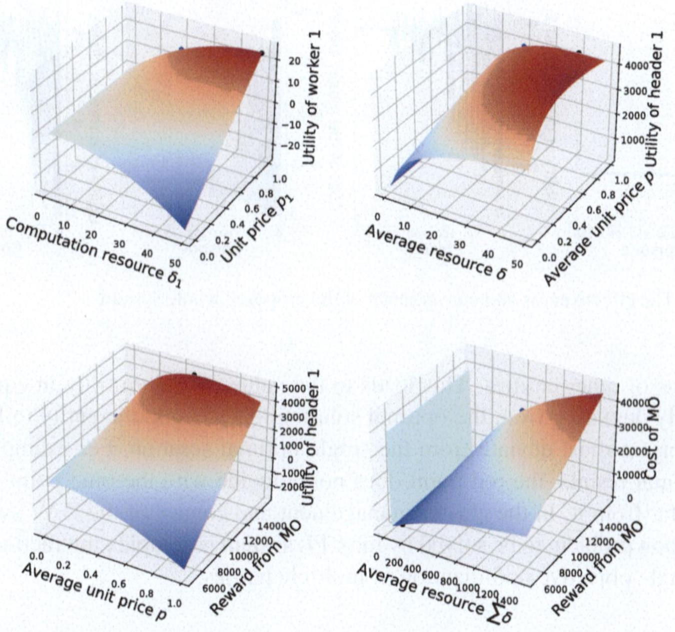

Fig. 5.16 Performance of the system model

5.2.4.3 Backward Induction Method Aided CFL

Accordingly, we focus on the economic analysis to explain the equilibrium performance of our proposed model, as shown in Fig. 5.16, where we highlight the theoretical optimal solution using black points considering only the unilateral model and the game equilibrium using blue points. Figure 5.16 clearly shows that the optimal solution of a single entity cannot be achieved under the game situation, and the equilibrium solution is the situation where multiple entities achieve stability. First, under the unified pricing mechanism in the same cluster, the optimal resource strategies of UAVs are relatively similar, for which we only observe one of the workers. Since the resources that can be sold by a UAV are limited, its optimal revenue is relatively close to the equilibrium revenue. Then, we observe the benefit of a certain cluster head, which is also affected by the resource pricing strategy, the number of resources traded, and the model owner's reward. In the fourth subgraph, when there are no resources, the training time approaches infinity, resulting in an infinite cost. Additionally, we can observe that the convergence order of the three entities is consistent with the action order of the SG, which is more specifically manifested by the fact that the leader can quickly converge to its own solution. In a master-slave game, in order to achieve a stable equilibrium, the model owner, the cluster head, and the worker, in turn, makes their own decisions to deal with

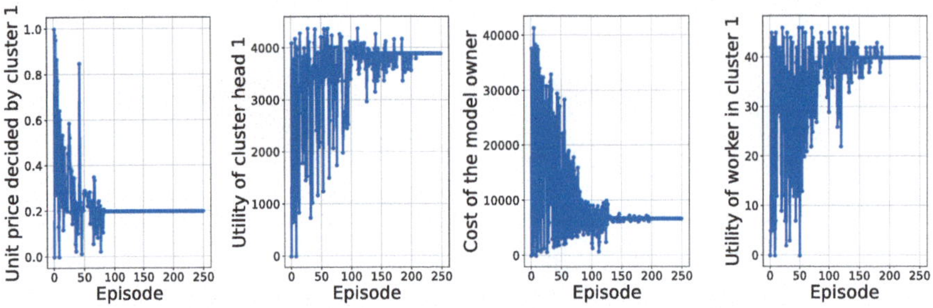

Fig. 5.17 The effectiveness and convergence of the proposed whole system

the actions of other entities. This leads to the value of entity utility in equilibrium necessarily deviating from the optimal solution. As a result, the entity utility value in equilibrium must deviate from the single optimal solution. For example, in the experimental results, the red point does not coincide with the blue point and even deviates far from it. In the cluster management and control of the UAV swarm, the optimization problem to be solved by the CFL system is complex interactions, rather than a single objective optimization of multiple parties.

5.2.4.4 Hierarchical Multi-Agent Reinforcement Learning Algorithm Aided CFL

System Evaluation

To evaluate the effectiveness and convergence of the proposed whole system, we observe the trend of variables and utilities in the system. In the process of distributed training of UAV swarms, the three variables of resource pricing in the cluster, the number of resources that nodes in the cluster are willing to sell, and the time spent in each round will affect each other. As shown in Fig. 5.17, we first observe the price strategy decided by the cluster head with randomly initial settings. As the training period increases, the unit price of the computation resource decided by cluster heads converges to the equilibrium state. Through experiments, we found that the pricing strategies of different clusters after the game are similar, this is because of the relatively balanced cluster size and the total number of samples in the cluster. The impact of the inconsistent training pace caused by the amount of data on the overall performance is mainly reflected in potentially important equipment falling behind, but it is far less than the negative effect of data structure heterogeneity on system performance.

In order to prevent the delay caused by an unbalanced cluster structure, the size of the five clusters is relatively balanced in our CFL architecture. Then, we record the utility of different clusters, which is affected by both pricing intra-cluster and selling inter-cluster. We first observe the unit price decided by cluster head 1 that the price reaches an equilibrium state after 80 episodes and the CPU frequency cycle

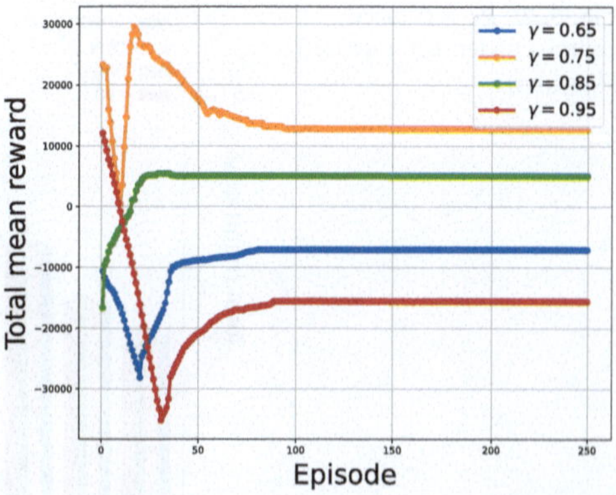

Fig. 5.18 The total mean reward under different discount factors γ

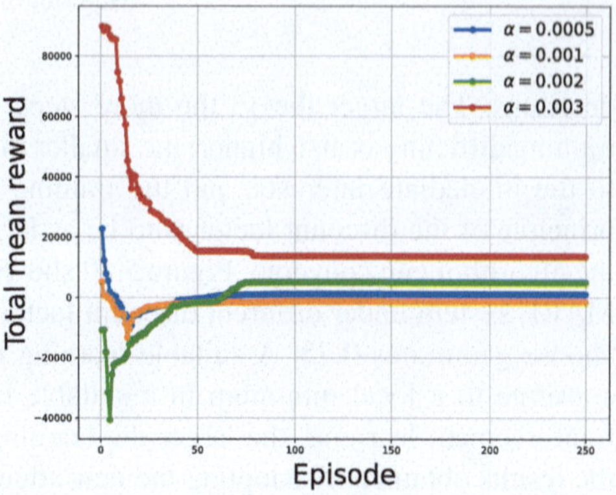

Fig. 5.19 The total mean reward under different learning rates α

price per cycle is approximately 0.2. Then, we evaluate the utility of cluster head 1, which is affected by the number of workers in cluster 1 and the payment from the model owner. Thirdly, the cost of the model owner throughout the three-party game tends to steady state after 150 episodes. The payment of the model owner to the cluster head is convergent, and the higher the reward, the higher the income of the cluster head. The payoffs for different clusters end up in a stable interval. Finally, we randomly select a worker node in each cluster to observe the change in its utility.

The Performance of RL Algorithm

The experimental results in Fig. 5.18 show that the clustered FL architecture under the three-stage Stackelberg game as an incentive mechanism, using the hierarchical MARL algorithm, has good convergence, and can achieve multi-party stability and balance. At the same time, it also shows that with the assistance of the MARL algorithm, the utility function and the cost function maintain bounded rationality. We further find different values of learning rate and future discount factor have some influence on the convergence performance of the proposed algorithm, shown in Fig. 5.19. The discount factor γ is used to adjust the short-term and long-term effects in reinforcement learning, that is, how long the agent considers when making

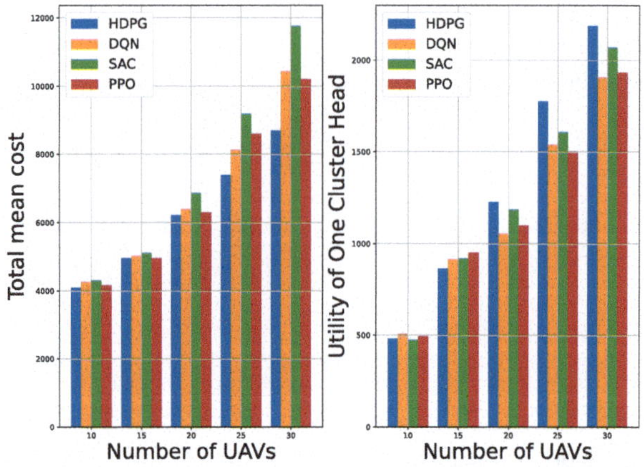

Fig. 5.20 Performance contrast under different DRL algorithms

decisions. The larger the γ, the more steps the agent considers forward, but the training difficulty is also higher; the smaller the γ, the more the agent pays attention to the immediate interests, and the training difficulty is also smaller. The value principle of the discount factor is to be as large as possible under the premise that the algorithm can converge. Figure 5.19 shows the convergence performance of the MARL system under different discount factors. In the final system experiment, we choose gamma as 0.75. A suitable learning rate can make the objective function converge to a local minimum in a suitable time. Theoretically, in the process of reinforcement learning, the larger the learning rate α, the larger the proportion of the results obtained by adopting the new attempt, and the smaller the proportion of keeping the old results. We evaluated the effect of different sizes of learning rates on system performance and chose 0.003.

The feasibility and effectiveness of the proposed hierarchical deep policy gradient (HDPG) algorithm are demonstrated by comparing it with popular DRL methods. As a model-free and deterministic policy gradient algorithm, HDPG is compared to deep Q-network (DQN), soft actor-critic (SAC), and proximal policy optimization (PPO). As shown in Fig. 5.20, HDPG can achieve higher rewards for the CH while consuming less cost of the MO compared to the other algorithms. The different algorithms have their own strengths and weaknesses. For example, while SAC may have poor performance in reducing MO cost, it has advantages in terms of CH's reward. The design of the hierarchical multi-agent behavior and learning process in HDPG helps agents learn their own strategies based on the strategies of competitors, which may help improve their performance. HDPG can perform better in these environments because it directly learns the policy, which can allow it to adapt more quickly to changing conditions like UAV swarms. In contrast, SAC and PPO algorithms learn the value function, which can make them more sensitive to changes in the environment. As the number of UAVs increases, HDPG has outstanding advantages in terms of system performance.

5.2.4.5 Robustness of the Hierarchical MARL Aided CFL System

The dynamic nature of distributed learning in UAV swarm scenarios cannot be ignored, as the clustering results of heterogeneous devices are not always static. The above analysis is based on a relatively static FL scenario. To evaluate the robustness and universality of the entire system, we introduce disturbance in each round of training. When the difference between nodes in the accumulated cluster becomes too large, the cluster must be reconstructed. In the extreme case, each round of training may require workers to re-select clusters, which is a significant issue due to the high dynamics of UAV swarms and the constant updating of local data.

Figure 5.21 shows the node ambiguity within cluster 1 in a static and dynamic distributed learning system. The difference in heterogeneous data caused by environmental noise increases, resulting in a larger variance in the data within dynamic clustering. In some rounds, the clustering results are adjusted in real-time according to the cosine similarity of the local learning direction, leading to very low intra-cluster data divergence. In most rounds, the static clustering system is in a more stable state after convergence. However, even in highly dynamic scenarios, the dynamic clustering system can still achieve convergence within a certain range. Interestingly, in the case of frequent re-clustering, the proposed dynamic system still performs well. Overall, based on fine-grained clustering, dynamic strategies result in lower loss values than static systems.

The challenge of using incentive mechanisms in highly dynamic scenarios is that clustering based on heterogeneous data does not necessarily result in higher local and system utility. The time cost involved cannot be considered as the cost of the model owner, otherwise, it will introduce a large error. We further observe the performance of the entire collaborative filtering (CFL) system under dynamic clustering, as shown in Fig. 5.22. It can be observed that the loss function of the dynamic CFL system fluctuates greatly at the beginning of training. This is due to the fact that environmental noise affects the difference of data within the cluster, and the learning difference within the cluster would be improved if there was no re-clustering. On the other hand, re-clustering also introduces a jitter to the whole system.

Based on the above experiments and analysis of the overall performance of the system, we can conclude that the hierarchical intensive mechanism based on CFL has strong robustness and can be extended to the dynamic UAV swarm scene. Taking into account the high dynamics and equipment heterogeneity of UAV swarms for distributed learning, the clustering system reflects high robustness and is highly suitable for such an FL model. Combining multiple entities in the CFL system to model the game mechanism can bring higher benefits and better system stability.

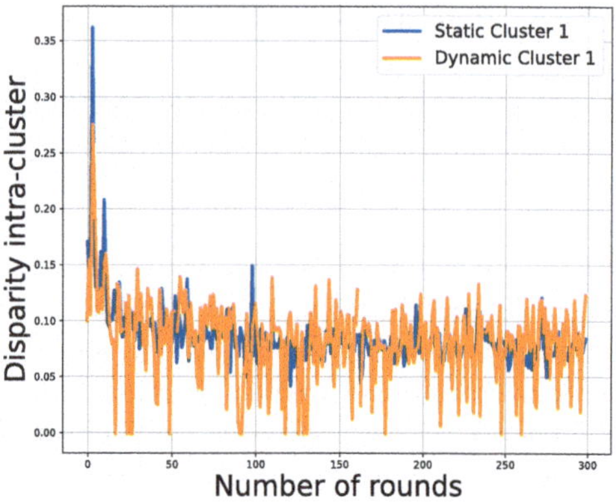

Fig. 5.21 Disparity intra-cluster under static and dynamic systems

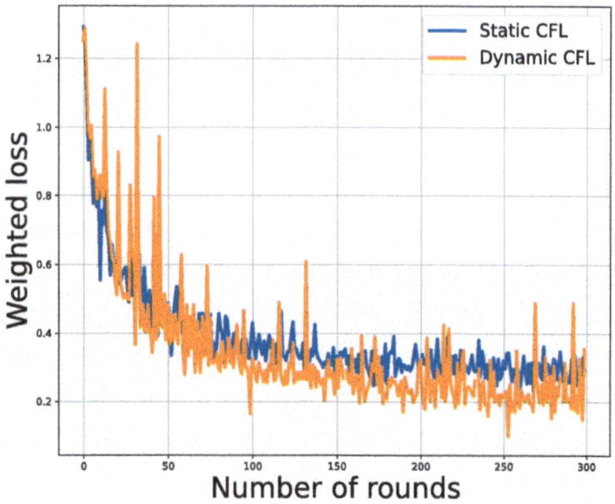

Fig. 5.22 Weighted loss under static and dynamic systems

5.3 Summary

This chapter discusses distributed learning empowered UAV cooperation methods. We first propose a clustered FL architecture, where learning and interaction are modeled as a three-stage Stackelberg game. A multi-agent reinforcement learning algorithm is introduced to search for the game equilibrium. The proposed method demonstrates good performance in resource-constrained dynamic scenarios. Additionally, this chapter employs a double auction mechanism to facilitate the trading of training models between data owners and FL tasks. To maximize individual utility and social welfare, an IDA mechanism is proposed. This mechanism uses a centralized controller to collect latent information, enabling rapid convergence to the optimal solution. To overcome the centralized reliance of IDA, we further propose an EWA-DA mechanism, which optimizes strategies through continuous interactions.

References

1. T. Mai, H. Yao, J. Xu, N. Zhang, Q. Liu, S. Guo, Automatic double-auction mechanism for federated learning service market in internet of things. IEEE Trans. Netw. Sci. Eng. **9**(5), 3123–3135 (2022)
2. W. He, H. Yao, T. Mai, F. Wang, M. Guizani, Three-stage stackelberg game enabled clustered federated learning in heterogeneous uav swarms. IEEE Trans. Veh. Technol. **72**(7), 9366–9380 (2023)
3. A. Ghosh, D. Chakraborty, A. Law, Artificial intelligence in internet of things. CAAI Trans. Intell. Technol. **3**(4), 208–218 (2018)
4. C. Qiu, H. Yao, C. Jiang, S. Guo, F. Xu, Cloud computing assisted blockchain-enabled Internet of Things. IEEE Trans. Cloud Comput. **10**, 247–257 (2022)
5. Q. Yang, Y. Liu, T. Chen, Y. Tong, Federated machine learning: concept and applications. ACM Trans. Intell. Syst. Technol. **10**(2), 1–19 (2019)
6. T. Li, A.K. Sahu, A. Talwalkar, V. Smith, Federated learning: challenges, methods, and future directions. IEEE Signal Process. Mag. **37**(3), 50–60 (2020)
7. B.P. Majumder, M.N. Faqiry, S. Das, A. Pahwa, An efficient iterative double auction for energy trading in microgrids, in *2014 IEEE Symposium on Computational Intelligence Applications in Smart Grid (CIASG)* (IEEE, Piscataway, 2014), pp. 1–7
8. Q. Yu, Y. Liu, D. Xia, L. Martínez, The strategy evolution in double auction based on the experience-weighted attraction learning model. IEEE Access **7**, 16730–16738 (2019)
9. S.R. Pandey, N.H. Tran, M. Bennis, Y.K. Tun, A. Manzoor, C.S. Hong, A crowdsourcing framework for on-device federated learning. IEEE Trans. Wireless Commun. **19**(5), 3241–3256 (2020)
10. J. Konečnỳ, Z. Qu, P. Richtárik, Semi-stochastic coordinate descent. Optim. Methods Softw. **32**(5), 993–1005 (2017)
11. J. Konečnỳ, H.B. McMahan, D. Ramage, P. Richtárik, Federated optimization: distributed machine learning for on-device intelligence (2016, preprint). arXiv:1610.02527
12. T. Mai, H. Yao, N. Zhang, L. Xu, M. Guizani, S. Guo, Cloud mining pool aided blockchain-enabled internet of things: an evolutionary game approach. IEEE Trans. Cloud Comput. **11**, 692–703 (2023). https://dx.doi.org/10.1109/TCC.2021.3110965
13. G. Iosifidis, L. Gao, J. Huang, L. Tassiulas, A double-auction mechanism for mobile data-offloading markets. IEEE/ACM Trans. Netw. **23**(5), 1634–1647 (2014)
14. J. Du, C. Jiang, E. Gelenbe, H. Zhang, Y. Ren, T.Q. Quek, Double auction mechanism design for video caching in heterogeneous ultra-dense networks. IEEE Trans. Wireless Commun. **18**(3), 1669–1683 (2019)
15. F. Li, H. Yao, J. Du, C. Jiang, Z. Han, Y. Liu, Auction design for edge computation ofloading in SDN-based ultra dense networks. IEEE Trans. Mobile Comput. **21**, 1580–1595 (2022)
16. L.P. Kaelbling, M.L. Littman, A.W. Moore, Reinforcement learning: a survey. J. Artif. Intel. Res. **4**, 237–285 (1996)
17. T. Mai, H. Yao, N. Zhang, W. He, D. Guo, M. Guizani, Transfer reinforcement learning aided distributed network slicing resource optimization in industrial IoT. IEEE Trans. Ind. Inf. **18**, 4308–4316 (2022)
18. M. Pipattanasomporn, H. Feroze, S. Rahman, Multi-agent systems in a distributed smart grid: design and implementation, in *2009 IEEE/PES Power Systems Conference and Exposition* (IEEE, Piscataway, 2009), pp. 1–8
19. J. Wang, C. Jiang, H. Zhang, Y. Ren, K.C. Chen, L. Hanzo, Thirty years of machine learning: the road to pareto-optimal wireless networks. IEEE Commun. Surv. Tuts. **22**(3), 1472–1514 (2020)
20. L. Matignon, G.J. Laurent, N. Le Fort-Piat, Independent reinforcement learners in cooperative markov games: a survey regarding coordination problems. Knowl. Eng. Rev. **27**(1), 1–31 (2012)

21. J. Wang, C. Jiang, K. Zhang, X. Hou, Y. Ren, Y. Qian, Distributed Q-learning aided heterogeneous network association for energy-efficient IIoT. IEEE Trans. Ind. Inf. **16**(4), 2756–2764 (2019)
22. D. Silver, G. Lever, N. Heess, T. Degris, D. Wierstra, M. Riedmiller, Deterministic policy gradient algorithms, in *International Conference on Machine Learning* (PMLR, New York, 2014), pp. 387–395
23. Y. Wang, Z. Su, T. Luan, R. Li, K. Zhang, Federated learning with fair incentives and robust aggregation for UAV-aided crowdsensing. IEEE Trans. Netw. Sci. Eng. **9**, 3179–3196 (2022)
24. Y. Wang, Z. Su, N. Zhang, A. Benslimane, Learning in the air: secure federated learning for UAV-assisted crowdsensing. IEEE Trans. Netw. Sci. Eng. **8**(2), 1055–1069 (2020)
25. E. Li, L. Zeng, Z. Zhou, X. Chen, Edge AI: on-demand accelerating deep neural network inference via edge computing. IEEE Trans. Wireless Commun. **19**(1), 447–457 (2019)
26. H. Zhanga, W. Quana, Networking automation and intelligence: a new era of network innovation. Education **2095**, 8099 (2021)
27. W.Y.B. Lim, J. Huang, Z. Xiong, J. Kang, D. Niyato, X.S. Hua, C. Leung, C. Miao, Towards federated learning in UAV-enabled internet of vehicles: a multi-dimensional contract-matching approach. IEEE Trans. Intel. Transp. Syst. **22**(8), 5140–5154 (2021)
28. X. Wang, Y. Han, C. Wang, Q. Zhao, X. Chen, M. Chen, In-edge AI: intelligentizing mobile edge computing, caching and communication by federated learning. IEEE Netw. **33**(5), 156–165 (2019)
29. C. Xu, Y. Qu, Y. Xiang, L. Gao, Asynchronous federated learning on heterogeneous devices: a survey (2021, preprint). arXiv:2109.04269
30. F. Sattler, K.R. Müller, W. Samek, Clustered federated learning: model-agnostic distributed multitask optimization under privacy constraints. IEEE Trans. Neural Netw. Learn. Syst. **32**(8), 3710–3722 (2021)
31. R. Jiang, S. Zhou, Cluster-based cooperative digital over-the-air aggregation for wireless federated edge learning, in *2020 IEEE/CIC International Conference on Communications in China (ICCC)* (IEEE, Piscataway, 2020), pp. 887–892
32. A. Ghosh, J. Hong, D. Yin, K. Ramchandran, Robust federated learning in a heterogeneous environment (2019, preprint). arXiv:1906.06629
33. X. Tu, K. Zhu, N.C. Luong, D. Niyato, Y. Zhang, J. Li, Incentive mechanisms for federated learning: from economic and game theoretic perspective. IEEE Trans. Cognit. Commun. Netw. **8**, 1566–1593 (2022)
34. Y. Gong, H. Yao, J. Wang, M. Li, S. Guo, Edge intelligence-driven joint offloading and resource allocation for future 6g industrial internet of things. IEEE Trans. Netw. Sci. Eng. **11**, 5644–5655 (2024)
35. Y. Zhao, M. Li, L. Lai, N. Suda, D. Civin, V. Chandra, Federated learning with non-iid data (2018, preprint). arXiv:1806.00582
36. Y. Cao, X. Ren, C. Qiu, X. Wang, Hierarchical reinforcement learning for blockchain-assisted software defined industrial energy market. IEEE Trans. Ind. Inf. **18**, 6100–6108 (2022)
37. P. Xia, L. Zhang, F. Li, Learning similarity with cosine similarity ensemble. Inf. Sci. **307**, 39–52 (2015)
38. J. Shao, Y. Xu, J. Liu, H. Takakura, Z. Li, X. Dong, On strategic interactions in blockchain markets: a three-stage stackelberg game approach, in *2021 IEEE Global Communications Conference (GLOBECOM)* (IEEE, Piscataway, 2021), pp. 1–6
39. L. Gu, W. Zhang, Z. Wang, D. Zeng, Service management and energy scheduling toward low-carbon edge computing. IEEE Trans. Sustain. Comput. **18**, 6100–6108 (2022)
40. C. Yu, W. Quan, D. Gao, Y. Zhang, K. Liu, W. Wu, H. Zhang, X. Shen, Reliable cybertwin-driven concurrent multipath transfer with deep reinforcement learning. IEEE Internet Things J. **8**(22), 16207–16218 (2021)
41. W. Quan, M. Liu, N. Cheng, X. Zhang, D. Gao, H. Zhang, Cybertwin-driven DRL-based adaptive transmission scheduling for software defined vehicular networks. IEEE Trans. Veh. Technol. **71**(5), 4607–4619 (2022)

42. Z. Qin, H. Yao, T. Mai, D. Wu, N. Zhang, S. Guo, Multi-agent reinforcement learning aided computation offloading in aerial computing for the internet-of-things. IEEE Trans. Serv. Comput. **16**, 1976–1986 (2023)
43. G. Cohen, S. Afshar, J. Tapson, A. Van Schaik, Emnist: extending mnist to handwritten letters, in *2017 International Joint Conference on Neural Networks (IJCNN)* (IEEE, Piscataway, 2017), pp. 2921–2926

Chapter 6
Semantic Communication Empowered UAV Swarm Networks

Traditional data-centric communication schemes are often constrained by the limited transmission bandwidth between UAVs. To address these challenges, this chapter introduces the semantic communication paradigm into UAV swarm networks. Such paradigm aims to reduce data transmission volumes and enhance communication efficiency by transmitting only the semantic information extracted from raw data, rather than transmitting the entire dataset. Despite the potential benefits of semantic communication, training accurate semantic models on resource-constrained UAV devices remains a significant challenge, primarily due to the dynamic topology and fluctuating communication links within UAV networks. To overcome this limitation, a hierarchical learning framework is proposed. In this framework, each UAV trains its own semantic model based on local data and uploads the model parameters to its corresponding cluster head. This architecture facilitates lightweight semantic communication among UAVs, thereby improving overall communication efficiency [1]. Furthermore, the chapter explores a semantic-aware environment interaction mechanism for UAV swarm networks. Real-world UAV swarm environments are characterized by uncertainty and randomness, which can lead to instability in information exchange and inefficiencies in data sampling–factors that become critical performance bottlenecks. To mitigate these challenges, a metaverse-based UAV swarm system is proposed. This system enables frequent data exchanges between the metaverse and the real world by utilizing semantic communication techniques to transmit only relevant semantic information, thereby significantly reducing communication latency [2].

6.1 Semantic Communication for UAV Swarm Cooperation

Recently, the UAV swarm has emerged as a transformative force to reshape our lives and works [3]. It refers to a coordinated group of multiple UAVs operating

H. Yao, T. Mai, *UAV Swarm Cooperation*, Wireless Networks,
https://doi.org/10.1007/978-3-031-96444-2_6

collaboratively to accomplish highly sophisticated and intelligent tasks. In this case, the UAV swarm is confronted with the need to integrate diverse capabilities including sensing, data collection, and communication services, which might lead to significant information exchange pressure. Due to the Shannon capacity limit, current data-oriented communication paradigms are not suitable for handling substantial volume of information exchange, thus restricting the advancement of UAV swarm intelligence. This sparked significant interest in semantic communication, a novel paradigm that transmits the semantics of the information into communication rather than the single symbol or bit [4]. Such a paradigm greatly reduces the amount of transmitted data while simultaneously strengthening the accuracy of communication. By integrating semantic communication into the UAV swarm network, accurate and real-time communication can be guaranteed, especially for tasks demanding extensive information exchange with low latency.

Generally, the effectiveness of semantic communication is mainly determined by the quality of semantic models, i.e., semantic encoder (decoder) and channel encoder (decoder), which are implemented in each communication node to extract the semantics of collected data. However, the limited computation and storage capabilities of UAVs hinder the onboard deployment of such semantic models. How to train these semantic models in the UAV swarm network is a prerequisite for guaranteeing accurate semantic communication. Traditionally, the training data collected by UAVs is sent back to a remote central server for model training. Such centralized learning paradigms introduce increased communication cost and communication delay under highly dynamic conditions, thus restricting the performance of the learning models. Besides that, directly transmitting raw data to third-party server raises data security and privacy issues.

In response, federated learning (FL) emerges as a promising solution to facilitate the model training of semantic communication [5]. Empowered by FL technology, the training process can be conducted locally on data owners, e.g., UAVs. Each UAV independently trains its own semantic model based on local data and then, periodically transmits model parameters to the server for reaching a global consensus. This paradigm accelerates the training process and reduces communication overhead by jointly training semantic models across UAV nodes. Another advantage is that FL enables data privacy by exchanging only model updates, rather than raw training data [6].

Nonetheless, there still exists several challenges which may impede the adaptive and sustainable implementation of FL [7]. FL paradigm can be classified into two categories, i.e., centralized FL (CFL) and decentralized FL (DFL). Conventionally, CFL employs the model owner, e.g., server, to coordinate training processes of data owners by aggregating and fusing local models [8]. This paradigm still relies on a central server, thus producing performance bottleneck and a single point of failure [9]. Specifically, DFL leverages data owners to act as parameter servers instead of a central server, in which the data owners broadcast local model updates to their neighbors so as to achieve the consistency of the global model [10]. Although the decentralized learning paradigm eliminates the single point of failure, it incurs high communication cost which limits network scalability.

Motivated by the above issues, we study hierarchical federated learning (HFL) framework, which capitalizes benefits of CFL and DFL, to support semantic communication in the UAV swarm. The main contributions of this article are summarized as follows:

- A HFL-powered semantic communication architecture is proposed for the UAV swarm, which is organized into clusters under cluster heads. We employ centralized learning for intra-cluster model training, whereas decentralized learning is employed for inter-cluster training. This design reduces the communication overhead while the system robustness is promised.
- To incentivize UAVs to participate in FL tasks, we formulate the cluster selection problem in intra-cluster training as an evolutionary game.Furthermore, we design a novel cluster management mechanism to mitigate the UAV dropout caused by the dynamic nature of the UAV swarm.
- We develop an inter-cluster decentralized training method to achieve decentralized model aggregation and updates, therefore avoiding the overall system failure arising from the single point of failure.

6.1.1 Semantic Communication

6.1.1.1 Semantic Communication

Semantic communication is the emerging communication paradigm beyond the Shannon limits, which focuses on the meaning or semantics transmission of the data rather than accurate bitstream transmission [4]. Compared to the data-oriented communication paradigm, semantic communication exhibits the following properties:

Improved Communication Efficiency: By facilitating the transmission of semantic information, semantic communication effectively reduces the data size, thereby minimizing the utilization of communication resources and enhancing communication efficiency.

Task-oriented: Semantic communication is task-oriented by only transmitting necessary information which is relevant to the specific task at the receiver.

The semantic communication system includes the transmitter, channel, and receiver.

Transmitter: The transmitter is responsible for extracting and encoding semantic features from the raw data. To achieve this effectively, advanced technologies, i.e., ML and AI are employed. For example, in natural language recognition, long short-term memory (LSTM) is commonly utilized, while convolutional neural networks (CNN) are adopted for image recognition [11].

Channel: The channel serves as the medium for propagating the semantic features from the transmitter to the receiver. The semantic features suffer distortion and channel noise during the propagation process.

Receiver: The receiver is in charge of decoding the received semantic features to reconstruct the original data.

6.1.1.2 Federated Learning

Federated learning is a kind of collaborative training technique that trains AI models over decentralized devices with local data samples. As shown in Fig. 6.1, two different paradigms of FL can be implemented.

In the CFL paradigm, model aggregation, updating, and coordination are performed through a central server. The server derives global model by fusion of local models and then transmits such model back to the data owners, i.e., devices. The CFL training process follows the subsequent steps [8].

(1) Task initialization: The server initiates the FL task and notifies task-related information the data owners. Based on their local strategies, data owners decide whether to participate in the task. The server then broadcasts the initial model parameters to all participating data owners.

(2) Local model training: Upon receiving the global model, the data owners commence the local model training by using their local data. The updated local parameters are subsequently uploaded to the server for further aggregation and synchronization.

(3) Global aggregation: The server receives and aggregates the local models from all data owners. Then, it derived the global model and sends back the updated global parameters.

Step *(2)* and *(3)* are repeated until convergence is reached.

Instead, without the aid of central server, the DFL paradigm allows each data owner to act as a parameter server [10]. In each iteration, data owners forward the model updates to their neighbors. Thereinafter, the data owner derives global model by conducting decentralized aggregation with its neighbors.

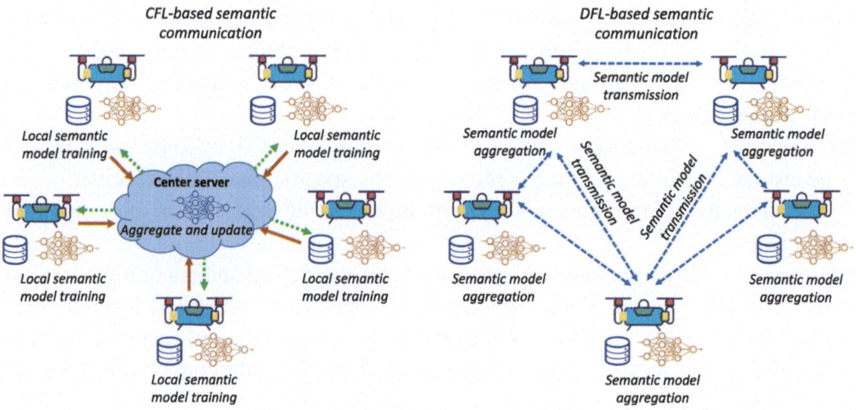

Fig. 6.1 CFL-based semantic communication vs DFL-based semantic communication

6.1.2 Hierarchical Learning Powered Semantic Communication Architecture

In this section, we design hierarchical federated learning-powered semantic communication architecture for the UAV swarm, which is organized into clusters under cluster heads.

Assume that the cluster heads are predetermined through the cluster head selection algorithm. The details of this algorithm are not the focus of this part. Figure 6.2 provides an overview of proposed architecture, including the swarm layer, the backbone layer, and the application layer. Besides, leveraged by DNNs, an auto-encoder is employed as the semantic model, in which the semantic encoder and decoder are jointly trained.

6.1.2.1 Swarm Layer

The swarm layer is constituted of a cooperative group of UAVs, which collect the training data and serve as data owners. In this layer, the intra-cluster model training adopts the centralized learning paradigm. First, each UAV determines its cluster selection so as to maximize derived utility. The initialized semantic model parameters are then disseminated to all UAVs within the cluster. After that, the UAV iteratively trains local semantic model based on its local data and sends the model parameters to the associated cluster head via the wireless channels. To address the issue of UAV dropout throughout this process, cluster management mechanism is employed, facilitating the adjustment of cluster selection by UAVs as needed.

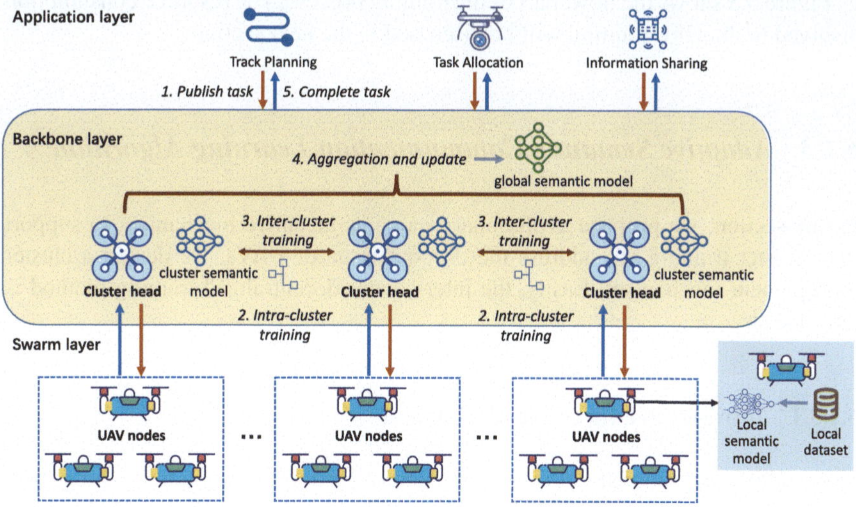

Fig. 6.2 HFL powered semantic communication architecture for UAV swarms

6.1.2.2 Backbone Layer

The backbone layer is formed by cluster heads, which serve as the model owners for aggregating updated parameters and facilitating inter-cluster training. We employ a decentralized method for inter-cluster training so as to eliminate the single point of failure and communication congestion. In each iteration, the cluster head receives local parameters from its cluster members and updates intermediate model accordingly. Next, through collaborating with neighbors, each cluster head performs intermediate model updates for global aggregation. Thereafter, the resulting global model is disseminated back to all UAVs within the cluster for another iteration of updating. After a number of iterations, the semantic model converges to the desired accuracy level.

6.1.2.3 Application Layer

The application layer is responsible for task punishment. Given the diverse communication demands (e.g., communication delay, accuracy) and data types (e.g., text, image, video) associated with various tasks, task-oriented semantic models (e.g., with different compression rates) are required. The task publisher in the application layer broadcasts the task to the UAV swarm, along with its requirements and corresponding reward. Upon task assignment, the semantic model is initialized for the specific task. Through cross-layer HFL training, the UAV swarm ultimately develops a well-trained semantic model. By utilizing this precise semantic model, low latency and stable communication is approached within the UAV swarm. As such, the UAV swarm becomes proficient in executing the task necessitating extensive communication and coordination with low latency.

Figure 6.3 shows the flowchart of the training process. The resource consumption involved in the HFL training will be discussed in the next section.

6.1.3 Adaptive Semantic Communication Learning Algorithm

In this section, we propose evolutionary game as incentive mechanism to support intra-cluster training. To address the dropout issue of UAVs, we design a cluster management mechanism. Lastly, the inter-cluster decentralized training method is introduced.

6.1.3.1 Intra-cluster Incentive Mechanism

Since selfish and rational UAVs are reluctant to participate in model training, how to stimulate the participation of the UAVs is a critical issue. As such, we apply the evolutionary game to formulate the cluster selection. The UAV utility and replicator

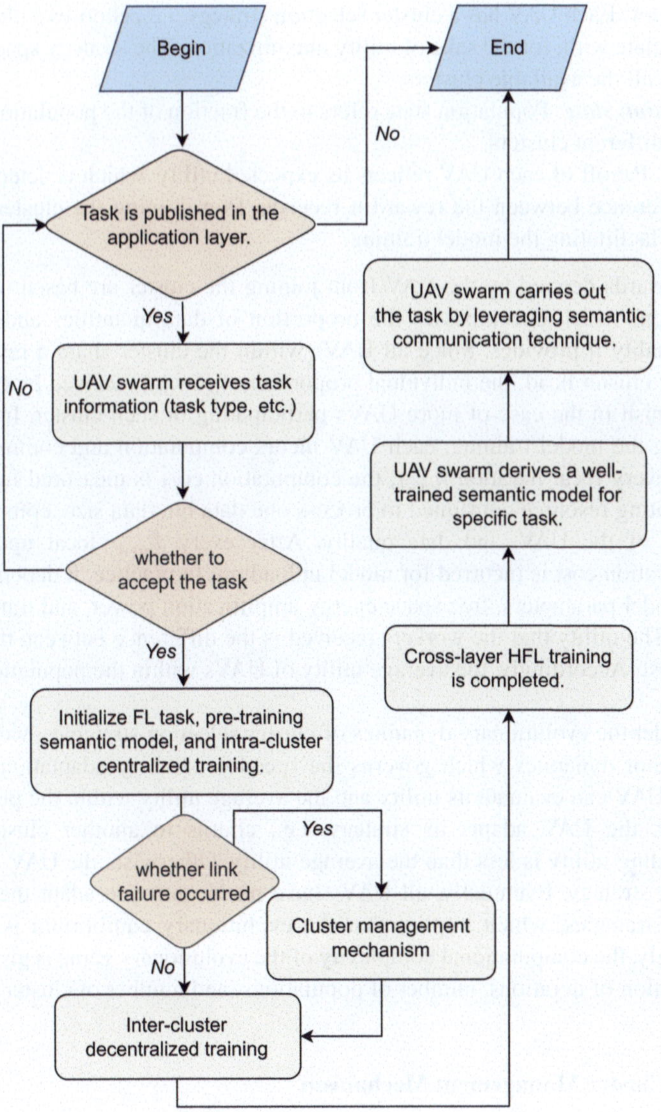

Fig. 6.3 Flowchart of the HFL powered semantic communication

dynamics are introduced to characterize the strategy adaptation of rational agents [12]. The evolutionary game based cluster selection can be formulated as:

- ***Players***: In the UAV swarm, each UAV is regarded as a player of the evolutionary game.
- ***Population***: UAVs are partitioned into several populations according to their local data quality and data quantities.

- ***Strategies***: Each UAV has a cluster selection strategy, i.e., chooses a cluster head to associate with, for the sake of utility maximization. The strategy space of each UAV is all the available clusters.
- ***Population state***: Population state refers to the fraction of the population that has joined different clusters.
- ***Payoff***: Payoff of each UAV reflects its expected utility, which is determined by the difference between the reward it receives from joining the cluster, and the cost of facilitating the model training.

The rewards derived by the UAV from joining the cluster are based on its data contribution, which encompasses the proportion of data quantities and the level of data quality it provides. Since all UAVs within the cluster share a reward pool offered by cluster head, the individual proportion of reward received by each UAV may diminish in the case of more UAVs participating in such cluster. In terms of facilitating the model training, each UAV incurs computation and communication costs. In every local iteration i_{local}, the computation cost is measured in terms of the computing resource consumed to process one data bit, data size, computational capability of the UAV, and data quality. After every I_{local} local updates, the communication cost is incurred for model uploading. In practice, it depends on the size of model parameters, free space energy amplification power, and transmission distance. The utility that the workers received is the difference between the reward and the cost. Accordingly, the average utility of UAVs within the population can be derived.

To model the evolutionary dynamics of cluster selection strategies, we consider the replicator dynamics which governs the speed of strategy adaptation. In each iteration, UAV can evaluate its utility and the average utility within the population. Thereafter, the UAV adapts its strategy, i.e., churns to another cluster if the corresponding utility is less than the average utility. Otherwise, the UAV will keep its current strategy. Eventually, all UAVs have no incentive to adapt their cluster selection strategies, which implies that the evolutionary equilibrium is reached. Accordingly, the computational complexity of the evolutionary game is given by the multiplication of iterations, number of populations, and number of cluster heads.

6.1.3.2 Cluster Management Mechanism

Due to the dynamic topology and volatile communication link in the UAV swarm, the FL training faces the issue of UAV dropout. In consideration of this problem, we design a cluster management mechanism to facilitate intra-cluster training.

We denote the link state between UAV and cluster head by link-state parameters, where a link-state parameter value of 0 refers to the link fault, whereas link-state parameter value of 1 indicates otherwise. For effective cluster management, we design a utility function to measure the marginal contribution of UAVs. We consider that the UAV's marginal contribution is the difference in intermediate model accuracy derived by cluster head between before and after the UAV joins this cluster. Specifically, the intermediate model accuracy of cluster head relates to the data quality and data size of all UAVs within the cluster. We hypothesized

Algorithm 17 Adaptive federated learning algorithm

Initialization: federated learning parameters, semantic model parameters, I_{local}, I_{global} and T
$i_{global} \leftarrow 1$
while $i_{global} < I_{global}$ **do**
  $i_{local} \leftarrow 1$
  for each UAV participating in FL task **do**
    chooses the cluster based on intra-cluster incentive mechanism.
  end for
  while $i_{local} < I_{local}$ **do**
    for each UAV **do**
      trains local semantic model
    end for
    $i_{local} \leftarrow i_{local} + 1$
  end while
  while link fault occurs **do**
    for each corresponding UAV **do**
      switch to another cluster based on cluster management mechanism.
    end for
  end while
  for each UAV **do**
    sends local parameters to associated cluster head.
  end for
  for each cluster head **do**
    aggregates an intermediate model based on received parameters.
  end for
  $i_{global} \leftarrow i_{global} + 1$
end while
$t \leftarrow 1$
while $t < T$ **do**
  for each cluster head **do**
    conducts intermediate model updates with neighbors for global aggregation.
  end for
  $t \leftarrow t + 1$
end while

that the rate of improvement in model accuracy should exhibit a diminishing trend as the data quality elevates, ultimately converging toward an upper threshold. As such, the UAV can potentially achieve higher utility by switching to a cluster exhibiting lower model accuracy. This design avoids the population imbalance across clusters which might deteriorate model quality imbalance, thus enhancing the system homogenization and learning efficiency. The detailed cluster management process is given as follows.

Step 1: Upon link fault occurring, the corresponding UAVs disassociate from their current cluster heads. To prevent these UAVs from dropping out of FL training, the cluster management mechanism is employed.

Step 2: The UAV sends the request bundle, containing data accuracy and data quantity, to connectable cluster heads according to the link-state parameters.

Step 3: Cluster head receives the request bundle as well as calculates the utility for UAV by considering its marginal contribution, and subsequently sends back the utility message.

Step 4: The UAV seamlessly switches to the cluster that maximizes the derived utility and proceeds to transmit local parameters to the cluster head.

6.1.3.3 Inter-cluster Decentralized Training Method

The inter-cluster decentralized training can be represented as a stochastic optimization problem, wherein each cluster head operates as a parameter server. Specifically, it requires constant model aggregations and updates between neighboring cluster heads until reaching convergence.

Such an updating process is also known as combinatorial adaption, in which the cluster head combines the model weights from neighbors and then, performs a simple step of stochastic gradient descent (SGD) based on its weights. In iteration t, the cluster head k transmits the model parameters $w_{t,k}$ and receives weights of its neighbors. With the receiving model parameters, the cluster head k conducts the aggregated model $\psi_{t,k}$. Then, gradient update $w_{t,k}$ is operated by running SGD according to the aggregated model $\psi_{t,k}$ and the loss associated with cluster head k. After T rounds of iterative update, each cluster head derives a global model, which is subsequently sent back to all UAVs within the cluster. Consequently, each UAV synchronizes its local model using the received model updates.

Note that the measurement of computation and communication costs involved in inter-cluster decentralized training is similar to that of intra-cluster training. Besides costs, the computational complexity of adaptive federated learning algorithm scales with the global model size and it is ruled by the number of iterations.

6.1.4 Simulation Results and Discussions

In this section, we present the performance evaluation of HFL-powered semantic communication architecture and the evolutionary game based cluster formation.

6.1.4.1 HFL Powered Semantic Communication Architecture

We consider a UAV swarm consisting of 20 UAVs and 5 cluster heads, where each UAV possesses a distinct amount of training data. In our experiment, the semantic model is trained by using the image dataset MNIST under the additive white Gaussian noise (AWGN) channel, with the signal-to-noise ratio (SNR) ranging from 0 to 10 dB. The local iterations are set to 100 steps, while the global iterations are set to 5 steps in the HFL paradigm. For comparison analysis, we adopt CFL paradigm as the baseline paradigm.

We first evaluate the performance of proposed HFL paradigm without considering the dynamics of the UAVs. In this scenario, the central server in the baseline sustains its uninterrupted operation. Figure 6.4a shows the training process of both the proposed HFL paradigm and the baseline. It is observed that the HFL paradigm converges after 150 rounds, which is slight slower than the baseline. From Fig. 6.4c, we can see that the performance gaps between the two paradigms remain within a reasonable range. This is because the decentralized training among clusters only aggregates the parameters from neighboring cluster heads, rather than all cluster heads. As a result, we can draw the conclusion that our HFL paradigm is capable of achieving nearly optimal performance.

Next, we conduct another simulation with consideration of the dynamics of the UAV swarm. Supported by the proposed HFL paradigm and the baseline, the performance evaluation of semantic communication is conducted. The accuracy of the semantic communication is measured by peak signal-to-noise ratio (PSNR) and mean square error (MSE). In Fig. 6.4b and c, we observe that the model trained with the HFL paradigm consistently outperforms the baseline as the channel signal-to-noise ratio increases. This is attributed to the cluster management mechanism, which avoids UAV dropout and ultimately ensures system robustness. This verifies the effectiveness of our proposed HFL paradigm in attaining accurate semantic communication within the UAV swarm.

6.1.4.2 Evolution Analysis for Cluster Selection

We evaluate the evolution behavior of the UAVs in the cluster formation process. UAVs are categorized into 4 populations based on their data quality and data quantity, with the population state being randomly initialized. The dynamics of population shares derived from associating with cluster heads are plotted in Fig. 6.5a. In each iteration, the UAV adjusts its cluster selection strategy according to the derived utility. After a certain number of iterations, the population shares converge to an equilibrium state. Our results demonstrate that the centralized controller consistently modifies the behavior of the UAVs.

As shown in Fig. 6.5b, the utilities of UAVs' strategies adaptation fluctuate over time. At the beginning, we can find that the utilities derived from associating with cluster header 2 are comparatively lower than those from others. Correspondingly, following the replicator dynamics, we have more workers churning to other clusters. This reflects that higher congestion in cluster header 2 results in diminished utility for UAVs associated with it. Eventually, the evolutionary equilibrium is achieved, in which no UAVs have the incentive to adapt their strategies.

6.1.5 Future Research Directions and Challenges

In this section, we discuss the challenges and future research directions of our semantic communication framework for the UAV swarm.

Fig. 6.4 Performance of proposed semantic communication architecture. (**a**) Convergence results of the proposed architecture. (**b**) Results in the high dynamic situation. (**c**) Results

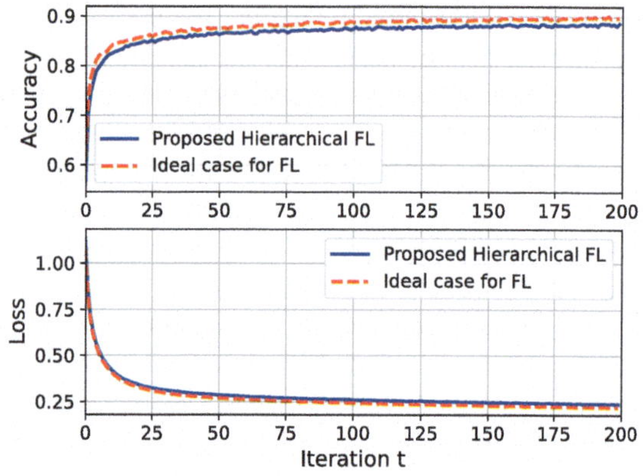

(a)

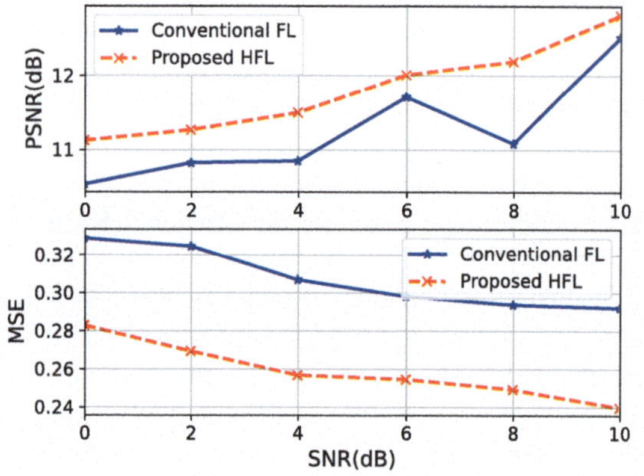

(b)

Setting	Metric	HFL	Baseline
Ideal case	Accuracy	0.8929	0.8936
	Loss	0.2362	0.2315
Dynamic condition	PSNR when SNR=4dB	11.5107	10.8573
	MSE when SNR=4dB	0.2572	0.3072

(c)

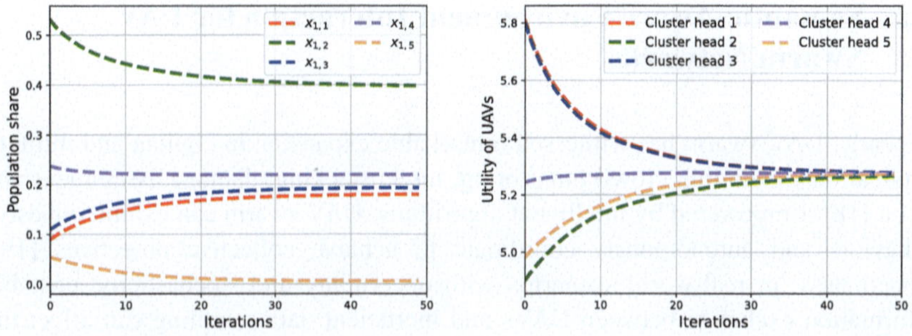

(a) (b)

Fig. 6.5 Simulation results of evolutionary game. (**a**) Population state of UAV. (**b**) Utility of choosing cluster head

Multi-modal Communication Since UAVs could collect environment information and generate data in multiple modalities, developing a multi-modal scheme has emerged as a crucial research area. In such a scheme, diverse data modalities, e.g., text, image, speech, and video, are transformed into a standardized format [13]. Therefore, to support the communication efficiency improvement within the UAV swarm, multi-modal semantic communication needs to be studied. Additionally, developing performance metrics to assess the effectiveness and efficiency of multi-modal semantic communication is an important direction for future research.

Resource Management Due to the energy and bandwidth constraints within the UAV swarm, an effective resource management scheme becomes crucial for guaranteeing successful task completion [14]. However, it encounters several challenges and limitations. On one hand, the dynamics in the UAV swarm make resource monitoring (e.g., location, speed, CPU utilization, network status, battery, and storage space) challenging. On the other hand, the heterogeneity in cross-layer resources adds complexity to resource management [15]. In our future works, the efficient and scalability resource management scheme should be explored.

Secure Semantic Communication In the face of widespread privacy threats, the integration of privacy protection technologies is imperative to establish secure semantic communication within the UAV swarm [16]. However, the existing researches rarely consider the trade-off between robustness and fairness, which hinders the efficient implementation of privacy-preserving mechanisms [17]. As research progresses, we should explore the potential of leveraging blockchain technology, i.e., a decentralized scheme that enhances privacy preservation, for robust and secure semantic communication.

6.2 Semantic-Aware Environment Interaction for UAV Swarm Networks

Recently, UAV swarm has witnessed remarkable expansion in civilian and military applications, including crowd monitoring, target tracking, damage assessment and so on [18]. Empowered by intelligent algorithms, UAV swarm can exhibit cohesive behaviors and autonomously coordinate to achieve collective objectives [19]. Nonetheless, in real-world scenarios with uncertainty and stochasticity, unstable information exchange between UAVs and inefficient data sampling can affect the model performance and training speed of intelligent algorithms [20]. Simultaneously, directly training the UAV swarm in a dynamic environment is confronted with undesirable factors such as physical obstacles, consequently resulting in additional costs and safety concerns [21]. The issues above become the bottleneck that impedes the enhancement of UAV swarm intelligence cooperation.

In particular, the metaverse can offer UAV swarm virtual services where monitoring, observation, analysis, and simulation can be realized collaboratively and virtually [22]. Utilizing sensor data collected by the UAV swarm, a Virtual Service Provider (VSP) creates and visualizes a virtual sub-world by replicating the physical environment and then provides a specific virtual service [23]. For an effective replication of the real world, we employ digital twin (DT) technology to capture the evolving states of the real-world entities, i.e., the UAVs, over time [24]. With robust virtual-real interaction capabilities and resilience to environmental factors, DT reliably generates high-fidelity digital representations of real-world entities from historical data, sensor data and physical objects [25], [26]. To this end, we use DT to establish a metaverse-based UAV swarm system, in which a learning model is trained with high-fidelity data to yield optimal decisions accurately and promptly [27].

To replicate the physical world to the finest details, the UAV swarm is required to guarantee real-time and continuous data synchronization from the physical world to the virtual world. In detail, the UAV swarm needs to effectively transmit relevant data, including basic information, status, and characteristics, to the VSPs within specified latency requirements [28]. However, when confronted with complex and dynamic environments, limited bandwidth and transmission unreliability may lead to the failure of the UAV swarm in guaranteeing real-time and large-scale data synchronization between these two worlds. The unreliability of the DT derived in this case would consequently affect the accuracy of the virtual services provided by the VSPs [29].

Fortunately, semantic communication (SemCom) as a promising technique, can help to reduce communication latency by transmitting only the semantic information while eliminating unnecessary information from the original data [30]. Moreover, it exhibits greater resistance to channel noise and interference compared to bit streams in conventional communication, thereby improving data transmission reliability [11]. In pursuit of achieving real-time synchronization, we employ the SemCom technique within the metaverse-based UAV swarm network to facilitate frequent data exchange and state synchronization between the two worlds.

In such a design, UAVs extract the semantic information of collected data based on its semantic model and then, transmit it to the VSP. Subsequently, VSP in the metaverse utilizes semantic information from corresponding UAVs to update its learning model. This design can significantly reduce communication latency and improve transmission dependability, equivalently alleviating communication pressure. However, there are still several issues that need to be addressed. On the one hand, the workers, i.e., UAVs, may be unwilling to participate in the data synchronization task without a reasonable incentive due to their self-interest. On the other hand, there may be malicious workers that perform intended operations such as forging, deleting, or replacing semantic information. Additionally, unstable wireless communication channel conditions and high mobility or energy constraints of UAVs may result in their unintentional behaviors of generating inaccurate semantic information. Both of these unreliable behaviors could deteriorate the decision-making performance of the VSP.

To tackle the aforementioned challenges, we introduce a novel two-layer architecture to ensure reliable worker selection and incentivize UAV participation in data synchronization tasks. In the lower layer, we consider a multi-armed bandit (MAB)-based worker selection scheme to assess the reliability of UAVs. Then each cluster head gathers semantic information from the selected workers within the cluster to realize low-cost and high-speed transmission. In the upper layer, we propose a deep learning (DL)-based auction model to attract workers to participate in data synchronization tasks in the metaverse.

The main contributions of this part are as follows:

- Leveraged by the SemCom technique, we propose a novel approach to enabling data synchronization in the metaverse-based UAV swarm system. Rather than transmitting raw data, UAVs within each cluster send extracted semantic information to these associated VSPs, thereby reducing the volume of transmitted data and consequently minimizing latency.
- We introduce a hierarchical architecture to investigate the reliability and sustainability of metaverse-based UAV swarm systems. In the lower layer, we propose an MAB-based online worker selection scheme in which reliable UAVs are selected for data synchronization tasks.
- In the upper layer of hierarchical architecture, we develop a DL-based auction game for incentivizing the UAVs' participation in semantic information trading, which is announced by VSPs.

6.2.1 System Model and Problem Formulation

6.2.1.1 Metaverse-based UAV Swarm Architecture

In this section, we propose a novel semantic-aware metaverse-based UAV swarm architecture to ensure UAV swarm cooperation.

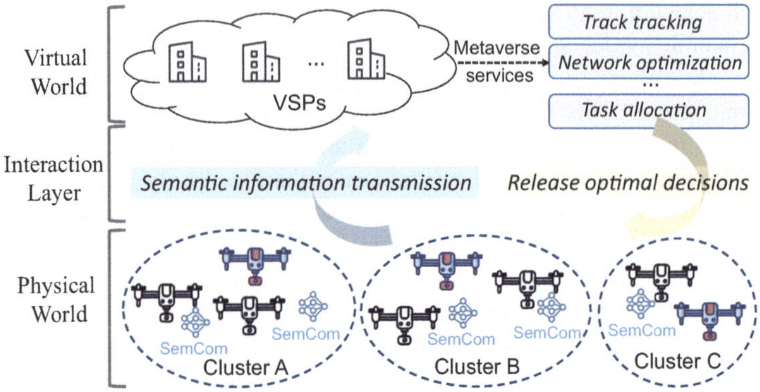

Fig. 6.6 Semantic-aware metaverse-based UAV swarm architecture

As depicted in Fig. 6.6, the presented architecture primarily has three components: the physical world which serves as the foundational underpinning, the virtual world which monitors the physical space, and the interaction layer which facilitates seamless connectivity between the two worlds.

Physical World The physical world consists of the UAV swarm and the task environment. Each UAV has a set of sensors to sense the surrounding environment state over time and extracts semantic information by equipped semantic extraction model.

Virtual World The virtual world is developed and maintained by VSPs in the metaverse, where a high-fidelity DT of the physical world is established with the recovery of received semantic information from the UAV swarm. Leveraged by the SemCom technique, the VSP updates the digital model in an effective and real-time manner. Empowered by ML and data analytics techniques, the VSP trains the learning model for providing optimal decisions in specific metaverse services, such as resource allocation, swarm network optimization, cooperative decisions, and so on. After each task execution by the UAV swarm, the VSP collects feedback information to improve the dataset and learning model.

Interaction Layer For the purpose of delivering better communication quality, the data transmission between the two worlds adopts the SemCom technique. The physical entity conveys semantic information to the VSP, which is equipped with a learning model to abstract environmental features, analyze cooperative tasks, derive optimal decisions, and subsequently release these decisions via the interaction layer to manage the UAV swarm.

6.2.1.2 System Model

We consider a semantic-aware metaverse-based UAV swarm with N UAVs and M VSPs, which are denoted by the sets $\mathcal{N} = \{1, \ldots, n, \ldots, N\}$ and $\mathcal{M} =$

$\{1, \ldots, m, \ldots, M\}$, respectively. To reduce the interruption probability caused by long-distance transmission, L UAV-BSs are introduced as relays to provide communication services between UAVs and VSPs, which are denoted as the set $\mathcal{L} = \{1, \ldots, l, \ldots, L\}$. The UAVs grouped into multiple clusters can collect sensing data and transmit extracted semantic information to its cluster head, i.e., UAV-BS, with the UAV-UAV channel. In the network, let n_l denote the number of UAVs served by UAV-BS l.

In practice, the VSP may not be able to identify reliable UAVs, which suffer the performance of the digital model and consequently degrade the service quality of the VSP [29]. We adopt a hierarchical framework, where initially in the lower layer, the reputation values and communication latency of UAVs are used to identify the trustable UAVs in the lower layer. In the upper layer, UAVs are encouraged to contribute their semantic information to VSPs in cluster units to facilitate data synchronization.

Figure 6.7 shows the data synchronization process for the metaverse-based UAV swarm, which mainly includes the reliable worker selection phase and semantic information trading phase. The details are given as follows.

Step 1: The VSPs send transaction requests to the UAV swarm for acquiring semantic information, which includes the semantic type and latency requirement.

Step 2: Upon receiving the request information from the VSPs, the cluster head informs its relevant UAVs and invites them to join in semantic information transactions.

Step 3: Through the worker selection scheme, the cluster heads select trustable UAVs, collect semantic information from all their cluster members, and then

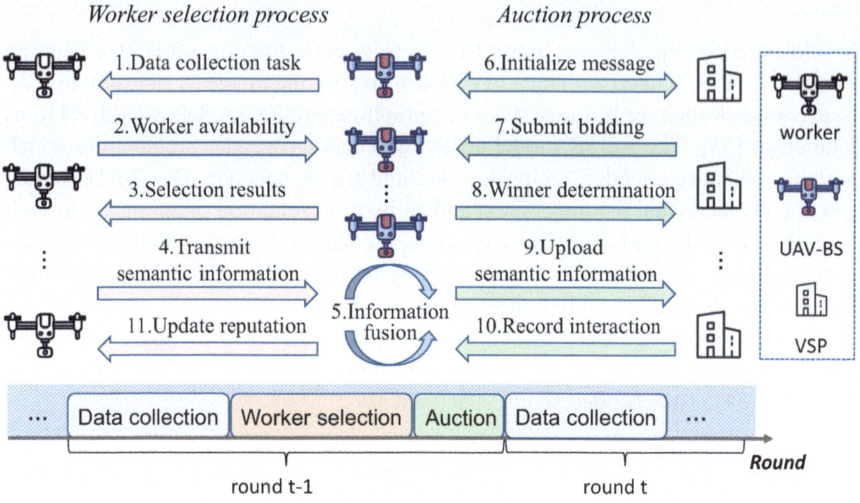

Fig. 6.7 The main procedures in the data synchronization

perform data fusion algorithms. The details about the worker selection scheme are given in Section IV.

Step 4: A multi-round DL-based auction is conducted to promote semantic information trading between the cluster head and VSPs. Typically, for each VSP, clusters with higher semantic information value can increase the performance of DT. More details about the auction design are presented in Section V.

Step 5: After the auction stage, the VSP assesses the quality of acquired semantic information through the attack detection scheme, e.g., the De-Pois scheme [31], to identify poison attacks and unreliable workers. The transactions with unreliable workers are recorded as negative interactions by the VSP and vice versa. Then, the cluster head receives the deserved reward from the VSP, where the reward is allocated to all the participating workers within the cluster in a fair manner.

Subsequently, the VSP receives the semantic information from the seller, i.e., cluster head, and proceeds to refresh its digital model through a series of operations which includes data cleaning, classification, and modeling. It is worth noting that the VSP should extract fundamental attributes from semantic information. For instance, in the collaborative search services provided by the VSP, semantic information is utilized to extract task areas and the number of UAVs to create a search information map, which can provide target appearance probability in the specific area. With the well-trained learning model at the VSP, optimal decisions for various tasks are derived and returned to the UAV swarm for the execution of corresponding tasks. The notations in this part are presented in Table 6.1.

6.2.1.3 Semantic Communication System

As illustrated in Fig. 6.8, we present our results on extracting semantics from an actual dataset associated with the UAV swarm and use the image as an example [32]. To accomplish this, we leverage the scene graph generation model called RelTR, as detailed in [33]. The RelTR model adopts an encoder-decoder architecture which contains a feature encoder, entity decoder, and triplet decoder. The RelTR model first obtains the visual feature context and entity representation of an image through the feature encoder and entity decoder. Then, the feature context of entities from the entity decoder layer is inferred by leveraging the triplet decoder with an attention mechanism, ultimately yielding the semantic triplet.

The transmitter, e.g., UAV, first conveys a set of images $D = [d_1, d_2, \ldots]$ over the physical channel, which is subject to transmission impairments including distortion and noise [34]. The encoded signal is defined as:

$$x = C_\alpha(S_\beta(d)), \tag{6.1}$$

where C_α and S_β are channel encoder with parameter α and semantic encoder with parameter β, respectively. The received signal is denoted as:

Table 6.1 Notations

Notation	Definition
$n \in \mathcal{N}$	Index of UAVs
$m \in \mathcal{M}$	Index of VSPs
$l \in \mathcal{L}$	Index of UAV-BSs
n_l	Number of UAVs served by UAV-BS l
(t_a, t_{ab}, t_b)	Semantic triplets
$R_{n,l}$	Data rate between UAV n and UAV-BS l
ω_n	Semantic information size of UAV n
$t_{n,l}$	Communication time of UAV n for UAV-BS l
I	Binary variable of workers' available status
W_t	Chosen worker set in the round t
F	Fixed setting of selected workers
k	Number of selected workers
$p_{m\to n}, q_{m\to n}$	Number of positive/negative interactions between VSP m and UAV n
β	Success probability of transmission
θ	Level of uncertainty
$\gamma_{m\to n}$	Reputation score of VSP m for UAV n
λ_{th}	Latency requirement between the UAV and its cluster head
r_n^t	Reputation value r_n^t of the UAV n in the round t
$a_{m,l}$	Semantic match level of VSP m for UAV-BS l
v_m	Valuation of VSP m

Fig. 6.8 Example of extracted semantic information for the metaverse [32]

$$y = hx + n, \tag{6.2}$$

where h is Rayleigh fading channel and $n \sim N(0, \sigma_n^2)$ is the Gaussian noise. Then, the received signal y, is decoded at the receiver as follows:

$$\hat{x} = S^{-1}(C^{-1}(y)), \tag{6.3}$$

where S^{-1} is the semantic decoder and C^{-1} is channel decoder at the receiver. The transmitter and receiver are jointly designed with DNNs.

The SemCom system outputs the triplet $< t_a, t_{ab}, t_b >$, which denotes the subject-relation-object relationship in text format. This representation effectively minimizes the data volume that needs to be transmitted. For example, a semantic triplet extracted by the UAV from Fig. 6.8 is < car on street >. This triplet is subsequently sent to the VSP for comprehensive computing and analysis. Each VSP providing traffic control services needs to extract semantic information about the conditions of physical entities such as roads, cars, traffic, and so on. Therefore, the semantic triplets < car on street > align with the information that VSP requires.

The data rate between UAV n and UAV-BS l can be obtained similar to that in [35]:

$$R_{n,l} = B_n \log_2(1 + \frac{p_n h_{n,l}}{B_n N_0}), \tag{6.4}$$

where p_n is transmit power, $h_{n,l}$ is the channel gain between UAV n and UAV-BS l, N_0 is the power spectral density of noise.

Accordingly, the communication latency from UAV n to UAV-BS l for transmitting the semantic information is:

$$t_{n,l} = \frac{\omega_n}{R_{n,l}}, \tag{6.5}$$

where ω_n is the size of semantic information.

6.2.2 *Lower Layer: MAB-Based Online Worker Selection*

In this section, we first introduce the worker selection scheme and the reputation model in the DT construction task. Then, we develop an MAB-based online worker selection algorithm.

6.2.2.1 Worker Selection for DT Construction Task

As the worker selection problem is investigated under a highly dynamic UAV swarm network, it becomes impractical to assume workers are perpetually prepared to offer their services [36]. In light of this, we use a binary variable I to capture workers' status which indicates whether they are available or not. Such information could be revealed to the cluster head at the beginning of each round t. On the other hand, due to the limited wireless channel resources, we focus on choosing a subset of the available workers for DT construction tasks. We denote the chosen subset in the round t as W_t. Thus, the constraint of worker availability is denoted as:

$$I_{t,n} = 1, \forall n \in W_t. \tag{6.6}$$

The selection of workers is not subject to a predetermined maximum threshold. Given that the number of available workers may not consistently align with the predefined parameter F, we should utilize a min function to constrain the number of selected workers:

$$k = \min\{F, \sum_{n=1}^{n_l} I_{t,n}\}, \tag{6.7}$$

where k is the number of selected workers. Then we have $|W_t| = k$ where $|W_t|$ is the cardinality of W_t. Precisely, when the number of overall available workers could not meet the fixed setting F, we entail all currently available workers in the subsequent round of selection without any selection criteria.

6.2.2.2 Reputation Model

We design a reputation model for UAV swarm. Our reputation model considers the communication quality, transmission reliability, and latency in data transactions.

Reputation Score

Consider $n \in \mathcal{N}$, the reputation opinions of VSP m to UAV n is expressed as a vector $\Upsilon_{m\to n} = \{b_{m\to n}, d_{m\to n}, u_{m\to n}\}$. Here, $b_{m\to n}$ is belief, $d_{m\to n}$ is disbelief and $u_{m\to n}$ is uncertainty. $b_{m\to n}, d_{m\to n}, u_{m\to n} \in [0, 1]$ and $b_{m\to n}+d_{m\to n}+u_{m\to n} = 1$. According to the subjective logic model [37], belief, disbelief, and uncertainty can be represented as follows:

$$\begin{cases} b_{m\to n} = (1 - u_{m\to n})\dfrac{p_{m\to n}}{p_{m\to n} + q_{m\to n}}, \\ d_{m\to n} = (1 - u_{m\to n})\dfrac{q_{m\to n}}{p_{m\to n} + q_{m\to n}}, \\ u_{m\to n} = 1 - \beta_{m\to n}, \end{cases} \tag{6.8}$$

where $p_{m\to n}$ and $q_{m\to n}$ are the numbers of positive and negative interactions, respectively. A trading interaction between the VSP and the UAV is treated as a positive event if the UAV can provide correct and trusted semantic information for the VSP to support the efficient digital copy, and vice versa. Specifically, through a poison attack detection scheme (as outlined in Sect. 6.2.1.2, Step 5), if the VSP judges the semantic information from the worker to be trustworthy, it categorizes the transaction as a positive interaction event. Conversely, interactions with malicious workers are considered negative and result in a decrement in the worker's reputation. $\beta_{m\to n}$, as the uncertainty in computing the reputation score, represents the success probability of transmission. The reputation score $\gamma_{m\to n}$ is denoted as follows:

$$\gamma_{m\to n} = b_{m\to n} + \theta u_{m\to n}, \tag{6.9}$$

where $\theta \in [0, 1]$ is the coefficient that indicates the level of uncertainty for reputation.

Latency Calculation

Since stale semantic information may lead to an inaccurate model, low latency is crucial for DT construction. The reputation valuation of the UAVs also depends on the transmission latency, which refers to the time spent on transmitting the semantic information from the worker to its cluster head.

Given that the cluster head has the lowest latency requirement λ_{th} of collecting semantic information from its corresponding UAV, it is imperative to establish a stringent criterion to guarantee the latency requirements. When the UAV's transmission delay of semantic symbols exceeds the threshold of λ_{th}, the corresponding UAV's reputation value is unilaterally reduced to zero. The delay indicator function is described as:

$$f(t_n, \lambda_{th}) = \begin{cases} 1, & \text{if } t_n \leq \lambda_{th}, \\ 0, & \text{if } t_n > \lambda_{th}, \end{cases} \tag{6.10}$$

where t_n is the communication latency incurred by UAV n.

Therefore, the reputation value r_n^t of the UAV n in the round t can be calculated as follows:

$$r_n^t = \gamma_{m \to n} \cdot f(t_n, \lambda_{th}). \tag{6.11}$$

6.2.2.3 MAB-based Worker Selection Scheme

The inherent uncertainty in UAV reputation during semantic information trading poses challenges for designing effective recruitment mechanisms. An MAB framework has been conceptualized as the learning strategy to navigate the balance between exploitation and exploration [38]. Compared with blockchain technologies, the flexibility and adaptability of the MAB algorithm make it a well-suited scheme for accommodating real-time adjustments in reputation values within the context of dynamic UAV swarm network [39].

To address the challenge of UAV reputations being non-stationary on a long-term basis, we adopt MAB which utilizes an estimator to adapt to the time-varying environment. Within the MAB framework, a slot machine holds multiple arms, each corresponding to an action. Pulling an arm generates a stochastic reward drawn from a probability distribution. Players decide which set of arms to pull to maximize their cumulative rewards.

We model the worker selection problem as a non-stationary MAB problem, where the cluster head and the UAVs are regarded as the player and the arms, respectively. The chosen subset of the arms W_t is regarded as a super arm and the corresponding reputation values of UAVs are seen as the reward for pulling the arm. The cluster head pulls k arms in every round and learns each selected UAV's reward, i.e., reputation value. We estimate reputation value $\bar{r}_n^t$ by computing weighted averages from their historical snapshots, which is represented as follows:

$$\bar{r}_n^t = \frac{1}{N_n^t} \sum_{s=1}^{t} \zeta^{t-s} r_n^s \mathbb{I}_{\{n \in W_s\}}, \tag{6.12}$$

where

$$N_n^t = \sum_{s=1}^{t} \zeta^{t-s} \mathbb{I}_{\{n \in W_s\}}, \tag{6.13}$$

$$\mathbb{I}_{\{n \in W_s\}} = \begin{cases} 1, & \text{if } n \in W_s, \\ 0, & \text{otherwise,} \end{cases} \tag{6.14}$$

where N_n^t is a normalization coefficient that indicates the number of times that cluster head selects UAV n until round t, and $\zeta \in (0, 1)$ is the discount factor. It allows us to progressively reduce the impact of past observations exponentially to aid in the tracking of changes in mean rewards. Essentially, a higher ζ leads to smoother estimation and a slower response to breakpoint.

To strike a balance between exploitation and exploration, we employ Upper Confidence Bound (UCB) policies [40]. By introducing an additional bonus term, the UCB policy incentivizes the cluster head to collect more observations from under-sampled UAVs. We denote UCB indexes of workers, where the uncertainty of estimation is taken into consideration, as follows:

$$\hat{r}_n^t = \bar{r}_n^t + \sqrt{\frac{\eta \ln p_n^t}{N_n^t}}, \tag{6.15}$$

where

$$p_n^t = \sum_{n=1}^{n_l} N_n^t. \tag{6.16}$$

Here, η is an exploration constant that controls the adaptability into our policy, and p_n^t is the summation of N_n^t across time steps t.

Leveraging the UCB criterion, the behavioral decision of selecting k workers for the data synchronization task is accordingly determined by

$$W_t = \arg\max_{|W_t|=k} \sum_{n \in W_t,} \hat{r}_n^t. \tag{6.17}$$

6.2.2.4 The Proposed Algorithm

Algorithm 18 MAB-based online worker selection algorithm

Input: The fixed setting about selected workers F; Workers' availability I.
Output: Selected worker set W_t.

```
1: Initialization: Success probability of transmission β and the probability of positive interactions p
2: Exploration phase:
3: while t < n_l do
4:     Recruit worker t for DT construction task
5:     Update r̄_n^t and N_n^t according to (6.12)-(6.14)
6:     t = t + 1
7: end while
8: Exploitation phase:
9: while t < T do
10:    Obtain k = min{m, Σ_{n∈N} I_{t,n}}
11:    Calculate r̂^t according to (6.15) and sort the workers in a descending order
12:    Choose k workers with the largest reputation values into W_t
13:    t = t + 1
14:    for each worker w in W_t do
15:        Update r̄_w^t according to (6.12)-(6.14)
16:        Update p_w = p_w + 1 or q_w = q_w + 1
17:    end for
18: end while
```

As shown in Algorithm 18, we propose a worker selection algorithm based on the Discounted-UCB policy which considers both the reputation score and the communication latency of each worker. We first initialize β and p. In the exploration phase, we observe each worker's reputation value at least once to update the values of $\bar{r}_n^t$ and N_n^t. In the exploitation phase, we make use of the acquired reputations from the preceding exploration phase to make decisions about the selection of workers. In line 10, we first obtain the number of clients that should be picked. In line 11, we calculate the reward, i.e., the reputation value of choosing each worker, and sort the workers in descending order of their UCB scores. Here we adopt a greedy strategy to choose the best k workers into the winning set W_t. The selected workers are employed to engage in data synchronization and their positive or negative interactions will be updated and recorded by their respective cluster heads at the end of each round.

Given that the other computational procedures only involve elementary numerical operations, the computational complexity of the algorithm depends on the choice of the search algorithm employed for the identification of the highest UCB index. This complexity inherently varies with the number of arms involved. Specifically, if we use merge sort to arrange the UCB scores of n_l arms, the computational complexity of the algorithm will be characterized by $O(n_l \log n_l)$.

6.2.3 Upper Layer: DL-Based Auction Design

After the online worker selection description, the cluster head, i.e., UAV-BS, receives the semantic information from its chosen workers and conducts information fusion. In this section, we design a continuous dynamic multi-round auction, enabling the successive semantic information trading between UAV-BSs and VSPs based on auction results. Note that the situation of auction interaction is recorded and returned to the cluster head for the purpose of updating workers' reputations.

6.2.3.1 The Semantic Information Market

We consider a semantic information market with L UAV-BSs and M VSPs. The objective of the market is to trade semantic information from the UAV swarm, which will facilitate the generation of digital copies within the VSPs. In this market, we consider a multi-round single-item auction to simplify the problem, where the VSP acts as the buyer and UAV-BS acts as the seller [41].

VSP (Buyer) The buyers provide payment to the sellers for acquiring the semantic information. Before each round of auction, each VSP submits the bid to its auctioneer. Given that all UAVs do not have the same match level and reputation value caused by their location, hardware specifications, and semantic model quality, the VSP needs to bid according to data relatedness and effectiveness of receiving semantic information.

UAV-BS (Seller) In this market, the sellers are the UAV-BSs that offer VSP semantic symbols. Upon receiving the bid profile, the UAV-BS selects the winning VSP for semantic information trading and decides the corresponding payment from the winning VSP.

Information exchange including bids and auction-related variables takes place between the cluster heads and VSPs via wireless links. The UAV swarm initializes the auction process after the fusion of semantic information by all the cluster heads. In every round of the multi-round single-item auction, the VSPs make their own private and independent valuation for the semantic information of the seller upon reception of the announcement. Thereinafter, the auctioneer collects bidding profiles $(b_1, b_2, \ldots, b_m, \ldots, b_M)$ from all VSPs and determines the winning VSP m^* along with the corresponding payment price p_m^*. The valuation, i.e., the private value, of VSP m for semantic information is denoted as v_m. The utility of the VSP is considered as:

$$u_m = \begin{cases} v_m - p_m^*, & \text{if VSP } m \text{ is the winner.} \\ 0, & \text{otherwise.} \end{cases} \tag{6.18}$$

To ensure system stability and guarantee the truthfulness of participants, the auction mechanism should possess desirable attributes, namely, Incentive Compatibility (IC) and Individual Rationality (IR) [42]. IC indicates the workers can maximize their utility by submitting truthful bids without considering the bids of other participants, i.e., $b_m = v_m$. IR guarantees that involvement in the auction results in non-negative utility for all participants, thereby motivating their participation in the semantic information trading market, denoted as $u_m \geq 0$.

Nonetheless, conventional single-item auction mechanisms, e.g., First-Price Auction (FPA) and Second-Price Auction (SPA), do not invariably guarantee the IC and IR properties. In the FPA, although revenue-optimal is ensured, the IC property is not satisfied. In the SPA, the second highest valuation determines the final payment even though the IC property is satisfied. Thus, we propose suitable allocation and pricing schemes to ensure the fulfillment of these two properties in our auction mechanism.

6.2.3.2 Valuation Model

Here, we propose a valuation model for VSPs to better acquire semantic information.

Given the multiple services offered by VSPs, it is evident that each VSP shows different interests and objectives when participating in the bidding process for semantic information acquisition. We consider the VSP's preference for semantic information as well as the reputation level of UAVs. According to the RelTR model introduced in Section III-C for extracting semantic information, we denote the count of semantic triplets for the UAV-BS l as T_l and the number of objects acquired by VSP m as O_m, respectively. Similar to [43], the preference, i.e., semantic match level, between VSP m and UAV-BS l is given by:

$$a_{m,l} = (\sum_{t=0}^{T_l}(\sum_{o=1}^{O_m} i + \sum_{o=1}^{O_m} j))/2T_l, \tag{6.19}$$

where i and j are binary variables. Specifically, $i = 1$ when $o = t_{l,a}$, $j = 1$ when $o = t_{l,b}$.

Meanwhile, the valuation of the semantic information also depends on the transmission latency, which represents how quickly the semantic information is conveyed to the VSP. The valuation function of VSP m receiving semantic information from cluster head l can be defined as $v_m = a_{m,l}\bar{r}_m$.

6.2.3.3 Optimal Auction Design

In this section, we adopt a deep learning-based approach [44] for the single-item auction.

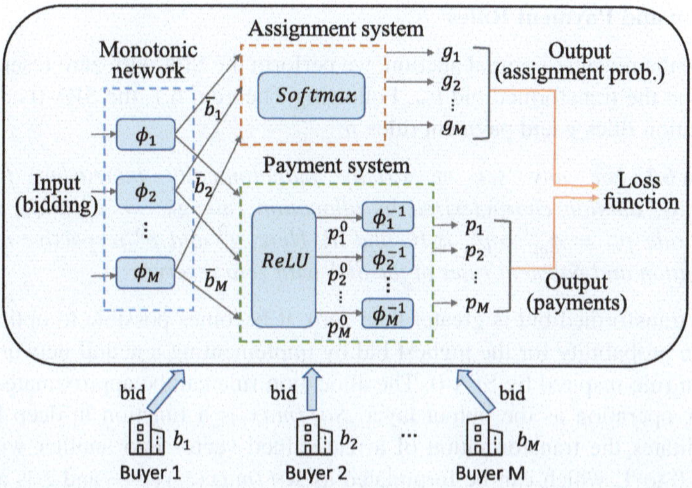

Fig. 6.9 Illustration of the DL-based optimal auction

We elaborate on the DL-based optimal auction, which maximizes the expected revenue for the sellers, while concurrently ensuring IR and IC constraints. Noted that the auctioneer operates without prior knowledge regarding the bidders and the determination of the winner. The cluster head can optimize the auction decision through a neural network architecture that contains three key components: a monotone transform function ϕ representing virtual valuation, an allocation rule g_i, and a payment rule p_i. In Fig. 6.9, we illustrate the DL-based optimal auction. Next, we describe the details of the three functions.

Monotone Transform Function

The proposed optimal auction mechanism utilizes virtual valuation transform functions, denoted as $\phi_m, m = 1, \ldots, M$, to derive allocation rule and payment rule in the neural network. Virtual valuation transform functions are strictly monotonically increasing, satisfying the IC and IR characteristics of the auction.

First, the input bid b_m of VSP is transform into virtual bids $\bar{b}_m = \phi_m(b_m)$. Each ϕ_m is modeled as a two-layer feedforward network, featuring K sets of J linear functions. In the network, we make use of operators such as min and max across multiple linear functions. Accordingly, the transform function ϕ_m is given by:

$$\phi_m(b_m) = \min_{k=1,\ldots,K} \max_{j=1,\ldots,J} (\omega_{kj}^m b_m + \beta_{kj}^m). \tag{6.20}$$

The inverse transform $\phi_m^{-1}(y)$ can be derived from the transform function as follows:

$$\phi_m^{-1}(y) = \max_{k=1,\ldots,K} \min_{j=1,\ldots,J} (\omega_{kj}^m)^{-1}(y - \beta_{kj}^m). \tag{6.21}$$

Allocation and Payment Rules

To ensure the revenue-optimal auction, we perform the SPA with zero reserve price (SPA-0) on the transformed bid $\bar{b}_m$. Following Theorem 6.1, the SPA-0 constrains the allocation rules g and payment rules p.

Theorem 6.1 *For any set of strictly monotonically increasing functions* $\phi_1, \ldots, \phi_M$*, auction characterized by allocation rule* $g_m = g_m^0 \circ \phi_m$ *and the payment rule* $p_m = \phi_m^{-1} \circ p_m^0$ *is IC and IR. Here,* g^0 *and* p^0 *respective represent the allocation and payment rules of the SPA with zero reserve.*

If the transformed bid is greater than zero, it becomes possible to optimize the allocation probability for the highest bid by implementing a neural network-based allocation rule inspired by SPA-0. The allocation rule can be approximated by the $softmax$ operation as the output layer. $Softmax$ is a function in deep learning that facilitates the transformation of a real-valued vector into another within the range of 0 to 1, which can be formulated as $softmax(z_i)\frac{z_i}{\sum_j z_j}$ and z is a vector. In the allocation network, the network's inputs consist of the transformed bids $\bar{\mathbf{b}} = (\bar{b}_1, \ldots, \bar{b}_M)$ and an additional dummy input, while the output is a vector representing assignment probabilities $\mathbf{g} = (g_1, ..., g_M)$.

$$g_m(\bar{\mathbf{b}}) = \text{softmax}_i\left(\bar{b}_1, \ldots, \bar{b}_{M+1}; \kappa\right) = \frac{e^{\kappa \bar{b}_m}}{\sum_{i=1}^{M+1} e^{\kappa \bar{b}_i}}, \forall m \in M \tag{6.22}$$

where $\bar{b}_{M+1}$ denotes the additional auxiliary input while $\kappa > 0$ impacts the quality of the approximation. A higher value of κ leads to an increase in the quality of the approximation.

The conditional payment rule determines the price p_m for the winning VSP m. In this process, the network calculates SPA-0 payment p_m^0 using an activation function, i.e., $ReLU$ function.

Then, it derives the conditional payment p_m through Theorem 1. Specifically, the input $\bar{b}_i$ is the larger value between transformed bids from other VSPs and zero. The payment rule network is expressed as :

$$p_m^0(\bar{\mathbf{b}}) = \mathrm{Re}\,LU\left(\max_{i \neq m} \bar{b}_i\right), \forall m \in M. \tag{6.23}$$

Here, we employ $ReLU(x) = \max(x, 0)$ to guarantee non-negativity for the payment. For the payment calculation, we apply a inverse transformation function to the SPA-0 price associated with $\bar{\mathbf{b}}$. The conditional payment to VSP m is then calculated as

$$p_m = \phi_m^{-1}(p_m^0(\bar{\mathbf{b}})). \tag{6.24}$$

Algorithm 19 DL-based auction algorithm

Input: Bids from VSPs $\mathbf{b} = (b_1, \ldots, b_M)$
Output: Assignment probabilities $\mathbf{g}$ and conditional payments $\mathbf{p}$;
1: Initialization: $\mathbf{w} = [\omega_{kj}^m] \in R_+^{M \times JK}$, $\boldsymbol{\beta} = [\beta_{kj}^m] \in R^{M \times JK}$
2: **while** $\hat{R}(\mathbf{w}, \boldsymbol{\beta})$ is not minimized **do**
3: Calculate transformed bid $\bar{b}_m = \phi_m(b_m) = \min_{k=1,\ldots,K} \max_{j=1,\ldots,J}(\omega_{kj}^m b_m + \beta_{kj}^m)$.
4: Compute the allocation probabilities $g_m = \text{softmax}_i\left(\bar{b}_1, \ldots, \bar{b}_{M+1}; \kappa\right) = \frac{e^{\kappa \bar{b}_m}}{\sum_{i=1}^{M+1} e^{\kappa \bar{b}_i}}$.
5: Compute the SPA-0 payments $p_m^0 = ReLU\left(\max_{i \neq m} \bar{b}_m\right)$.
6: Compute the conditional payment $p_m = \phi_m^{-1}(p_m^0(\bar{\mathbf{b}}))$.
7: Compute the loss $\hat{R}(\mathbf{w}, \boldsymbol{\beta}) = -\sum_{m=1}^{M} g_m(\bar{\mathbf{b}}) p_m$.
8: Calculate $\mathbf{w}$, β using SGD solver
9: **end while**
10: **return** Winner m^*, payment $\mathbf{p}^*$

Neural Network Training

In the following, we define the loss function $\hat{R}(\mathbf{w}, \boldsymbol{\beta})$ to optimize the deep learning network training. More precisely, loss function is formulated by the negative expectation of revenue, which can be determined through the allocation probability g and payment p. The goal of the network is to achieve the minimization of the loss function, as well as optimize the weights and biases associated with the transformation function, consequently enabling the maximization of revenue for the UAV swarm.

Our dataset has L bidder valuation profiles from corresponding VSPs and v_m^l represents the valuation of VSP m. As previously mentioned, the valuation v_m^l is expressed by the corresponding match level and reputation value, both of which are independently and identically sampled from a well-defined distribution function [44]. Therefore, we can derive the distribution of v_m by calculating the Jacobian determinant [45]. The loss function can be calculated as:

$$\hat{R}(\mathbf{w}, \boldsymbol{\beta}) = -\sum_{m=1}^{M} g_m(\bar{\mathbf{b}}) p_m, \tag{6.25}$$

where $\mathbf{w}$ and $\boldsymbol{\beta}$ are the matrices that contain weights and biases, respectively. The optimization of the loss function is achieved by applying a SGD solver to adjust the parameters $(\mathbf{w}, \boldsymbol{\beta})$.

6.2.3.4 Algorithm Complexity Analysis

The DL-based auction algorithm is presented in Algorithm 19. The computation cost mainly depends on the DL model training and decision-making [41]. In the DL-based auction model training stage, the training time is mostly used to assess the computation complexity. In the decision-making stage, the matrix operation and the activation function computation are considered for computation complexity evaluation. Therefore, the computation complexity of Algorithm 19 is expressed as $(O_M(i_n) + O_A(i_n)) * i_l$, where i_n is the number of nodes within the network, i_n is the number of layers, and $O_M(.)$ and $O_A(.)$ are computation complexity associated with the matrix operation and the activation function computation for each layer, respectively.

6.2.4 *Simulation Results and Discussions*

In this section, we assess the effectiveness of proposed worker selection scheme and DL-based auction method for DT construction.

6.2.4.1 Simulation Setup

We consider a metaverse-based UAV swarm network with 10 UAV-BS as cluster heads, 100 UAVs, and 5 VSPs. Our experiments are conducted on the VisDrone2019 dataset, a remotely sensed image dataset collected by UAVs [32]. The VisDrone2019 dataset has 10 classes and 54,200 instances of objects, including diverse and complex scenes under various weather and lighting conditions. The training and validation subsets consist of 7018 and 1609 images, respectively. The pre-defined deep learning model, i.e., the RelTR model, is deployed in each UAV to extract the corresponding semantic symbols. The parameter settings are given in Table 6.2.

6.2.4.2 Results of Worker Selection Scheme

First, we conduct an analysis of the worker selection scheme. The performance evaluation of our scheme is based on an average of over 10,000 rounds. We compare the performance of our algorithm with the following baseline schemes:

Table 6.2 Parameter setting

Parameter	Value
Number of UAV-BSs, L	10
Number of UAVs, N	100
Success probability of transmission, β	[0.8,1]
Level of uncertainty for reputation, θ	0.5
Exploration constant, η	1.5
Semantic information size of UAV, ω	3–5 Mbit
Bandwidth of UAV, B	8–10 MHz
Transmit power of UAV, p	1 W
Channel Gain of UAV, h	5–25 dBm
Power spectral density of noise, N_0	−169 dBm/HZ
Lowest latency requirement, λ_{th}	0.5 s

- *Oracle Scheme:* Oracle is the theoretically optimal method and always selects the best k workers in each round based on prior knowledge.
- ε-*Greedy Scheme*: This scheme combines the exploration and exploitation by using a parameter ε. At each round t, it explores workers with probability ε and exploits the best workers with probability $1 - \varepsilon$. For a more comprehensive analysis of algorithm performance, two variants of the method were employed, with ε values set to 0.3 and 0.7, respectively.
- *Random Scheme*: The Random scheme randomly chooses k workers in each selection. This scheme ensures randomized exploration of available workers by the cluster head.

As shown in Fig. 6.10, we evaluate the average reward curves by varying the number of available workers when $k = 3$. It can be found that the average reward obtained under our proposed scheme is slightly lower than the Oracle scheme, primarily due to the exploration actions. Meanwhile, the 0.3-Greedy scheme and 0.7-Greedy scheme exhibit similar performances, but their average rewards are both lower than our proposed approach. Given the severely suboptimal worker selection, the Random scheme has the lowest reward. As the value of n_l increases, it illustrates an increasing trend in the average rewards for all schemes. This phenomenon can be attributed to the emergence of more workers with higher reputations and less time delay, thereby promoting a better selection process compared to the previous.

Figure 6.11 illustrates the average reward changes when selecting a different number of workers with $n_l = 16$ We can observe that even with variations in the recruitment quantity of workers, our scheme still achieves satisfactory performance. It is also noteworthy that an increase in the selected number of workers results in a decrease in average rewards. This phenomenon reveals that the smaller k diminishes the recruitment of suboptimal workers in each round.

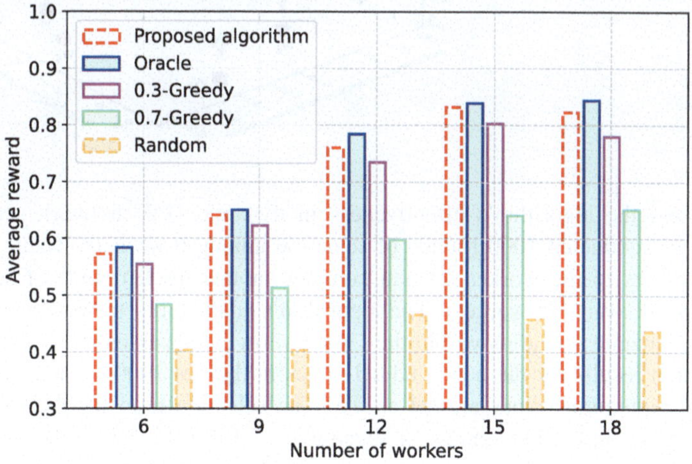

Fig. 6.10 Average reward versus the number of workers

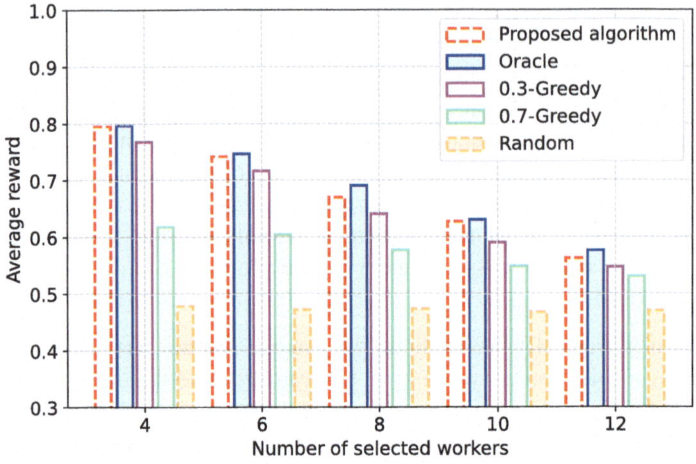

Fig. 6.11 Average reward versus the selected number of workers

Fig. 6.12 Ratio of selected original workers

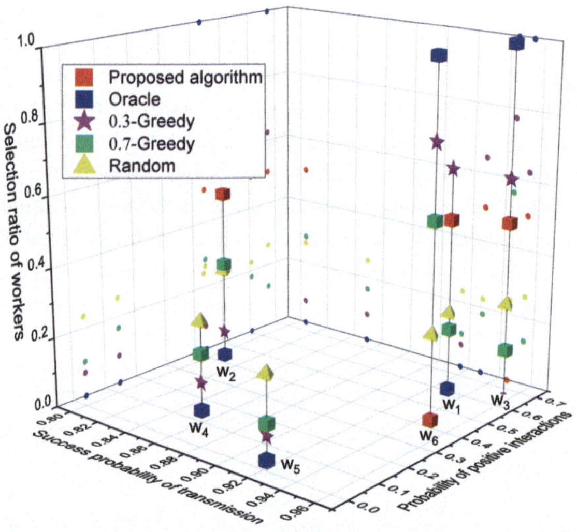

To assess the algorithm's effectiveness in dynamic UAV network scenarios, we initially conducted 1000 rounds of selection among 6 workers. Then, we add 6 new workers to the worker set and conduct another 1000 rounds of selection among all the available workers. The algorithm's adaptability to worker selection in highly dynamic scenarios within the UAV swarm can be assessed by examining the selection results for the newly added nodes in the last 1000 rounds. Note that the initial reputation values are set as 0.5. The selection ratio for the original 6 workers in the last 1000 rounds is illustrated in Fig. 6.12. The Oracle scheme consistently selects workers w_1, w_3 with the highest reputations based on known information. It is crucial to note that this approach is impractical in the real-

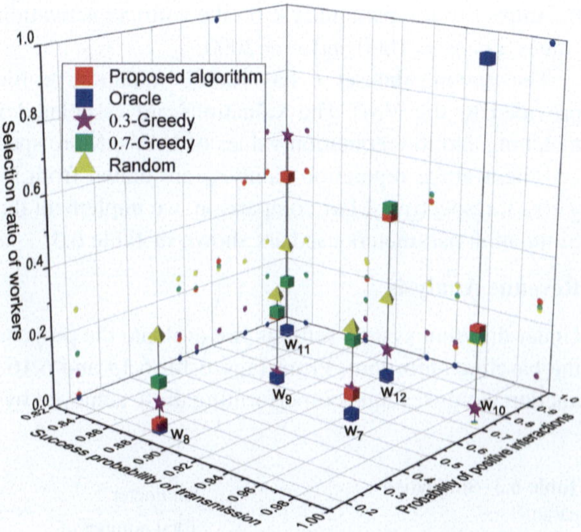

Fig. 6.13 Ratio of selected newly added workers

world scenario and is used merely to represent the theoretically optimal solution. In contrast, the random scheme uniformly selects each worker with almost equal probability. Our proposed method exhibits a tendency to select workers with the highest reputation values while also allocating certain selection opportunities to other workers. These observations verify the performance of our scheme when dealing with the exploration-exploitation trade-off.

Figure 6.13 shows the selection ratio of newly added workers. For the later participating workers with a high probability of positive interactions, the 0.3-Greedy scheme prioritizes the selection of existing workers and seldom chooses them. In contrast, the 0.7-Greedy scheme has the opportunity for exploration but with limited utilization, thus leading to unsatisfactory results. According to our proposed scheme, we observe that newly added workers have enough opportunities to update their true reputation values, and their chances are not less compared to existing workers. This implies that our scheme is effective and robust in the dynamic UAV swarm network compared to the baselines.

6.2.4.3 Evaluation of DL-based Auction

In this section, we experimentally validate the effectiveness of our DL-based optimal auction algorithm.

We employ the TensorFlow deep learning library to construct a neural network for the implementation of the proposed algorithm. The learning rate for the neural network is set to 0.001. In the neural network, each transformation function ϕ_m is structured by utilizing 5 groups of 10 linear functions. Furthermore, to investigate the impact of the parameter κ on the softmax activation function, we consider two values, i.e., $\kappa = 1000$ and $\kappa = 2000$.

The training dataset L has 10,000 valuation profiles of semantic information provided to the VSP. The valuations are associated with the level of semantic matching and the reputation values of UAVs. More specifically, the match level a_m and the average reputation value $\bar{r}_m$ are drawn from the distributions $f_A(a_m)$ and $f_R(\bar{r}_m)$, respectively. For comparison, we implement the SPA as a baseline scheme. Simulation parameters used are shown in Table 6.3.

Revenue Analysis

Under different system settings, we evaluate the proposed scheme in comparison to the baseline algorithm. From Figs. 6.14, 6.15 and 6.16, we find that the DL-based auction yields revenue outperforming those achieved by the baseline scheme for the

Table 6.3 Simulation parameters

Parameter	Value
Total number of VSPs, M	5,10,15
Training dataset size, L	10,000
Learning rate	0.001
Number of groups, Q	5
Number of linear functions, S	10
Quality of approximation κ	1000, 2000
Range of match level, a_m	$\sim U[0.3, 0.6]$, $U[0.6, 0.9]$
Range of reputation value, $\bar{r}_m$	$\sim U[0.5, 1]$

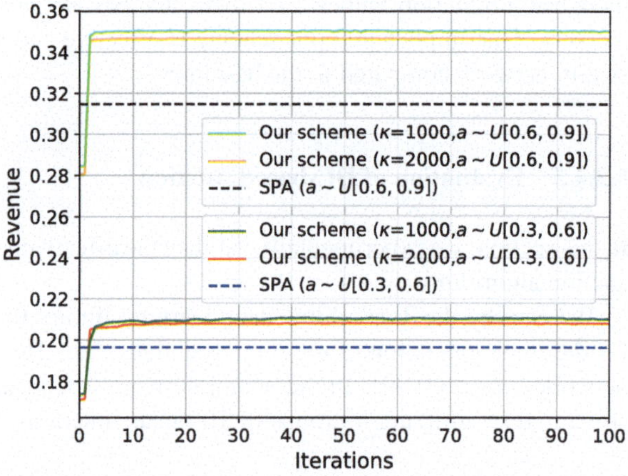

Fig. 6.14 Test revenue statistics for 5 VSPs

given number of VSPs and distributions. This indicates that the proposed scheme can maximize the expected revenue while achieving the IC and IR. In the case of $N = 5$ and $a_m \sim U[0.6, 0.9]$, we can observe that the revenue obtained by our scheme is 0.2082 with $\kappa = 2000$ whereas the SPA is 0.1966. It is clear that our scheme produces better performance compared to the baseline.

The Impact of the VSPs with Different Distributions

Then, we consider the impact of different uniform distributions on the cluster head's revenue. As depicted in Fig. 6.15, the expected revenue has increased with a_m following distribution $U[0.6, 0.9]$ while revenue is lower with $a_m \sim U[0.3, 0.6]$. The reason is that the bids of VSPs are positively proportional to the a_m. We can also find similar trends in other settings, i.e., as evidenced in Figs. 6.14 and 6.16.

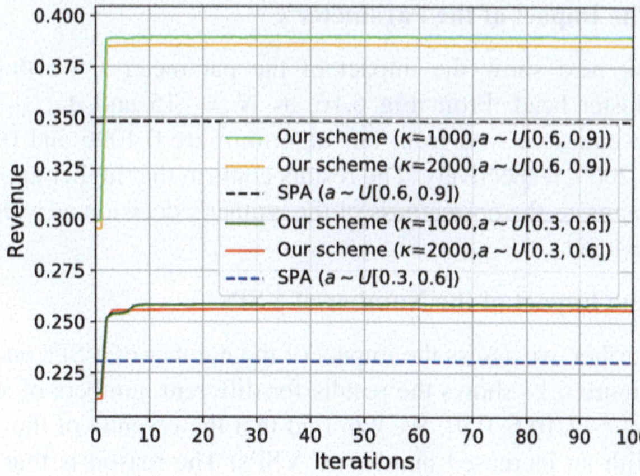

Fig. 6.15 Test revenue statistics for 10 VSPs

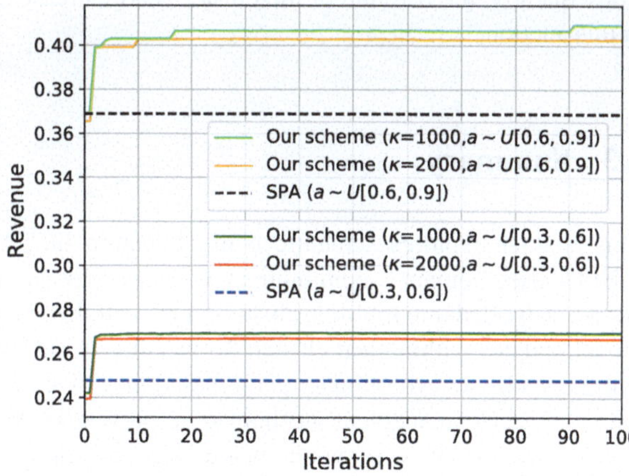

Fig. 6.16 Test revenue statistics for 15 VSPs

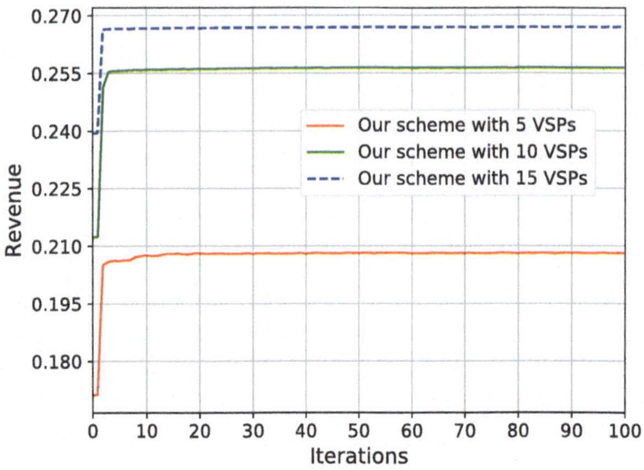

Fig. 6.17 Test revenue statistics under different VSPs

The Impact of the Parameter κ

We next show the impact of the parameter κ on the expected revenue of the cluster head. From Fig. 6.16, as $N = 15$ and $a_m \sim U[0.6, 0.9]$, the expected revenues derived from our algorithm are 0.4096 and 0.4031 for $\kappa = 1000$ and $\kappa = 2000$, respectively. The results confirm that higher approximation quality slightly increases the correctness of the winner's decision as well as decreases the expected revenue.

The Impact of the Number of VSPs

Further, we assess the impact of the number of VSPs on the cluster head's revenue. Figure 6.17 shows the results for different numbers of sellers with the distribution $a_m \sim U[0.6, 0.9]$. We can find that the revenue of the UAV swarm has improved with an increased number in VSPs. The reason is that more VSPs will intensify the competition, which potentially motivates VSPs to submit higher bidding prices. Consequently, the revenue of the cluster head increases as more buyers participate in this auction.

6.3 Summary

This chapter explores semantic communication in UAV swarm networks. First, a HFL-based semantic communication architecture is proposed, where the UAV swarm is divided into multiple clusters, and centralized federated learning is used for model updates within each cluster. To incentivize UAVs to participate in federated learning tasks, an evolutionary game-based incentive mechanism is designed, along with a cluster management mechanism to prevent UAV dropout. Additionally, this chapter investigates a semantic-aware UAV swarm system in the metaverse. The proposed hierarchical game-theoretic framework enables efficient and timely data synchronization. At the lower layer, a MAB-based UAV selection mechanism is employed, considering the reputation and communication latency of UAVs. At the upper layer, a DL-based auction algorithm is introduced for semantic information trading between UAV-BS and VSPs, ensuring individual rationality and incentive compatibility. Future research can further extend to the application of semantic communication for multimodal information fusion and other related areas.

References

1. J. Xu, H. Yao, R. Zhang, T. Mai, S. Huang, S. Guo, Federated learning powered semantic communication for uav swarm cooperation. IEEE Wirel. Communi. **31**(4), 140–146 (2024)
2. J. Xu, H. Yao, R. Zhang, T. Mai, S. Huang, Z. Xiong, D. Niyato, Semantic-aware uav swarm coordination in the metaverse: a reputation-based incentive mechanism. IEEE Trans. Mobile Comput. **23**(12), 13821–13833 (2024)
3. R. Shahzadi, M. Ali, H.Z. Khan, M. Naeem, Uav assisted 5g and beyond wireless networks: a survey. J. Netw. Comput. Appl. **189**, 103114 (2021)
4. H. Xie, Z. Qin, G.Y. Li, B.-H. Juang, Deep learning enabled semantic communication systems. IEEE Trans. Signal Process. **69**, 2663–2675 (2021)
5. H.B. McMahan et al., Advances and open problems in federated learning. Found. Trends® Mach. Learn. **14**(1) (2021)
6. T. Mai, H. Yao, J. Xu, N. Zhang, Q. Liu, S. Guo, Automatic double-auction mechanism for federated learning service market in internet of things. IEEE Trans. Netw. Sci. Eng. (2022)
7. Q.-V. Pham, M. Zeng, R. Ruby, T. Huynh-The, W.-J. Hwang, Uav communications for sustainable federated learning. IEEE Trans. Veh. Technol. **70**(4), 3944–3948 (2021)
8. W.Y.B. Lim, N.C. Luong, D.T. Hoang, Y. Jiao, Y.-C. Liang, Q. Yang, D. Niyato, C. Miao, Federated learning in mobile edge networks: a comprehensive survey. IEEE Commun. Surv. Tutor. **22**(3), 2031–2063 (2020)
9. W.Y.B. Lim, S. Garg, Z. Xiong, Y. Zhang, D. Niyato, C. Leung, C. Miao, Uav-assisted communication efficient federated learning in the era of the artificial intelligence of things. IEEE Netw. **35**(5), 188–195 (2021)
10. T. Sun, D. Li, B. Wang, Decentralized federated averaging. Preprint (2021). arXiv:2104.11375
11. X. Luo, H.-H. Chen, Q. Guo, Semantic communications: overview, open issues, and future research directions. IEEE Wirel. Commun. **29**(1), 210–219 (2022)
12. T. Mai, H. Yao, N. Zhang, L. Xu, M. Guizani, S. Guo, Cloud mining pool aided blockchain-enabled internet of things: an evolutionary game approach. IEEE Trans. Cloud Comput. **11**(1), 692–703 (2023)
13. S.R. Pokhrel, J. Choi, Understand-before-talk (ubt): A semantic communication approach to 6g networks. IEEE Trans. Veh. Technol. **72**(3), 3544–3556 (2023)
14. Y. Shen, Y. Qu, C. Dong, F. Zhou, Q. Wu, Joint training and resource allocation optimization for federated learning in uav swarm. IEEE Int. Things J. **10**(3), 2272–2284 (2023)
15. T. Li, C. Li, C. Yang, J. Shao, Y. Zhang, L. Pang, L. Chang, L. Yang, Z. Han, A mean field game-theoretic cross-layer optimization for multi-hop swarm uav communications. J. Commun. Netw. **24**(1), 68–82 (2021)
16. Z. Yang, D. Li, N. Zhao, Z. Wu, Y. Li, D. Niyato, Secure precoding optimization for noma-aided integrated sensing and communication. IEEE Trans. Commun. **70**(12), 8370–8382 (2022)
17. J. Chen, J. Wang, C. Jiang, Y. Ren, L. Hanzo, Trust-worthy semantic communications for the metaverse relying on federated learning. Preprint (2023). arXiv:2305.09255

18. S. Javaid, N. Saeed, Z. Qadir, H. Fahim, B. He, H. Song, M. Bilal, Communication and control in collaborative uavs: recent advances and future trends. IEEE Trans. Intell. Trans. Syst. **24**(6), 5719–5739 (2023)
19. Z. Chang, H. Deng, L. You, G. Min, S. Garg, G. Kaddoum, Trajectory design and resource allocation for multi-uav networks: deep reinforcement learning approaches. IEEE Trans. Netw. Sci. Eng. **10**(5), 2940–2951 (2023)
20. G. Shen, L. Lei, Z. Li, S. Cai, L. Zhang, P. Cao, X. Liu, Deep reinforcement learning for flocking motion of multi-uav systems: learn from a digital twin. IEEE Internet Things J. **9**(13), 11,141–11,153 (2021)
21. L. Zhou, S. Leng, Q. Wang, A federated digital twin framework for uavs-based mobile scenarios. IEEE Trans. Mobile Comput. **23**(6), 7377–7393 (2024)
22. L.U. Khan, Z. Han, D. Niyato, M. Guizani, C.S. Hong, Metaverse for wireless systems: vision, enablers, architecture, and future directions. Preprint (2022) arXiv:2207.00413
23. S. Cai, Y. Xu, A multi-objective optimization approach to resource allocation for edge-based digital twin, in *GLOBECOM 2022-2022 IEEE Global Communications Conference* (IEEE, 2022), pp. 2733–2738
24. W. Sun, N. Xu, L. Wang, H. Zhang, Y. Zhang, Dynamic digital twin and federated learning with incentives for air-ground networks. IEEE Trans. Netw. Sci. Eng. **9**(1), 321–333 (2020)
25. L.U. Khan, Z. Han, W. Saad, E. Hossain, M. Guizani, C.S. Hong, Digital twin of wireless systems: overview, taxonomy, challenges, and opportunities. IEEE Commun. Surv. Tutor. **24**(4), 2230–2254 (2022)
26. G. Ji, J.-g. Hao, J.-l. Gao, C.-Z. Lu, Digital twin modeling method for individual combat quadrotor uav, in *2021 IEEE 1st International Conference on Digital Twins and Parallel Intelligence (DTPI)* (IEEE, 2021), pp. 1–4
27. M.G. Kapteyn, J.V. Pretorius, K.E. Willcox, A probabilistic graphical model foundation for enabling predictive digital twins at scale. Nature Comput. Sci. **1**(5), 337–347 (2021)
28. B. Du, H. Du, H. Liu, D. Niyato, P. Xin, J. Yu, M. Qi, Y. Tang, Yolo-based semantic communication with generative ai-aided resource allocation for digital twins construction. IEEE Internet Things J. **11**(5), 7664–7678 (2024)
29. Y. Han, D. Niyato, C. Leung, C. Miao, D.I. Kim, A dynamic resource allocation framework for synchronizing metaverse with iot service and data, in *ICC 2022-IEEE International Conference on Communications* (IEEE, 2022), pp. 1196–1201
30. G. Shi, Y. Xiao, Y. Li, X. Xie, From semantic communication to semantic-aware networking: model, architecture, and open problems. IEEE Commun. Mag. **59**(8), 44–50 (2021)
31. J. Chen, X. Zhang, R. Zhang, C. Wang, L. Liu, De-pois: an attack-agnostic defense against data poisoning attacks. IEEE Trans. Inf. Forensics Secur. **16**, 3412–3425 (2021)
32. P. Zhu, L. Wen, D. Du, X. Bian, H. Fan, Q. Hu, H. Ling, Detection and tracking meet drones challenge. IEEE Trans. Pattern Anal. Mach. Intell. **44**(11), 7380–7399 (2021)
33. Y. Cong, M.Y. Yang, B. Rosenhahn, Reltr: relation transformer for scene graph generation. IEEE Trans. Pattern Anal. Mach. Intell. **45**(9), 11169–11183 (2023)
34. X. Kang, B. Song, J. Guo, Z. Qin, F.R. Yu, Task-oriented image transmission for scene classification in unmanned aerial systems. IEEE Trans. Commun. **70**(8), 5181–5192 (2022)
35. A.S. Matar, X. Shen, Joint subchannel allocation and power control in licensed and unlicensed spectrum for multi-cell uav-cellular network. IEEE J. Selected Areas Commun. **39**(11), 3542–3554 (2021)
36. T. Huang, W. Lin, W. Wu, L. He, K. Li, A.Y. Zomaya, An efficiency-boosting client selection scheme for federated learning with fairness guarantee. IEEE Trans. Parallel Distrib. Syst. **32**(7), 1552–1564 (2020)
37. T. Cheng, G. Liu, Q. Yang, J. Sun, Trust assessment in vehicular social network based on three-valued subjective logic. IEEE Trans. Multimedia **21**(3), 652–663 (2019)
38. W. Chen, Y. Wang, Y. Yuan, Combinatorial multi-armed bandit: general framework and applications, in *International Conference on Machine Learning*. PMLR (2013), pp. 151–159

39. H. Xiao, L. Cai, J. Feng, Q. Pei, W. Shi, Resource optimization of mab-based reputation management for data trading in vehicular edge computing. IEEE Trans. Wirel. Commun. **22**(8), 5278–5290 (2023)
40. A. Garivier, E. Moulines, On upper-confidence bound policies for switching bandit problems, in *International Conference on Algorithmic Learning Theory* (Springer, 2011), pp. 174–188
41. K. Zhu, Y. Xu, Q. Jun, D. Niyato, Revenue-optimal auction for resource allocation in wireless virtualization: a deep learning approach. IEEE Trans. Mobile Comput. **21**(4), 1374–1387 (2020)
42. M. Babaioff, W.E. Walsh, Incentive-compatible, budget-balanced, yet highly efficient auctions for supply chain formation, in *Proceedings of the 4th ACM Conference on Electronic Commerce* (2003), pp. 64–75
43. Z.Q. Liew, H. Du, W.Y.B. Lim, Z. Xiong, D. Niyato, H. Yu, Economics of semantic communication in metaverse: an auction approach, in *2023 IEEE 20th Consumer Communications & Networking Conference (CCNC)* (IEEE, 2023), pp. 398–403
44. P. Dütting, Z. Feng, H. Narasimhan, D. Parkes, S.S. Ravindranath, Optimal auctions through deep learning, in *International Conference on Machine Learning*. PMLR (2019), pp. 1706–1715
45. N.C. Luong, Z. Xiong, P. Wang, D. Niyato, Optimal auction for edge computing resource management in mobile blockchain networks: a deep learning approach, in *2018 IEEE International Conference on Communications (ICC)* (IEEE, 2018), pp. 1–6
46. J. Chen, C. Du, Y. Zhang, P. Han, W. Wei, A clustering-based coverage path planning method for autonomous heterogeneous uavs. IEEE Trans. Intell. Trans. Syst. **23**(12), 25546–25556 (2022)

Chapter 7
Deterministic Network Empowered UAV Swarm

Conventional networks that rely on best-effort (BE) transmission methods face significant challenges in delivering high-quality, differentiated end-to-end transmission, e.g., UAV control signal. In recent years, advancements in deterministic networking technologies have provided promising solutions to address this issue. By integrating deterministic networking technologies, it becomes feasible to provide low-latency, jitter-free, and highly reliable communication services, thereby ensuring precise real-time control and data transmission. Therefore, this chapter introduces a TSN-5G-UAV deterministic networking architecture to provide deterministic transmission assurance for UAV remote control. Building on this framework, we propose a scheduling model that integrates Genetic Algorithms (GA) with reinforcement learning. This model effectively minimizes end-to-end delay and queue occupancy by optimizing scheduling strategies within resource-constrained environments, thereby significantly enhancing overall network performance [1].

7.1 Differentiated Traffic Scheduling for TSN-5G-UAV Networks

In recent years, the emergence of new applications in UAV swarms, such as remote control and remote sensing, has introduced stringent demands on network performance. These time-sensitive services, particularly those requiring low latency and minimal jitter, are highly sensitive to time-related performance metrics. Conventional networks, which rely on best-effort (BE) transmission methods, struggle to deliver high-quality end-to-end communication, particularly in environments with such stringent requirements. Consequently, achieving reliable and efficient transmission in these settings necessitates the integration of advanced networking technologies capable of meeting these exacting performance standards.

H. Yao, T. Mai, *UAV Swarm Cooperation*, Wireless Networks,
https://doi.org/10.1007/978-3-031-96444-2_7

Recently, a large number of studies have investigated scheduling Time-sensitive traffic. Gate Control List (GCL) [2] is a scheduling mechanism used in TSN to control the transmission of time-critical frames in an Ethernet network. Craciunas et al. [3] conducted research on the effectiveness of GCL and employed satisfiability modulo theories (SMT) to solve traffic scheduling problems. Dürr and Nayak in [4] proposed a time-sensitive traffic scheduling method called "No-wait packet scheduling". End-to-end latency is greatly reduced because traffic enters the link for processing without waiting. Khoshnevisan et al. in [5] focused on the end-to-end design of 5G networks for industrial and use SPS to provide URLLC wireless communication. In [6], X. Wang et al. proposed a TSN-5G hybrid network architecture and an algorithm that combines particle swarm optimization and reinforcement learning to schedule industrial traffic.

In this part, we intend to solve the problem of deterministic end-to-end scheduling [7] of complex industrial traffic in TSN-5G-UAV networks [8]. The network traffic scheduling problem can be modeled as an ILP problem. However, it is difficult to obtain results from ILP solver in the wide-scale industrial Internet. When using heuristic algorithms to solve, it is hard to converge to the optimal solution. Recognizing the limitations of both ILP solver and heuristic algorithms, this study proposes a two-stage population generation algorithm to tackle the end-to-end scheduling problem arising from complex industrial traffic within large-scale industrial networks. This algorithm combines principles from the elite GA [9] and reinforcement learning [10], improved based on the traffic characteristics and specific business demands inherent within the industrial Internet. The contributions of this part are as follows.

- To address the heterogeneity challenge in the hybrid system, we incorporate the concept of the minimum time unit from the 5G system into our proposed traffic scheduling model. Differentiated traffic scheduling is carried out in response to the wide variety of industrial traffic and complex needs.
- We propose a GA and PPO based two-stage population generation algorithm to solve the differentiated scheduling problem of traffic. We use PPO to select elite individuals in the population to avoid local optimal solutions. The algorithm can obtain the global optimal solution while ensuring the convergence speed.
- Experimental results show that our algorithm greatly improves the performance of end-to-end delay and the number of occupied queues compared with GA, particle swarm optimization (PSO) algorithm, and their improved algorithms.

7.1.1 System Model and Problem Formulation

In this section, we first introduce the network model and then elaborate on the constraints that TSN traffic and 5G traffic need to meet respectively.

7.1.1.1 Network Model

As shown in Fig. 7.1, in order to be able to use mathematical language to describe the traffic scheduling process, we model the TSN-5G-UAV industrial network as an undirected graph $G \equiv (V, E)$. Among them, $V = \{v_i | i \in (1, 2, \ldots, n)\}$ represents the network switches and other network equipment in the factory. $E = \{e_i | i \in (1, 2, \ldots, n)\}$ represents the network link that connects network devices and switches. If there is a flow propagating on a link with v_i as the starting point and v_j as the ending point, it can be represented as $[v_i, v_j]$.

We use F to represent the set of flows and $f_i \in F$ to represent any stream in the set. We use $f_i.src$ to represent the sending node of the flow, which is the starting point, and $f_i.dst$ to represent the receiving node of the flow, which is the endpoint. To represent the flow load, period, and deadline, we use $f_i.pld$,$f_i.prd$ and $f_i.ddl$ respectively. Therefore, we can represent a flow as follows:

$$f_i = \{f_i \cdot src, f_i \cdot dst, f_i \cdot pld, f_i \cdot prd, f_i \cdot ddl\}. \tag{7.1}$$

Since most of the traffic in the Industrial Internet is periodic, when considering problems with traffic scheduling, we also take the periods of different flows into consideration. Due to the cyclic nature of flow, the transmission of all flows will repeat after the common cycle time. So we use T_{com} [11] to represent the least common multiple (lcm) of the period of all flows. When solving a scheduling problem, the solutions we need are as follows: the transmission offset of the first frame of the data flow on the link and the queue ID in the switch. Therefore we use $f_i^{[v_i,v_j]} \cdot o$ and $f_i^{[v_i,v_j]} \cdot q$ to represent them. The parameters appearing are shown in the Table 7.1.

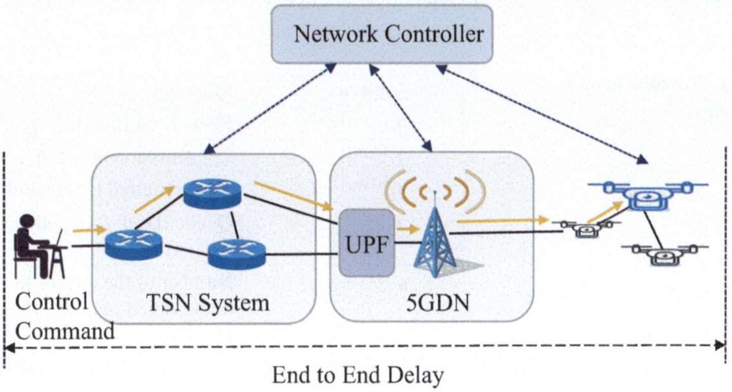

Fig. 7.1 Network model

7.1.1.2 Network Constraints

Network constraints ensure the deterministic transmission of TT traffic in industrial networks and define the space of feasible solutions when we solve the scheduling problem. The constraints are as follows.

1. Frame constraints: We set the start point of the link to v_a and the endpoint to v_b, Frame constraints can be expressed like this:

$$f_i{}^{[v_a,v_b]} \cdot o \geq 0, \tag{7.2}$$

$$f_i{}^{[v_a,v_b]} \cdot o + f_i{}^{[v_a,v_b]} \cdot d + d^{[v_a,v_b]} \leq f_i \cdot prd. \tag{7.3}$$

2. Link constraints: We set the start point of the link to v_a and the endpoint to v_b, Link constraints can be expressed like this:

$$\begin{aligned}(f_i{}^{[v_a,v_b]} \cdot o + \alpha f_i \cdot prd \geq f_j{}^{[v_a,v_b]} \cdot o \\ + \beta f_j \cdot prd + f_j{}^{[v_a,v_b]} \cdot d),\end{aligned} \tag{7.4}$$

or

$$\begin{aligned}(f_j{}^{[v_a,v_b]} \cdot o + \beta f_j \cdot prd \geq f_i{}^{[v_a,v_b]} \cdot o \\ + \alpha f_i \cdot prd + f_i{}^{[v_a,v_b]} \cdot d),\end{aligned} \tag{7.5}$$

In order to make the scheduling results applicable to different T_{com}, we introduce variables α and β, which can be represented by the number of flows in T_{com}. For example, $\alpha = \{0, 1, 2, \ldots, \frac{T_{com}}{f_i \cdot prd} - 1\}$.

Table 7.1 Variable names and meaings

Symbol	Meaning
$f_i{}^{[v_i,v_j]}$	Flow f_i on link $[v_i, v_j]$
$f_i{}^{[v_i,v_j]} \cdot o$	Transmission offset of f_i
$f_i{}^{[v_i,v_j]} \cdot d$	Time required to transmit f_i
$f_i^{[v_i,v_j]} \cdot q$	Queue ID of f_i in node v_i, in the port to v_j
$q^{[v_i,v_j]} \cdot_{sum}$	Number of the queues that can be occupied in node v_i, in the port to v_j
$q^{[v_i,v_j]} \cdot_{num}$	Number of the queues that has been occupied in node v_i, in the port to v_j
θ	Time synchronization error

3. Flow constraints: We set the links to $[v_a, v_b]$, $[v_b, v_c]$, flow constraints can be expressed like this:

$$f_i^{[v_b,v_c]} \cdot o \geq f_i^{[v_a,v_b]} \cdot o + f_i^{[v_a,v_b]} \cdot d + d^{[v_a,v_b]} + d^{v_b} + \theta. \tag{7.6}$$

4. Queue constraints: We set the link to $[v_a, v_b]$, queue constraints can be expressed like this:

$$0 \leq f_i^{[v_a,v_b]} \cdot q \leq q^{[v_a,v_b]} \cdot sum. \tag{7.7}$$

5. 5G constraints: In 5G networks, flows cannot start transmission at arbitrary times. Instead, the start transmission time must be an integer multiple of the transmission time interval (T_{TTI}) [11]. Therefore, 5G constraints can be expressed like this:

$$f_i^{[v_a,v_b]} \cdot o = XT_{TTI}, \tag{7.8}$$

X and Y, as positive integers, can be expressed as $X, Y = \{0, 1, 2, \ldots, \frac{T_{com}}{T_{TTI}} - 1\}$, $X < Y$.

6. Differentiated scheduling constraints: In the industrial network, there are many types of traffic, and the size and transmission quality requirements of different types of traffic are different [12]. We assume that there is a network link composed of the following nodes in order: v_a, v_b, v_c, v_d. Differentiated scheduling constraints on-time performance metrics can be expressed like this:

$$(f_i^{[v_c,v_d]} \cdot o + f_i^{[v_c,v_d]} \cdot d + d^{[v_c,v_d]} - f_i^{[v_a,v_b]}.o \leq latency), \tag{7.9}$$

or

$$(f_i^{[v_c,v_d]} \cdot o + f_i^{[v_c,v_d]} \cdot d + d^{[v_c,v_d]} \leq Deadline). \tag{7.10}$$

7.1.1.3 Objective Function

1. Number of queues that are occupied: This optimization enables the allocation of additional queues to BE flows, thereby enhancing their performance. The formula articulating the minimum count of occupied queues is as follows:

$$\min \sum_{[v_a,v_b] \in E} q^{[v_a,v_b]} \cdot num, \tag{7.11}$$

Since the number of used queues is less than its maximum value:

$$\forall v_a \in V : q^{[v_a,v_b]} \cdot num \in (0, 1, \ldots, q^{[v_a,v_b]} \cdot sum).$$

2. Total end-to-end delay: To control network delay, it is necessary to minimize queuing delay through traffic scheduling. We assume that there is a network link composed of the following nodes in order: v_a, v_b, v_c, v_d. The formula aimed to minimize end-to-end delay is as follows:

$$\min \sum f_i{}^{[v_c,v_d]} \cdot o + f_i{}^{[v_c,v_d]} \cdot d + d^{[v_c,v_d]} - f_i{}^{[v_a,v_b]} \cdot o. \tag{7.12}$$

7.1.2 Reinforcement Learning-Based Genetic Algorithm for Traffic Scheduling

In this article, we propose a genetic algorithm (GA) that leverages reinforcement learning (RL) for traffic scheduling. This algorithm harnesses the global search capability of the elite GA to obtain feasible solutions. Proximal Policy Optimization (PPO) is employed to determine the number of elite individuals and enhance the performance of the elite GA. In this section, we will introduce the algorithm architecture and the process of determining the number of elite individuals by the PPO algorithm.

7.1.2.1 Algorithm Architecture

The effectiveness of genetic algorithms heavily relies on the quality of the initial population, which significantly impacts both the convergence speed and the solution quality. Consequently, randomly generated binary sequences prove unsuitable as initial population genomes. To address this, an elite GA, devoid of an objective function, is employed as the first stage of the algorithm to generate the initial population [13]. Nevertheless, to maintain that all solutions generated by the algorithm adhere to feasibility standards, the penalty function will enforce a "death penalty" to remove solutions that fail to meet the scheduling requirements.

In the second stage, we also employ the elitist GA. However, our approach differs: to avert local optima and enhance convergence towards the global optimal solution, we integrate PPO to determine the count of elite individuals within each generation. To comprehensively explore the solution space for attaining the global optimal solution, we implement a penalty function based on the degree of constraint violation. Additionally, as the number of iterations nears the upper limit, we employ a "death penalty" mechanism as same as stage one.

Taking the algorithm aimed at minimizing end-to-end delay as an example, the integration of GA with the PPO algorithm involves feeding the output of the GA and the action of the previous generation of the PPO algorithm into the neural network as the state for this iteration. In this context, reward is assigned a value related to the delay. The action of the PPO algorithm in this iteration becomes the determinant for the number of elites in GA. Importantly, the action determined by the PPO algorithm plays a crucial role in influencing the number of elites selected in this iteration of the genetic algorithm. This integrated approach leverages the strengths of both algorithms to optimize the end-to-end delay, using the neural network as a key intermediary.

7.1.2.2 Reinforcement Learning For Elites Number Control

The process of the PPO algorithm is shown in Algorithm 20. When the PPO algorithm iteratively updates, it observes the probability of behavior taken by the current strategy agent and the probability of behavior taken by the previous strategy. It calculates the ratio of the probabilities to control the update amplitude of the new strategy. To address the problem of repeated sampling in each iteration, the PPO algorithm introduces importance sampling by using PPO-Clip.

The PPO algorithm comprises two main components: the Actor and the Critic. The Actor is responsible for determining the policy or strategy, while the Critic evaluates the value function associated with the current policy. The update method for the Critic is similar to other Actor-Critic algorithms and typically involves calculating the Timing Difference (TD) error. For the update method of the Actor, the PPO-clip method is used. The objective function of PPO-clip is as follows:

$$\begin{aligned} \mathcal{L}(\theta) = \mathbb{E}_t[\min &\left(\frac{\pi_\theta(a_t|s_t)}{\pi_{\theta_{\text{old}}}(a_t|s_t)} A_t \right), \\ \text{clip} &\left(\frac{\pi_\theta(a_t|s_t)}{\pi_{\theta_{\text{old}}}(a_t|s_t)}, 1-\epsilon, 1+\epsilon \right) A_t], \end{aligned} \tag{7.13}$$

where $\pi_\theta(a_t|s_t)$ is the probability of action a_t in s_t under policy θ, A_t is the advantage function at time t, θ_{old} is the parameters of the policy at the previous iteration, ϵ is the PPO clipping parameter.

Through the above process, the PPO algorithm can control the number of elite individuals.

7.1.3 Simulation Results and Discussions

In this section, we assess the efficacy of our proposed two-stage population generation algorithm for time-sensitive traffic scheduling in TSN-5G-UAV hybrid industrial networks.

Algorithm 20 The elite number algorithm

Input: loss function $losses$, policy network $actor$, value network $critic$, discount factor $gamma$, number of training iterations per sequence $epoch$, total number of iterations $num_episodes$
Output: action $actions$

```
Initialize policy network actor ,value network critic
Calculate action probabilities using actor
sampled action
Get states, actions, rewards, next_states
Calculate the target value for the next state using value network critic
Calculate current state value using value network critic
Compute advantages
for each num_episodes do
    Update advantage
    Calculate old log probabilities of the selected actions
    for each training epoch do
        Calculate current log probabilities of the selected actions
        Calculate the ratio and surrogate losses
        Clip the surrogate losses
        Calculate actor and critic losses
        Update actor and critic networks
    end for
end for
```

7.1.3.1 Experiment Setting

In this part, our main research problem is to calculate the traffic schedule table. The network topology required for our experiments is shown in Table 7.2. When both the sending and receiving ends of the traffic are static, the traffic's path is determined using the Dijkstra shortest path algorithm. The network's link delay is set to 0.1 microseconds, and the switch's processing delay is configured as 5 microseconds. Utilizing a 5G network featuring a 200 MHz carrier bandwidth and a 30 kHz subcarrier spacing, the time slot length is established at 70 microseconds.

According to [14], we select two traffic cycle sets as shown in Table 7.3. The size and business requirements of different types of traffic are shown in Table 7.4. In a factory environment, the proportions of different flows are variable, the proportions of various types of factory traffic in the experiment are shown in Table 7.5.

The fitness function of the algorithm is as follows:

Table 7.2 Network topology

Component name	Number of components
TSN link	7
5G link	1
devices connected to TSN link	15
devices connected to 5G link	10

Table 7.3 TSN-5G-UAV network traffic cycle

Cycle 1	{1000,1500,2000,2500}
Cycle 2	{2000,3000,4000,5000}

Table 7.4 Traffic types and transmission requirements

	Data size	Transmission requirements
Best effort	30 ~1500 bytes	n.a.
Video	1000 ~1500 bytes	End-to-end delay<10 ms
Audio/Voice	1000 ~1500 bytes	End-to-end delay<50 ms
Configuration& diagnostics	500 ~1500 bytes	End-to-end delay<100 ms
Cycle	50 ~1000 bytes	End-to-end delay<2 ms
Network control	50 ~500 bytes	n.a.
Isochronous	30 ~100 bytes	Deadline<period

Table 7.5 Traffic ratio

Type of traffic	Proportion
Best effort	50%
Video	15%
Audio/Voice	10%
Configuration&diagnostics	2%
Cycle	15%
Network control	3%
Isochronous	5%

$$\min \sum_{i=1}^{N} R_i C_i + F, \tag{7.14}$$

R_i is the degree of violation of constraint, if no constraints are violated, R_i=0. C_i is penalty coefficient. N is the constraint quantity. F is the objective function value we want to minimize, which can be total number of queues occupied by time-sensitive traffic or the end-to-end delay depending on the situation.

As each flow passes through a switch, we need to confirm the values of two variables: the time begin to transmit and the occupied queue number, in order to facilitate crossover, mutation and other operations, we use 10 bits of binary number to represent them. Consequently, the length of each individual in the randomly generated population equals 20 times the quantity of switches traversed. The designed population consists of 150 individuals, subject to 1000 iterations. In the second stage, we set the crossover probability at 0.95 and the mutation probability stands at 0.01. In the second stage, the initial mutation probability is set at 0.02. To enhance population quality and expedite optimal solution acquisition, we configure the crossover and mutation probabilities as functions that gradually diminish as the iteration count increases.

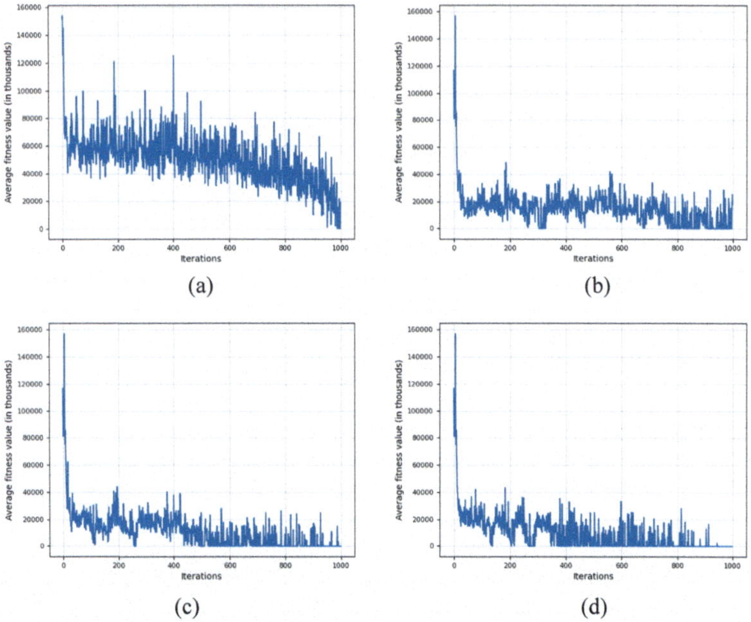

Fig. 7.2 Comparison of algorithm convergence. (**a**) GA. (**b**) GA(elite). (**c**) Q-learning. (**d**) PTPG

7.1.3.2 Convergence Analysis

When the average fitness remains stable, the algorithm can be considered to have converged. If the algorithm attains a stable fitness function value with a less number of iterations, it suggests better convergence of the algorithm. As shown in Fig. 7.2, the four pictures show the convergence status of GA, elite GA, elite GA with Q-learning and PTPG algorithm. When the basic versions of GA are used, the algorithm is difficult to converge, and when the improved versions of GA are used, although the algorithm converges, the convergence speed is slow. When using our proposed PTPG algorithm, the convergence speed is greatly improved. Experimental results show that compared with GA and its elite version, our proposed PTPG algorithm has better performance in terms of convergence.

7.1.3.3 Number of Occupied Queues Analysis

We investigate the effect of the number of flows on the number of queues occupied by time-sensitive traffic. The period is selected from cycle 1. The queue occupancy rate changes with the number of flows, as shown in Fig. 7.3. We notice that as the traffic increases, the queue occupancy rate also increases. This is because when the number of traffic increases, in order to ensure deterministic transmission and reduce queuing delay, traffic that cannot be transmitted in time will be transmitted in other

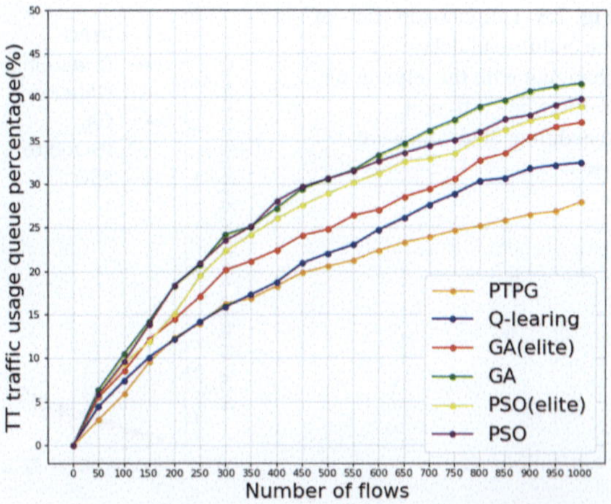

Fig. 7.3 Comparison chart of the percentage of occupied queues changing with the amount of traffic when different scheduling algorithms are used

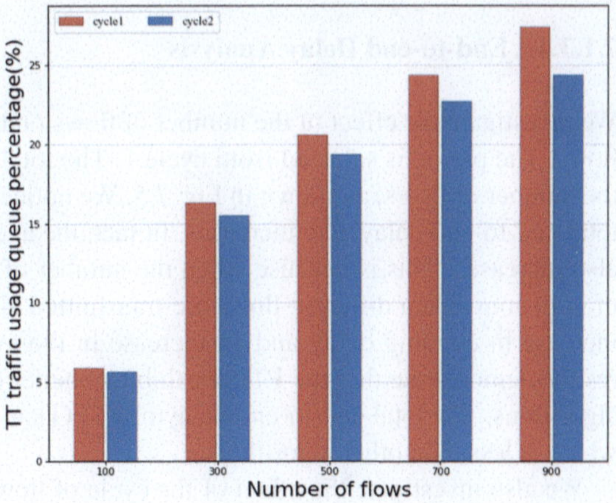

Fig. 7.4 Comparison of the percentage of occupied queues of PTPG algorithm for different period sets with 100–900 flows

queues. The results demonstrate that the PTPG exhibits a better number of queues than the other algorithms. The PTPG can obtain a 27.9% queue occupancy rate with 1000 flows, which is less than other algorithms.

We also investigate the effect of the cycle of flows on the number of queues occupied by time-sensitive traffic. The number of flows is set to 100, 300, 500, 700, 900 respectively. The number of queues occupied by time-sensitive traffic changes with the period, as shown in Fig. 7.4. We know that when the traffic period increases, the queue occupancy rate decreases. This is because an increase in the period of traffic means that the probability of collision, when different flows are transmitted, is reduced, so the number of queues that need to be occupied is also reduced. Experimental results show that when the traffic cycle is selected from cycle 2, the queue occupancy rate is approximately 10% lower than when the traffic cycle is selected from cycle 1.

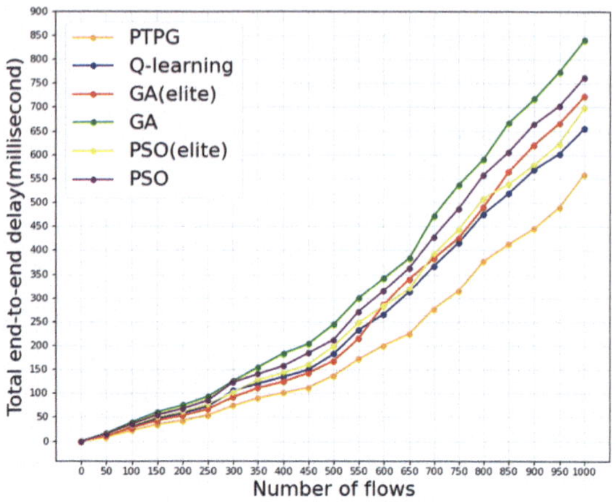

Fig. 7.5 Comparison chart of the end-to-end delay changing with the amount of traffic when different scheduling algorithms are used

7.1.3.4 End-to-end Delay Analysis

We investigate the effect of the number of flows on the total end-to-end delay of TT flows. The period is selected from cycle 1. The total end-to-end delay changes with the number of flows, as shown in Fig. 7.5. We notice that as the traffic increases, the total end-to-end delay also increases. In fact, the average end-to-end delay per flow also increases. This is because when the number of flows increases, the possibility of collisions when different flows are transmitted also increases, which leads to an increase in queuing delay and an increase in the average delay of each flow. The results demonstrate that the PTPG exhibits a better end-to-end delay than the other algorithms. The total end-to-end delay of PTPG can obtain 560ms with 1000 flows, which is less than other algorithms.

We also investigate the effect of the cycle of flows on the total end-to-end delay of TT flows. The number of flows is also set to 100, 300, 500, 700, 900 respectively. The total end-to-end delay of TT flows changes with the period, as shown in Fig. 7.6. We know that when the traffic period increases, the queue occupancy rate also increases. This is because as the traffic cycle increases, the frame constraints of TT flows are also relaxed, so the total end-to-end delay also increases. Experimental results show that when the traffic cycle is selected from cycle 2, the total end-to-end delay of TT flows is approximately 10%~40% higher than when the traffic cycle is selected from cycle 1. Moreover, when the number of flows increases, the difference in delay between cycle 1 and cycle 2 will increase accordingly.

7.2 Summary

This chapter investigates the control problem of UAV networks assisted by deterministic networking. In response to the diverse traffic types and complex demands

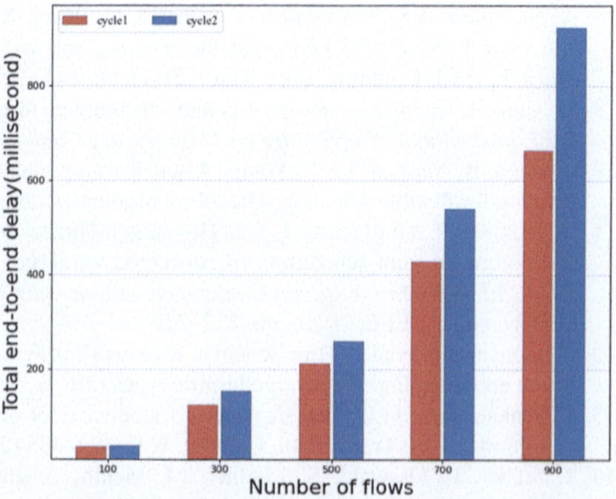

Fig. 7.6 Comparison of the end-to-end delay of PTPG algorithm for different period sets with 100–900 flows

within the network, a scheduling model tailored to different transmission requirements is proposed. Additionally, a two-stage PTPG based on PPO is developed to address the scheduling problem, resulting in performance improvements in terms of latency and queue management. Compared to other scheduling methods, the proposed approach significantly enhances the feasibility of traffic scheduling and optimizes network resource utilization while minimizing jitter impacts on networks. Future research will focus on further investigating differentiated traffic scheduling based on varying latency and jitter requirements.

References

1. J. Guo, H. Yao, W. He, T. Mai, T. Ouyang, F. Wang, Reinforcement learning-based genetic algorithm for differentiated traffic scheduling in industrial tsn-5g networks, in *2024 International Wireless Communications and Mobile Computing (IWCMC)* (IEEE, 2024), pp. 1283–1289
2. R. Nakayama, Y. Ito, Study on affect of gcl of time-aware shaper on qos for ieee802. 1 tsn networks, in *2020 IEEE 9th Global Conference on Consumer Electronics (GCCE)* (IEEE, 2020), pp. 946–947
3. S.S. Craciunas, R.S. Oliver, W. Steiner, Formal scheduling constraints for time-sensitive networks. Preprint (2017). arXiv:1712.02246
4. F. Dürr, N.G. Nayak, No-wait packet scheduling for ieee time-sensitive networks (tsn), in *Proceedings of the 24th International Conference on Real-Time Networks and Systems* (2016), pp. 203–212
5. M. Khoshnevisan, V. Joseph, P. Gupta, F. Meshkati, R. Prakash, P. Tinnakornsrisuphap, 5g industrial networks with comp for urllc and time sensitive network architecture. IEEE J. Selected Areas Commun. **37**(4), 947–959 (2019)
6. X. Wang, H. Yao, T. Mai, S. Guo, Y. Liu, Reinforcement learning-based particle swarm optimization for end-to-end traffic scheduling in tsn-5g networks. IEEE/ACM Trans. Netw. **31**(6), 3254–3268 (2023)
7. J. Yang, G. Yu, Traffic scheduling for 5g-tsn integrated systems, in *2022 International Symposium on Wireless Communication Systems (ISWCS)* (IEEE, 2022), pp. 1–6

8. A. Nasrallah, A.S. Thyagaturu, Z. Alharbi, C. Wang, X. Shao, M. Reisslein, H. ElBakoury, Ultra-low latency (ull) networks: the ieee tsn and ietf detnet standards and related 5g ull research. IEEE Commun. Surv. Tutor. **21**(1), 88–145 (2018)
9. M. Gen, R. Cheng, A survey of penalty techniques in genetic algorithms, in *Proceedings of IEEE International Conference on Evolutionary Computation* (IEEE, 1996), pp. 804–809
10. R. Chen, B. Yang, S. Li, S. Wang, A self-learning genetic algorithm based on reinforcement learning for flexible job-shop scheduling problem. Comput. Ind. Eng. **149**, 106778 (2020)
11. D. Ginthör, R. Guillaume, J. von Hoyningen-Huene, M. Schüngel, H.D. Schotten, End-to-end optimized joint scheduling of converged wireless and wired time-sensitive networks, in *2020 25th IEEE International Conference on Emerging Technologies and Factory Automation (ETFA)*, vol. 1 (IEEE, 2020), pp. 222–229
12. I.I. Consortium et al., "Time sensitive networks for flexible manufacturing testbed characterization and mapping of converged traffic types (2019)
13. S. Venkatraman, G.G. Yen, A generic framework for constrained optimization using genetic algorithms. IEEE Trans. Evol. Comput. **9**(4), 424–435 (2005)
14. T. Stüber, L. Osswald, S. Lindner, M. Menth, A survey of scheduling in time-sensitive networking (tsn). Preprint (2022). arXiv:2211.10954

Chapter 8
Conclusion

This book systematically discusses UAV swarm networks from a networking perspective, including adaptive clustering and networking, efficient routing algorithms, and resource scheduling strategies. As shown in Fig. 8.1, we also analyze the opportunities brought by emerging technologies such as distributed learning, semantic communication, and deterministic networks for UAV swarm networks. In Chap. 2, we propose intelligent clustering and networking methods to address the dynamic topology, unstable links, and self-organization characteristics of UAV swarm networks. By organizing the network efficiently and using optimization algorithms, communication efficiency and network stability are enhanced. Among them, the FSC strategy generates clusters via fission and dynamically adjusts links to improve communication performance. Additionally, to address the limited coverage of UAVs, a RL architecture with centralized training and distributed execution is proposed to solve the problems of "credit allocation" and "false reward signals," significantly optimizing cooperative communication performance in dynamic environments.

Due to the dynamic and resource-constrained nature of UAV swarm networks, routing algorithm design faces numerous challenges. Traditional algorithms such as OLSR and AODV rely on flooding mechanisms, leading to high communication overhead. While geographic routing is simple, it often falls into inefficient local optima. To tackle these challenges, in Chap. 3, we propose a series of intelligent routing optimization methods that combine MARL and evolutionary game theory to achieve efficient routing in dynamic environments. By incentivizing UAV nodes to cooperate in forwarding data packets, network performance and stability are significantly enhanced. Furthermore, by using MLP to filter key routing information, we design a routing algorithm based on a multi-agent actor-critic framework, achieving joint optimization of lifetime, data transmission rate, and throughput.

In Chap. 4, we focus on the resource scheduling problem in UAV swarm networks. To address the limited computational resources and communication bandwidth, intelligent solutions are proposed to improve task execution efficiency.

H. Yao, T. Mai, *UAV Swarm Cooperation*, Wireless Networks,
https://doi.org/10.1007/978-3-031-96444-2_8

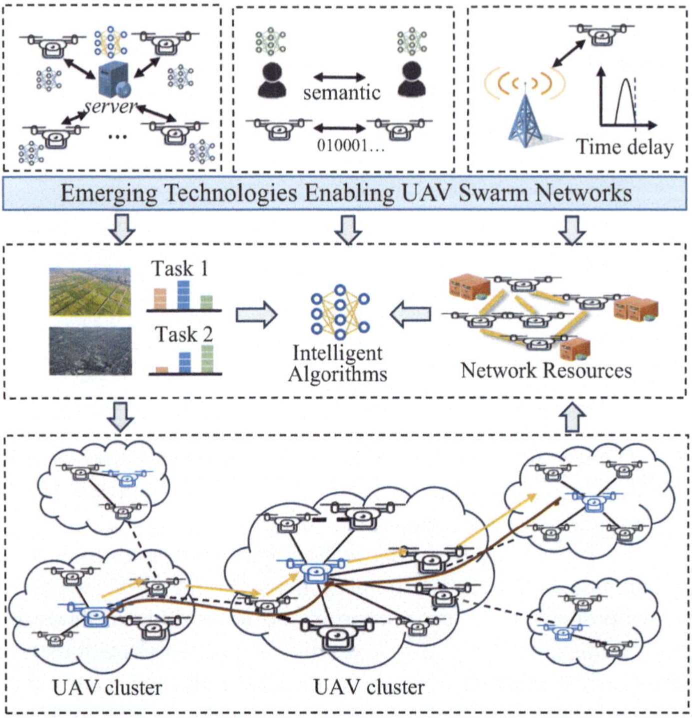

Fig. 8.1 Book organization

This chapter utilizes virtualization technology to abstract physical resources into virtual resources, enabling flexible allocation of network resources to meet node demands. To address the challenges of insufficient energy resources in UNDN, we propose a resource-efficient content-sharing mechanism that incentivizes content producers to share content and reveal private information. Building on this, we introduce a contract-based approach and use the GS algorithm to achieve a stable matching between content consumers and producers, thereby enhancing the overall system efficiency.

To solve the inefficiency of traditional centralized learning due to resource constraints and dynamic network topologies, we explore the application of distributed learning in UAV swarm collaboration in Chap. 5. Traditional FL allows for collaborative training across devices while ensuring data privacy, but it struggles to adapt to the heterogeneity of UAV swarms. To address this, a CFL architecture is proposed, where UAVs are divided into clusters based on optimization of directional similarity, and models are updated within the clusters. Additionally, a three-phase Stackelberg game model is designed to optimize resource allocation, and a hierarchical reinforcement learning algorithm is used to enhance resource utilization efficiency. To address the issue of node incentives, this chapter introduces a dual-auction mechanism and an IDA algorithm, allowing participants to optimize strategies, and uses a reinforcement learning-based EWA-DA algorithm for autonomous optimization.

Traditional communication schemes fail to meet the needs of the intelligent development of UAV swarms. In Chap. 6, we explore the application of semantic communication in UAV swarm networks to address the issues of large data exchange demands and low communication efficiency. This chapter proposes a HFL framework, enabling UAVs to train semantic models based on local data, achieving global parameter consistency through aggregation at cluster heads, and designing a cluster management mechanism to cope with network dynamics, thereby enhancing reliability. Furthermore, to address uncertainty and inefficient data sampling, semantic communication is combined to reduce communication delays. At the lower level, a working UAV selection mechanism ensures data synchronization, while the upper layer optimizes semantic information trading through a deep learning-driven auction mechanism, significantly enhancing communication efficiency and network response in large-scale data transmission and complex task scenarios.

In Chap. 7, we discuss the application of deterministic networks in UAV swarm control to achieve precise real-time control and data transmission in environments with large task transmission demand differences, limited resources, and dynamic complexity. We propose a scheduling model based on GA. By optimizing scheduling strategies, end-to-end delay and queue occupancy are effectively reduced, improving network performance.

The manufacturer's authorised representative in the EU is Springer Nature Customer Service Centre GmbH, Europaplatz 3, 69115 Heidelberg, Germany. If you have any concerns regarding our products, please contact ProductSafety@springernature.com

Printed and bound by CPI Group (UK) Ltd, Croydon, CR0 4YY
01/10/2026
02225971-0007